Micromechanics of
Composite Materials

Micromechanics of Composite Materials

Contributors

Alexander L. Kalamkarov, Marcelo A. Savi et al.

AURIS
Reference

www.aurisreference.com

Micromechanics of Composite Materials

Contributors: Alexander L. Kalamkarov, Marcelo A. Savi et al.

Published by Auris Reference Limited
www.aurisreference.com

United Kingdom

Micromechanics of Composite Materials

ISBN: 978-1-78154-946-9

British Library Cataloguing in Publication Data
A CIP record for this book is available from the British Library

Printed in the United Kingdom

Exclusively distributed by CBS Publishers & Distributors Pvt. Ltd.

Sales & Distribution Rights only for India, Pakistan, Bangladesh, Sri Lanka, Nepal and Bhutan. This book is not to be sold outside these territories.

Contents

List of Abbreviations

AFOSR	Air Force Office of Scientific Research
BF	Bright field
CFC	Carbon fiber reinforced epoxy composites
CNT	carbon nanotube
CrN	Coatings. Chromium nitride
DOE	Design of Experiment ,
DLC	Diamond-like carbon
FRPCs	Fiber-reinforced polymer matrix composites
FE	finite element
GMC	Generalized method of cells
INCT-EIE	Institute of Science and Technology on Smart Structures for Engineering
ISS	Interfacial shear stress
MRS	Micro-Raman spectroscopy
MC	Monte Carlo
MWCNT	Multiple walled carbon nanotube
NSERC	Natural Sciences and Engineering Research Council of Canada
PBCs	Periodic boundary conditions
PMCs	Polymer matrix composite materials
RVE	Representative volume element
SEM	scanning electron microscope
SWCNT	Single walled carbon nanotube
TEM	Transmission electron microscopy

List of Contributors

Alexander L. Kalamkarov
Dalhousie University Department of Mechanical Engineering B3H 4R2 Halifax, Nova Scotia, Canada

Marcelo A. Savi
Universidade Federal do Rio de Janeiro COPPE – Department of Mechanical Engineering 68503 Rio de Janeiro, RJ, Brazil

Zhenkun Lei
State Key Laboratory of Structural Analysis for Industrial Equipment, Dalian University of Technology, Dalian 116024, China

Xuan Li
State Key Laboratory of Structural Analysis for Industrial Equipment, Dalian University of Technology, Dalian 116024, China

Fuyong Qin
State Key Laboratory of Structural Analysis for Industrial Equipment, Dalian University of Technology, Dalian 116024, China

Wei Qiu
Department of Mechanics, Tianjin University, Tianjin 300072, China

Juergen M. Lackner
JOANNEUM RESEARCH Forschungsgesellschaft mbH, Institute of Surface Technologies and Photonics, Functional Surfaces, Leobner Strasse 94, A-8712 Niklasdorf, Austria

Wolfgang Waldhauser
JOANNEUM RESEARCH Forschungsgesellschaft mbH, Institute of Surface Technologies and Photonics, Functional Surfaces, Leobner Strasse 94, A-8712 Niklasdorf, Austria

Lukasz Major
Polish Academy of Sciences, Institute of Metallurgy and Materials Sciences, Ul. Reymonta 25, 30-059 Krakow, Poland

Marcin Kot
AGH University of Science and Technology, Faculty of Mechanical Engineering and Robotics, Laboratory of Tribology and Surface Engineering, A. Mickiewicza Ave. 30, 30-059 Krakow, Poland

Pierre Ladevèze
LMT-Cachan (ENS Cachan, CNRS-UMR8535, UPMC, PRES UniverSud Paris)

Federica Daghia
LMT-Cachan (ENS Cachan, CNRS-UMR8535, UPMC, PRES UniverSud Paris)

Emmanuelle Abisse
LMT-Cachan (ENS Cachan, CNRS-UMR8535, UPMC, PRES UniverSud Paris)

Camille Le Mauff
LMT-Cachan (ENS Cachan, CNRS-UMR8535, UPMC, PRES UniverSud Paris)

Leandro José da SilvaI
Department of Mechanical Engineering, Federal University of São João Del Rei -
UFSJ, Brazil, Praça Frei Orlando, 170, São João Del Rei, MG, Brazil

Túlio Hallak PanzeraI
Department of Mechanical Engineering, Federal University of São João Del Rei -
UFSJ, Brazil, Praça Frei Orlando, 170, São João Del Rei, MG, Brazil

André Luis ChristoforoI
Department of Mechanical Engineering, Federal University of São João Del Rei -
UFSJ, Brazil, Praça Frei Orlando, 170, São João Del Rei, MG, Brazil

Juan Carlos Campos RubioI
Department of Mechanical Engineering, Federal University of Minas Gerais - UFMG,
Belo Horizonte, MG, Brazil

Fabrizio Scarpa
Advanced Composites Centre for Innovation and Science, University of Bristol, Bristol, UK

Xiaojun Zhu
State Key Lab for Manufacturing Systems Engineering, Xi'an Jiaotong University,
Xi'an 710049, China

Xuefeng Chen
State Key Lab for Manufacturing Systems Engineering, Xi'an Jiaotong University,
Xi'an 710049, China

Zhi Zhai
State Key Lab for Manufacturing Systems Engineering, Xi'an Jiaotong University,
Xi'an 710049, China

Zhibo Yang
State Key Lab for Manufacturing Systems Engineering, Xi'an Jiaotong University, Xi'an 710049, China

Xiang Li
State Key Lab for Manufacturing Systems Engineering, Xi'an Jiaotong University, Xi'an 710049, China

Zhengjia He
State Key Lab for Manufacturing Systems Engineering, Xi'an Jiaotong University, Xi'an 710049, China

Andrew Ritchey
Purdue University

Joshua Dustin
Purdue University

Jonathan Gosse
The Boeing Company USA

R. Byron Pipes
Purdue University

Royan J D'Mello
Composite Structures Laboratory, Department of Aerospace Engineering, University of Michigan
William E. Boeing Department of Aeronautics and Astronautics, University of Washington

Marianna Maiarù
Composite Structures Laboratory, Department of Aerospace Engineering, University of Michigan
William E. Boeing Department of Aeronautics and Astronautics, University of Washington

Anthony M Waas
Composite Structures Laboratory, Department of Aerospace Engineering, University of Michigan
William E. Boeing Department of Aeronautics and Astronautics, University of Washington

Preface

The text *Micromechanics of Composite Materials* presents a broad exposition of analytical and numerical methods for modeling composite materials, laminates, polycrystals and other heterogeneous solids, with emphasis on connections between material properties and responses on several length scales, ranging from the nano and microscales to the macroscale. Micromechanical modeling and effective properties of the smart grid-reinforced composites have been focused in first chapter. Second chapter discusses recent advances of interfacial micromechanics in fiber reinforced composites using micro-Raman spectroscopy. Third chapter focuses on multilayered coating structures on soft, flexible CFC composites substrates in comparison to comparatively hard and stiff austenite steel. In fourth chapter, we present a new and relatively simple micromechanics-based interface model which takes into account the interaction between delamination and microcracking. Fifth chapter describes the analytical and experimental characterization of a class of polymeric composites made from epoxy matrix reinforced with unidirectional natural sisal and banana fibers with silica microparticles and maleic anhydride fabricated by manual molding. The aim of sixth chapter is to investigate the comprehensive influence of three microstructure parameters (fiber cross-section shape, fiber volume fraction, and fiber off-axis orientation) and strain rate on the macroscopic property of a polymer matrix composite. Self-consistent micromechanical enhancement of continuous fiber composites has been outlined in seventh chapter. Micromechanics modeling on electrical conductivity of CNT-polymer composites has been discussed in eighth chapter. Effect of the curing process on the transverse tensile strength of fiber-reinforced polymer matrix lamina using micromechanics computations has been investigated in ninth chapter. Micromechanics modeling of the electrical conductivity of carbon nanotude (CNT)-polymer nanocomposites has been proposed in tenth chapter.

Chapter 1

MICROMECHANICAL MODELING AND EFFECTIVE PROPERTIES OF THE SMART GRID-REINFORCED COMPOSITES

Alexander L. Kalamkarov[I]; Marcelo A. Savi[II]

[I]Dalhousie University Department of Mechanical Engineering B3H 4R2 Halifax, Nova Scotia, Canada

[II]Universidade Federal do Rio de Janeiro COPPE – Department of Mechanical Engineering 68503 Rio de Janeiro, RJ, Brazil

ABSTRACT

Smart composite structures reinforced with a periodic grid of generally orthotropic cylindrical reinforcements that also exhibit piezoelectric behavior are analyzed using the asymptotic homogenization method. The analytical expressions for the effective elastic and piezoelectric coefficients are derived. In particular, the smart orthotropic composite structures with cubic, conical and diagonal actuator and reinforcement orientations are investigated.

INTRODUCTION

The mechanical modeling of composite structures made of reinforcements embedded in a matrix has been the focus of investigation for some time. Noteworthy among the earlier models is the composite cylinders model proposed by Hashin and Rosen (1964). Budiansky (1965) developed a model which predicted the elastic moduli of multiphase composites with isotropic constituents. Other work can be found in Mori and Tanaka (1973), Sendeckyj (1974), Vinson and Sierokowski (1986), Christensen (1990), Drugan and Willis (1996), Kalamkarov and Liu (1998), Andrianov et al. (2006) and others.

Micromechanical models for the smart composites must take into consideration both local and global properties. Accordingly, the developed models should be rigorous enough to enable the consideration of the spatial distribution, mechanical properties, and behavior of the different constituents (reinforcing elements, matrix and actuators) at the local level, but not too complex to be described and used via straightforward analytical and numerical approaches.

Effective technique that can be used for the analysis of smart composites with regular structures is the multi-scale asymptotic homogenization method. The mathematical framework of this method can be found in Bensoussan et al. (1978), Sanchez-Palencia (1980), Bakhvalov and Panasenko (1984), Kalamkarov (1992). This method is mathematically rigorous and it enables the prediction of both the local and global effective properties of the periodic composite structure. Many problems in the framework of elasticity and thermoelasticity have been solved using this approach. For example, Kalamkarov and Georgiades developed general micromechanical models pertaining to smart composite structures with homogeneous (2002a) and non-homogeneous (2002b) structural boundary conditions, the later resulting in a boundary-layer type solution. Kalamkarov (1992) developed comprehensive micromechanical model for a thin composite layer with wavy upper and lower surfaces. This model was subsequently used to analyze the wafer and rib-reinforced smart composite plates as well as the sandwich composites with honeycomb fillers, see, e.g., Kalamkarov and Kolpakov (1997, 2001), Kalamkarov et al. (2009a). More recently, Kalamkarov et al. (2006), Georgiades et al. (2006) and Challagulla et al. (2007, 2008) have determined effective coefficients for the network-reinforced composite plates and shells. Saha et al. (2007a,b) investigated the smart composite sandwich shells made of generally orthotropic materials. The objective of these studies was to transform a general anisotropic composite material with a periodic array of reinforcements and actuators into a simpler one that is characterized by some effective coefficients. It is implicit of course that the physical problem based on these effective coefficients should give predictions differing as little as possible from those of the original problem.

The micromechanical models for the composite structures reinforced with a periodic grid of generally orthotropic cylindrical reinforcements have been developed in Kalamkarov et al. (2009b, 2010). Hassan et al. (2011), Hassan et al. (2009) and Hassan (2011) investigated smart grid-reinforced composite structures.

The review of micromechanical modeling of smart grid-reinforced structures based on the application of the asymptotic homogenization method is presented in the present paper. The formulated model is subsequently used to evaluate the effective elastic and piezoelectric coefficients of such structures.

Following this introduction the rest of the paper is organized as follows. The basic problem formulation is presented in the next section. That is then followed by the general piezoelastic model pertaining to smart 3D grid-reinforced composite structures with generally orthotropic reinforcements and actuators. The micromechanical model is further illustrated by means of several practically-important examples.

Nomenclature

σ_{ij}	= *stress tensor*
e_{kl}	= *strain tensor*
u_i,	= *displacement field*
C_{ijkl}	= *tensor of elastic coefficients*
P_{ijk}	= *tensor of piezoelectric coefficients*
R	= *control signal*
f_i	= *body forces*
ε	= *small parameter characterizing dimension of a unit cell Y, dimensionless*
$\tilde{C}_{ijkl}$	= *effective elastic coefficients*
$\tilde{P}_{ijk}$	= *effective piezoelectric coefficients*

ASYMPTOTIC HOMOGENIZATION MODEL FOR 3D COMPOSITE STRUCTURES

Consider a smart composite structure in a form of an inhomogeneous solid occupying domain Ω with boundary $\partial\Omega$ that contains a large number of periodically arranged reinforcements and actuators, see Fig. 1(a). It can be observed that this periodic structure is obtained by repeating a small unit cell Y in the domain Ω, see Fig. 1(b).

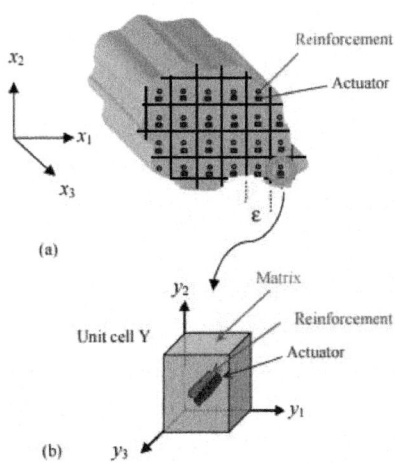

(a)

(b)

Figure 1: (a) 3D smart composite solid of a regular structure and (b) unit cell Y.

The elastic deformation of this structure can be described by the following boundary-value problem:

$$\frac{\partial \sigma_{ij}}{\partial x_j} = f_i \quad in\ \Omega, \quad u_i(\boldsymbol{x}) = 0 \quad on\ \partial\Omega,$$

(1)

$$\sigma_{ij} = C_{ijkl}e_{kl} - P_{ijk}R_k(\boldsymbol{x}), \quad e_{ij} = \frac{1}{2}\left(\frac{\partial u_i}{\partial x_j} + \frac{\partial u_j}{\partial x_i}\right)$$

(2)

In Eqs. (1) and (2) and in the sequel all indices assume values 1,2,3 and the summation convention is adopted, C_{ijkl} is the tensor of elastic coefficients, e_{kl} is the strain tensor which is a function of the displacement field u_i, P_{ijk} is a tensor of piezoelectric coefficients describing the effect of a control signal R on the stress field σ_{ij}. Finally, f_i represent body forces. Note that the present analysis is limited by considering only converse piezoelectric effect. This limitation does not affect the derived formulae for the effective elastic and piezoelectric coefficients of the smart grid-reinforced composite structure.

It is assumed in Eq. (1) that all the elastic and piezoelectric coefficients are periodic functions of spatial coordinates with a unit cell Y of characteristic dimension ε. Small parameter ε is made non-dimensional by dividing the size of the unit cell by a certain characteristic dimension of the overall structure.

The development of asymptotic homogenization model for the 3D smart composite structures can be found in Kalamkarov (1992), Kalamkarov et al. (2009a,b), Hassan et al. (2009). Here a brief review of the steps involved in the development of the model is given.

The first step is to define the so-called «fast» or microscopic variables according to:

$$y_1 = x_1/\varepsilon, \quad y_2 = x_2/\varepsilon, \quad y_3 = x_3/\varepsilon$$

(3)

As a consequence of introducing **y** coordinates, the derivatives are also transformed according to

$$\frac{\partial}{\partial x_1} \rightarrow \frac{\partial}{\partial x_1} + \frac{1}{\varepsilon}\frac{\partial}{\partial y_1}, \quad \frac{\partial}{\partial x_2} \rightarrow \frac{\partial}{\partial x_2} + \frac{1}{\varepsilon}\frac{\partial}{\partial y_2},$$

$$\frac{\partial}{\partial x_3} \rightarrow \frac{\partial}{\partial x_3} + \frac{1}{\varepsilon}\frac{\partial}{\partial y_3}$$

(4)

The boundary value problem and corresponding stress field defined in Eqs. (1) and (2) are thus transformed into the following expressions:

$$\frac{\partial \sigma_{ij}(x,y)}{\partial x_j} + \frac{1}{\varepsilon}\frac{\partial \sigma_{ij}(x,y)}{\partial y_i} = f_i \ in \ \Omega,$$

(5)

$$u_i(x,y) = 0 \ on \ \partial\Omega$$

$$\sigma_{ij}(x,y) = C_{ijkl}(y)\frac{\partial u_k}{\partial x_l}(x,y) - P_{ijk}(y)R_k(x)$$

(6)

The next step is to consider the following asymptotic expansions in terms of powers of the small parameter ε:

$$u_i(x,y) = u_i^{(0)}(x,y) + \varepsilon u_i^{(1)}(x,y) + \varepsilon^2 u_i^{(2)}(x,y) + \dots$$

(7)

$$\sigma_i(x,y) = \sigma_i^{(0)}(x,y) + \varepsilon \sigma_i^{(1)}(x,y) + \varepsilon^2 \sigma_i^{(2)}(x,y) + \dots$$

(8)

By substituting Eqs. (7) and (8) into Eqs. (5) and (6) and considering at the same time the periodicity of $u^{(i)}$ in y one can readily eliminate the microscopic variable y from the first term $u^{(0)}$ in the asymptotic displacement field expansion thus showing that it depends only on the macroscopic variable x. Subsequently, by separating terms with like powers of ε one obtains a series of differential equations, the first two of which are:

$$\frac{\partial \sigma_{ij}^{(0)}}{\partial y_j} = 0$$

(9a)

$$\frac{\partial \sigma_{ij}^{(1)}}{\partial y_j} + \frac{\partial \sigma_{ij}^{(0)}}{\partial x_j} = f_i$$

(9b)

Where

$$\sigma_{ij}^{(0)} = C_{ijkl}\left(\frac{\partial u_k^{(0)}}{\partial x_l} + \frac{\partial u_k^{(1)}}{\partial y_l}\right) - P_{ijk}R_k$$

(10a)

$$\sigma_{ij}^{(1)} = C_{ijkl}\left(\frac{\partial u_k^{(1)}}{\partial x_l} + \frac{\partial u_k^{(2)}}{\partial y_l}\right)$$

(10b)

Combination of Eqs. (9a) and (10a) leads to the following expression:

$$\frac{\partial}{\partial y_j}\left(C_{ijkl}\frac{\partial u_k^{(1)}(x,y)}{\partial y_l}\right) =$$

$$\frac{\partial P_{ijk}(y)}{\partial y_j}R_k(x) - \frac{\partial C_{ijkl}(y)}{\partial y_j}\frac{\partial u_k^{(0)}(x)}{\partial x_l}$$

(11)

The separation of variables in the right-hand-side of Eq. (11) prompts to represent the solution for $u^{(1)}$ as:

$$u_m^{(1)}(x,y) = \frac{\partial u_k^{(1)}(x)}{\partial x_l}N_m^{kl}(y) + R_k(x)M_m^k(y)$$

(12)

where the auxiliary functions N_m^{kl} and M_m^k are periodic in y and they satisfy the following problems:

$$\frac{\partial}{\partial y_j}\left(C_{ijmn}(y)\frac{\partial N_m^{kl}(y)}{\partial y_n}\right) = -\frac{\partial C_{ijkl}}{\partial y_j}$$

(13)

$$\frac{\partial}{\partial y_j}\left(C_{ijmn}(y)\frac{\partial M_m^k(y)}{\partial y_n}\right) = \frac{\partial P_{ijk}}{\partial y_j}$$

(14)

One observes that Eqs. (13) and (14) depend only on the fast variable **y** and thus are formulated in the domain Y of the unit cell, remembering at the same time that all C_{ijkl}, P_{ijk} and N_m^{kl} and M_m^k are Y-periodic in **y**. Consequently, Eqs. (13) and (14) are appropriately called the unit-cell problems.

The next important step in the model development is the homogenization procedure. This is carried out by first substituting Eq. (12) into Eq. (10a), and combining the result with Eq. (9b). The resulting expressions are then integrated over the unit cell Y (with the volume |Y|) remembering to treat **x** as a parameter as far as integration with respect to **y** is concerned. After cancelling the terms that vanish due to periodicity, this yields the following equation:

$$\tilde{C}_{ijkl}\frac{\partial^2 u_k^{(0)}(x)}{\partial x_j\partial x_l} - \tilde{P}_{ijk}\frac{\partial R_k(x)}{\partial x_j} = f_i$$

(15)

where the following definitions are introduced:

$$\tilde{C}_{ijkl} = \frac{1}{|Y|} \int_Y \left(C_{ijkl}(y) + C_{ijmn}(y) \frac{\partial N_m^{kl}(y)}{\partial y_n} \right) dv \tag{16}$$

$$\tilde{P}_{ijk} = \frac{1}{|Y|} \int_Y \left(P_{ijk}(y) - C_{ijmn}(y) \frac{\partial M_m^k(y)}{\partial y_n} \right) dv \tag{17}$$

Coefficients $\tilde{C}_{ijkl}, \tilde{P}_{ijk}$, defined by Eqs. (16) and (17) are the effective elastic and piezoelectric coefficients respectively. It is noticed that they are constant unlike their original rapidly varying material counterparts C_{ijkl}, P_{ijk} and therefore problem (15) is much simpler than the original problem given by Eqs. (5) and (6). It is worth mentioning that although the present work pertains to piezoelectric actuators, the model derived applies equally well if the smart composite structure is associated with some general transduction properties that can be used to induce residual strains and stresses. In that case, the coefficients $\tilde{P}_{ijk}$ represent the appropriate effective actuation coefficients (rather than the piezoelectric ones).

3D Smart Grid-Reinforced Composite Structures

In the subsequent Sections we will consider the problem of a smart 3D composite structure reinforced with Nfamilies of reinforcements/actuators, see for example Fig. 2 where an explicit case of multiple families of reinforcements is shown.

(a)   (b)

Figure 2: (a) 3D grid-reinforced smart composite structure and (b) its unit cell.

We assume that the members of each family are made of different generally orthotropic materials that may exhibit piezoelectric characteristics and that the reinforcements of each family make angles $\varphi_1^n, \varphi_2^n, \varphi_3^n$ $(n = 1, 2, ..., N)$ with the y_1,

y_2, y_3 axes respectively. It is further assumed that the orthotropic reinforcements/ actuators have significantly larger elastic moduli than the matrix material, so we are justified in neglecting the contribution of the matrix phase in the ensuing analytical treatment. The error resulting from this simplifying assumption is discussed below. Clearly, for the case of the lattice grid structures there is no surrounding matrix and assumption of zero matrix rigidity is exact.

The nature of the grid-reinforced composite structure of Fig. 2 is such that it would be more efficient if we first considered a simpler type of unit cell made of only a single reinforcement/actuator as shown in Fig. 3. Having dealt with this situation, the effective elastic and piezoelectric coefficients of more general structures with multiple families of reinforcements/actuators can be determined by superposition of the solution for each of them found separately. In following this procedure, one must naturally accept the error incurred at the regions of intersection between the reinforcements. However, our approximation will be quite accurate since these regions of intersection are highly localized and do not contribute significantly to the integral over the entire volume of the unit cell. Essentially, the error incurred will be negligible if the dimensions of the actuators/reinforcements are much smaller than the spacing between them. The mathematical justification for this argument in the form of the so-called principle of the split homogenized operator can be found in Bakhvalov and Panasenko (1989).

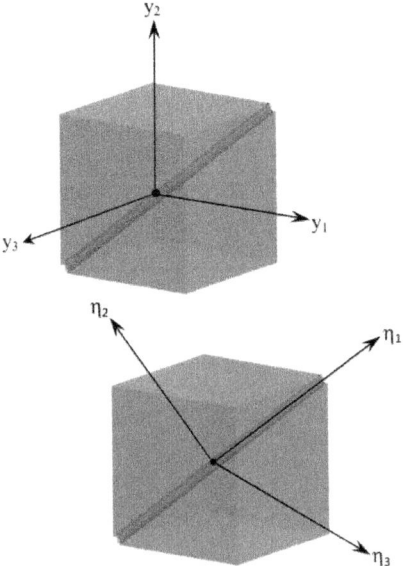

Figure 3: Unit cell in case of a single reinforcement family in the original and rotated microscopic coordinates.

In order to evaluate the accuracy of the above two key assumptions pertaining to the asymptotic model, the finite element analysis was carried out in Hassan et al. (2011). It is shown that errors in the values of the effective properties are negligibly small for a large mismatch between the stiffness of the reinforcements and the matrix. The finite element results have also indicated that the error from ignoring the regions of overlap of reinforcements will only be significant for the cases of grid-reinforced structures with more than three different reinforcement families; if the unit cell consists of up to three different reinforcements the associated error is negligibly small.

In order to calculate the effective coefficients for the simpler smart structure of Fig. 3, unit cell problems given by Eqs. (13) and (14) should be solved and, subsequently, Eqs. (16) and (17) should be applied.

The problem formulation for the structure shown in Fig. 3 begins with the introduction of the following local functions:

$$b_{ij}^{kl} = C_{ijkl}(y) + C_{ijmn}(y)\frac{\partial N_m^{kl}(y)}{\partial y_n}$$
(18)

$$p_{ij}^{k} = P_{ijk}(y) - C_{ijmn}(y)\frac{\partial M_m^{k}(y)}{\partial y_n}$$
(19)

The unit cell problems in Eqs. (13) and (14) can be then written as follows:

$$\frac{\partial}{\partial y_j} b_{ij}^{kl} = 0$$
(20)

$$\frac{\partial}{\partial y_j} p_{ij}^{k} = 0$$
(21)

Perfect bonding conditions are assumed at the interfaces between the actuators/reinforcements and the matrix. This assumption yields the following interface conditions:

$$N_n^{kl}(r)\big|_s = N_n^{kl}(m)\big|_s, \quad b_{ij}^{kl}(r)n_j\big|_s = b_{ij}^{kl}(m)n_j\big|_s$$
(22)

$$M_n^{k}(r)\big|_s = M_n^{k}(m)\big|_s, \quad p_{ij}^{k}(r)n_j\big|_s = p_{ij}^{k}(m)n_j\big|_s$$
(23)

In Eqs. (22) and (23) "r", "m", and "s" denote the actuator/reinforcement, matrix, and reinforcement/matrix interface, respectively; while n_j denote the components of the unit normal vector at the interface. As was mentioned earlier, we will further assume that $C_{ijkl}(m) = 0$, which implies from Eqs. (18)

and (19) that $b_{ij}^{kl}(m) = p_{ij}^{k}(m) = 0$. Therefore, the interface conditions in Eqs. (22) and (23) become

$$b_{ij}^{kl}(r)n_j\Big|_s = 0$$

(24)

$$p_{ij}^{k}(r)n_j\Big|_s = 0$$

(25)

In summary, the unit cell problems that must be solved for the 3D grid-reinforced smart composite structure with a single family of orthotropic reinforcements/actuators are given by Eqs. (20) and (21) in conjunction with Eqs. (22)-(25).

In order to solve the pertinent unit cell problems we perform a coordinate transformation of the global coordinate system $\{y_1, y_2, y_3\}$ into the new coordinate system $\{\eta_1, \eta_2, \eta_3\}$ shown in Fig. 3. With the new coordinate system we note that since the reinforcement is oriented along the η_1 coordinate axis, the problem at hand becomes independent of η_1 and depend only on η_2 and η_3. As a result, the ensuing analysis becomes much easier.

Effective Elastic and Piezoelectric Coefficients

A scheme for the determination of the effective elastic and piezoelectric coefficients for 3D grid-reinforced composite structures with generally orthotropic reinforcements is given in detail in Kalamkarov et al. (2009b) and Hassan et al. (2009). It is noteworthy to mention that in the limiting particular case of 2D grid-reinforced structure with isotropic reinforcements the developed expressions for the effective elastic coefficients converge to those obtained earlier by Kalamkarov (1992).

With reference to Fig. 3, we begin by rewriting Eqs. (18), (22) and (24) in the $\{\eta_1, \eta_2, \eta_3\}$ coordinates to get:

$$b_{ij}^{kl} = C_{ijkl}(\mathbf{y}) + C_{ijmn}q_{pn}\frac{\partial N_m^{kl}(\mathbf{y})}{\partial \eta_p},$$

$$\left(b_{ij}^{kl}q_{2j}n_2'(r) + b_{ij}^{kl}q_{3j}n_3'(r)\right)\Big|_s = 0$$

(26)

Here, q_{ij} are the direction cosines characterizing the axes rotation (see Fig. 3); n_2' and n_3' are the components of the unit normal vector in the new coordinate system. Expanding Eq. (26) and keeping in mind the independency of the unit cell problem on η_1 yields:

$$b_{ij}^{kl} = C_{ijkl} + C_{ijm1}q_{21}\frac{\partial N_m^{kl}}{\partial \eta_2} + C_{ijm2}q_{22}\frac{\partial N_m^{kl}}{\partial \eta_2} + C_{ijm3}q_{23}\frac{\partial N_m^{kl}}{\partial \eta_2} +$$

$$C_{ijm1}q_{31}\frac{\partial N_m^{kl}}{\partial \eta_3} + C_{ijm2}q_{32}\frac{\partial N_m^{kl}}{\partial \eta_3} + C_{ijm3}q_{33}\frac{\partial N_m^{kl}}{\partial \eta_3}$$

(27)

Apparently, Eqs. (26) and (27) can be solved by assuming a linear variation of the local functions N_m^{kl} with respect to η_2 and η_3:

$$N_1^{kl} = \lambda_1^{kl}\eta_2 + \lambda_2^{kl}\eta_3, \quad N_2^{kl} = \lambda_3^{kl}\eta_2 + \lambda_4^{kl}\eta_3,$$

$$N_3^{kl} = \lambda_5^{kl}\eta_2 + \lambda_6^{kl}\eta_3,$$

(28)

where λ_i^{kl} are constants to be determined from the boundary conditions. Once these coefficients are determined, the coefficients b_{ij}^{kl} are found from the Eq. (27). In turn, these are used to calculate the effective elastic coefficients of the structure of Fig. 3 by integrating over the volume of the unit cell

$$\tilde{C}_{ijkl} = \frac{1}{|Y|}\int_Y b_{ij}^{kl}dv$$

(29)

Noting that b_{ij}^{kl} are constants, the effective elastic coefiicients become

$$\tilde{C}_{ijkl} = \mu_f b_{ij}^{kl}$$

(30)

where μ_f is the volume fraction of the reinforcement within the unit cell. It can be proved in general case that the effective elastic coefficients $\tilde{C}_{ijkl}$ maintain the same symmetry and convexity properites as their actual material counterparts C_{ijkl}, see Kalamkarov (1992).

The above derived effective moduli pertain to grid-reinforced structures with a single family of reinforcements. For structures with more than one family of reinforcements the effective moduli can be obtained by superposition. The effective elastic coefficients of a grid-reinforced structure with N families of generally orthotropic reinforcements will be given by:

$$\tilde{C}_{ijkl} = \sum_{n=1}^{N} V \mu_f^{(n)} b_{ij}^{(n)kl}$$

(31)

where the superscript (n) represents the n-th reinforcement family with the reinforcement volume fraction $\mu_f^{(n)}$.

Let us now proceed to calculation of the effective piezoelectric coefficients from the unit cell problem given by Eqs. (19), (23) and (25) which in coordinates $\{\eta_1, \eta_2, \eta_3\}$ becomes

$$p_{ij}^k = P_{ijk} - C_{ijmn} q_{pn} \frac{\partial M_m^k}{\partial \eta_p},$$

$$\left(p_{ij}^k q_{2j} n_2'(r) + p_{ij}^k q_{3j} n_3'(r) \right)\Big|_s = 0$$

$$(32)$$

Keeping in mind independency on η_1, Eq. (32) yields:

$$p_{ij}^k = P_{ijk} - (C_{ijm1} q_{21} \frac{\partial M_m^k}{\partial \eta_2} + C_{ijm2} q_{22} \frac{\partial M_m^k}{\partial \eta_2} + C_{ijm3} q_{23} \frac{\partial M_m^k}{\partial \eta_2} +$$

$$+ C_{ijm1} q_{31} \frac{\partial M_m^k}{\partial \eta_3} + C_{ijm2} q_{32} \frac{\partial M_m^k}{\partial \eta_3} + C_{ijm3} q_{33} \frac{\partial M_m^k}{\partial \eta_3})$$

$$(33)$$

It can be shown that Eq. (33) in conjunction with Eq. (32) can be solved by assuming a linearity of functions Σ_i^k (**y**) in η_2 and η_3:

$$M_1^k = \Sigma_1^k \eta_2 + \Sigma_2^k \eta_3, \quad M_2^k = \Sigma_3^k \eta_2 + \Sigma_4^k \eta_3,$$

$$M_3^k = \Sigma_5^k \eta_2 + \Sigma_6^k \eta_3,$$

$$(34)$$

where Σ_i^k are constants that can be determined from the boundary conditions. The functions given by Eqs. (33) and (34) are used to calculate the effective piezoelectric coefficients of the smart composite structure of Fig. 3 by integrating over the volume of the unit cell, which on account of Eqs. (17) and (19) yields

$$\tilde{P}_{ijk} = = \frac{1}{|Y|} \int_Y p_{ij}^k dv$$

$$(35)$$

Since the local functions p_{ij}^k a are constant, the effective piezoelectric coefficients become

$$\tilde{P}_{ijk} = \mu_f p_{ij}^k$$

$$(36)$$

where μ_f is the volume fraction of the actuators/reinforcement within the unit cell.

The effective piezoelectric coefficients derived above pertain to grid-reinforced smart composite structures with a single family of actuators/

reinforcements. For structures with multiple families of inlusions the effective actuation coefficients can be obtained by superimposition. For instance, pertaining to a grid-reinforced smart composite structure with N families of actuators/reinforcements the effective coefficients will be given by

$$\tilde{P}_{ijk} = \sum_{n=1}^{N} \mu_f^{(n)} p_{ij}^{(n)k},$$

(37)

where the superscript (n) represents the nth reinforcement/actuator family, as in the above Eq. (31).

Examples of Smart Grid-Reinforced Composite Structures

The developed micromechanical model will now be used to analyze three different practically important examples of smart 3D grid-reinforced composite structures with orthotropic actuators/reinforcements, see Hassan et al. (2009, 2011). The first example, structure S_1 is shown in Fig. 2. It has three families of orthotropic actuators/reinforcements, each family oriented along one of the coordinate axes. The second example, structure S_2 is shown in Fig. 4. It is formed by a conical array of orthotropic reinforcements/actuators.

Spatial arrangement of reinforcements/ actuators as viewed from the top

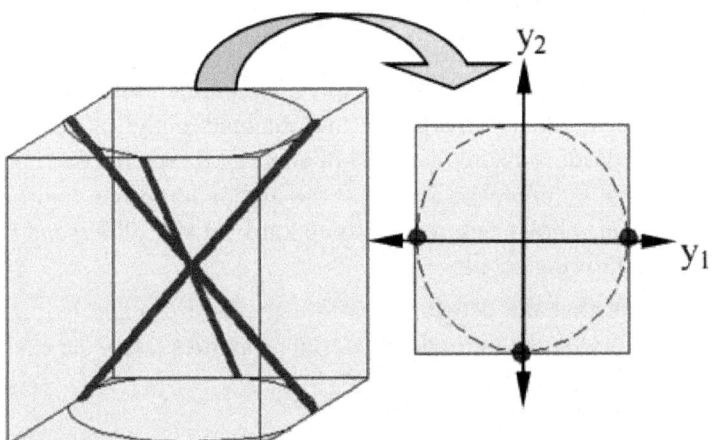

Figure 4: Unit cell of smart composite structure S_2 with conical arrangement of orthotropic reinforcements/actuators.

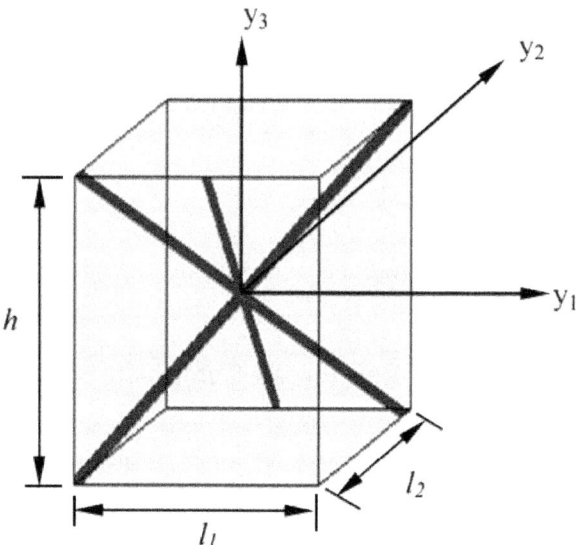

Figure 5: Unit cell of smart composite structure S_3 with diagonally arranged orthotropic actuators/reinforcements.

The third example structure S_3 is shown in Fig. 5. It has a unit cell formed by three actuators/reinforcements, two of them extended diagonally across the unit cell between two diametrically opposite vertices while the third reinforcement is spun between the middle of the bottom edge and the middle of the top edge on the opposite face. The effective elastic and piezoelectric coefficients for the above introduced three structures are calculated on the basis of Eqs. (31) and (37). Although the obtained analytical results are too lengthy to be reproduced here, the plots of some of these effective coefficients vs. reinforcement volume fraction or vs. the inclination of the reinforcements with the y_3 axis are shown below, see Kalamkarov et al. (2009b) and Hassan et al. (2009, 2011) for the details.

We assume that the actuators/reinforcements are made of piezoelastic material PZT-5A with the following material properties (see Cote et al., 2002):

$$C_{11}^{(p)} = C_{22}^{(p)} = 121.0\,\text{GPa},\ C_{33}^{(p)} = 111.0\,\text{GPa},$$

$$C_{12}^{(p)} = 75.4\,\text{GPa},\ C_{13}^{(p)} = C_{23}^{(p)} = 75.2\,\text{GPa},$$

$$C_{44}^{(p)} = 22.6\,\text{GPa},\ C_{55}^{(p)} = C_{66}^{(p)} = 21.1\ \text{GPa},$$

$$P_{13}^{(p)} = P_{23}^{(p)} = -5.45\times10^{-6}\,\text{C}/\text{mm}^2,$$

$$P_{33}^{(p)} = 1.56\times10^{-5}\,\text{C}/\text{mm}^2, P_{42}^{(p)} = P_{51}^{(p)} = 2.46\times10^{-5}\,\text{C}/\text{mm}^2.$$

We start by providing numerical results for the effective coefficients of structure S_1 shown in Fig. 2. Typical piezoelectric coefficients are plotted vs. volume fraction in Fig. 6. As expected, these coefficients increase in magnitude as the volume fraction increases. One also observes from Fig. 6 that the values of $\tilde{P}_{333}$ are larger than $\tilde{P}_{113}$ for a given volume fraction, which is to be expected because the former refers to the stress response in the direction y_3 in which electric field is applied.

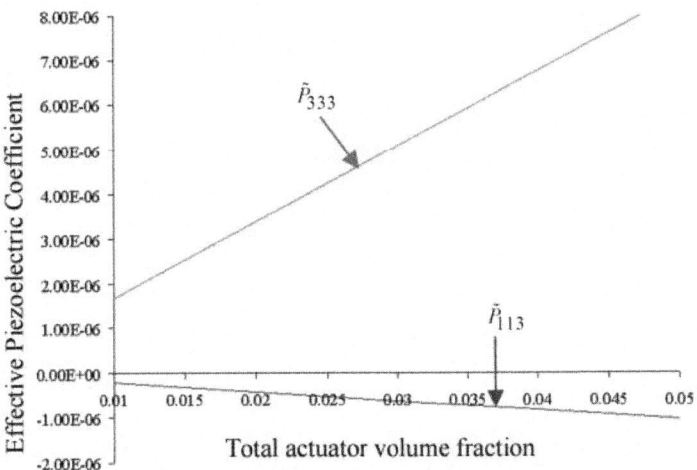

Figure 6: Plot of 113 $\tilde{P}_{113}$ **and** $\tilde{P}_{333}$ ' effective piezoelectric coefficients vs. actuator volume fraction for structure S_1

We now turn our attention to structure S_2 of Fig. 4. Typical effective piezoelectric coefficients are plotted vs. the total volume fraction of the actuators/reinforcements within the unit cell in Figs. 7 and 8. As expected, the plots show an increase in the effective piezoelectric coefficients as the overall volume fraction increases. And it is seen from Figs. 6 and 7 that the magnitude of the coefficient $\tilde{P}_{333}$ (which refers to the stress response of the structure in the y_1-direction when an external field is applied in the y_3-direction) is larger for Structure S_1 than for Structure S_2. This is expected and is attributed to the geometry of the unit cells. Figure 6 refers to structure S_1 with some actuators/ reinforcements oriented entirely in the y_3 direction. Figure 7 refers to structure S_2 where none of the reinforcements are oriented in the y_3-direction (all 3 actuators/reinforcements are oriented at about 34°to the y_3 axis). Consequently, the stress response of structure S_2 in the y_1 direction when a voltage is applied in the y_3 direction is smaller and so is the corresponding effective coefficient $\tilde{P}_{333}$.

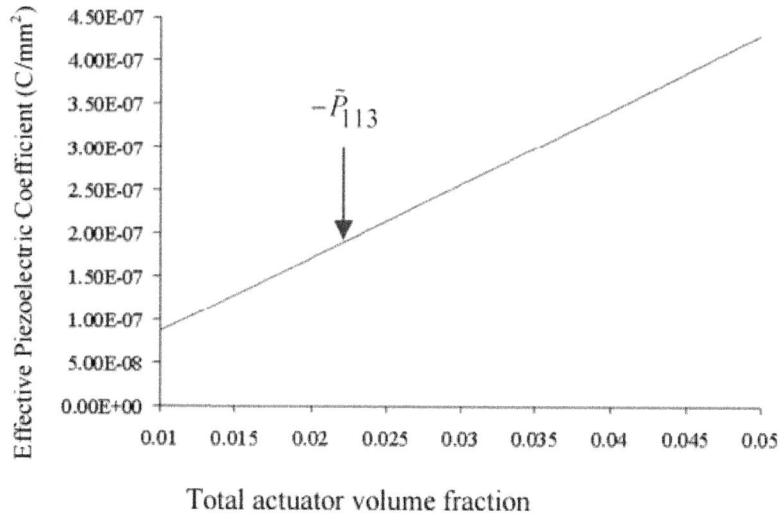

Figure 7: Plot of $^{-\tilde{P}_{113}}$ effective piezoelectric coefficient vs. actuator volume fraction for structure S2 (actuators oriented at 33.7o to the y_3 axis)

It is also of interest to analyze the variation of the effective coefficients of structure S_2 vs. the angle of inclination of the actuators/reinforcements to the y_3 axis. As this angle increases, the actuators/reinforcements are oriented progressively closer to y_1-and y_2-axes, and, consequently, further away from the y_3 axis. Thus, one expects a corresponding increase in the values of effective coefficients, as it is seen in the Fig. 8 plotting $\tilde{P}_{113}$ and $\tilde{P}_{223}$.

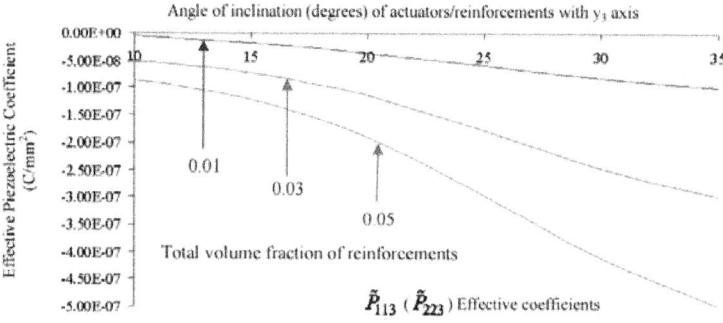

Figure 8: Plot of $\tilde{P}_{113}$ $(=\tilde{P}_{223})$ effective piezoelectric coefficient vs. inclination of actuators/reinforcements with the y_3 axis for different volume fractions (structure S_2).

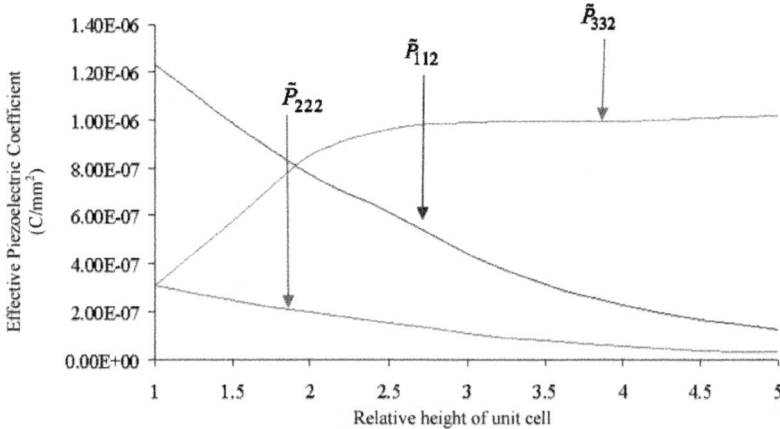

Figure 9: Plot of $\tilde{P}_{112}$, $\tilde{P}_{222}$, **and** $\tilde{P}_{332}$ effective piezoelectric coefficients vs. relative height of unit cell for structure S_3.

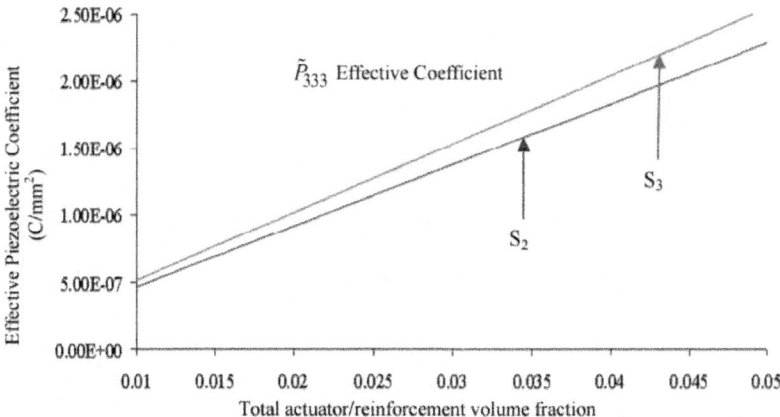

Figure 10: Plot of $\tilde{P}_{333}$ effective piezoelectric coefficient vs. total volume fraction for structures S_2 and S_3.

We now turn our attention to structure S_3 shown in Fig. 5 and we will present graphically some of the effective piezoelectric coefficients vs. the relative height of the unit cell, see Fig. 9. The relative height is defined as the ratio of the height to the length of the unit cell. The width of the unit cell and the cross-sectional area of the reinforcements/actuators stay the same. It is noted that when the relative height of the unit cell is increased the total volume fraction of the reinforcements/actuators as well as the orientation angle between these actuators and y_3-axis will decrease. This has very interesting consequences on the effective coefficients. In particular, decrease of the angle of inclination of the actuators with the y_3 axis will reduce the stiffness in y_1 and y_2 directions because the actuators are oriented further away from the y_1-y_2 plane. The simultaneous decrease in the overall actuator volume fraction makes this effect even more pronounced. These trends are clearly visible in Fig. 9 for the coefficients $\tilde{P}_{112}$ and $\tilde{P}_{222}$. However, as far as the stiffness in the y_3-direction is concerned the two factors that accompany the increase in the relative height of the unit cell are in direct competition with one another. That is, decreasing the angle of inclination of the actuators with y_3-axis increases such coefficients $\tilde{P}_{332}$ and $\tilde{P}_{333}$, but decreasing the overall actuator volume fraction naturally reduces the magnitude of these coefficients. As Fig. 9 shows, the former effect dominates the latter one, especially for low to moderate values of the relative height of the unit cell. However, after a certain point, the two factors tend to compensate each other, so that the value of $\tilde{P}_{332}$ increases at a modest rate.

Finally, we compare the typical effective coefficients of structures S_2 and S_3 by varying the total volume fraction of the actuators/reinforcements but keeping the same dimensions of the respective unit cells, see Fig. 10. Under these circumstances structure S_3 has higher effective piezoelectric coefficient $\tilde{P}_{333}$ than structure S_2. This is attributed to the different angles of inclination of the actuators/reinforcements to the y_3 axis.

The above discussed examples demonstrate that the derived micromechanical model allows developing a smart composite structure with the desirable combination of effective properties via selection of relevant material and geometric parameters such as number, type and cross-sectional dimensions of the actuators/reinforcements, relative dimensions of the unit cell, and the spatial orientation of the actuators/reinforcements.

We also note that the advantage of our model is that the effective coefficients can be computed easily without the need of time consuming numerical (such as finite element) calculations, see Hassan et al. (2011) where the analytical and finite element models are compared.

CONCLUSIONS

Smart composite structures reinforced with a periodic grid of generally orthotropic cylindrical reinforcements that also exhibit piezoelectric behavior are considered. The review of micromechanical modeling of these smart structures based on the application of the asymptotic homogenization method is presented. The micromechanical model decouples the original boundary-value problem into a simpler set of problems called the unit cell problems which describe the elastic and piezoelectric effective properties of the smart 3D grid-reinforced composite structures. By means of the solution of the unit cell problems the explicit expressions for the effective elastic and piezoelectric coefficients are obtained. The general orthotropy of the reinforcement material is important from the practical viewpoint, and makes the mathematical analysis much more complicated. It is worth mentioning that even though the analysis presented is applied to the piezoelectric material, the model derived accommodates equally well any smart composite structure exhibiting some general transduction characteristic that can be used to induce strains or stresses in some controlled manner. The developed micromechanical model is applied to different examples of orthotropic smart composite structures with cubic, conical and diagonal actuators/reinforcements arrangements. It is shown in these examples that the micromechanical model provides a complete flexibility in designing a 3D smart grid-reinforced composite structure with desirable piezoelastic characteristics to conform to a particular engineering application by tailoring certain material and/or geometric parameters. Examples of such parameters include the type, number, material and cross-sectional characteristics and relative orientations of the actuators and reinforcements.

ACKNOWLEDGEMENTS

The authors acknowledge the support of the Natural Sciences and Engineering Research Council of Canada (NSERC), the Brazilian Research Agencies CNPq, CAPES, FAPERJ and the National Institute of Science and Technology on Smart Structures for Engineering (INCT-EIE). The support of the Air Force Office of Scientific Research (AFOSR) is also acknowledged.

REFERENCES

1. Andrianov, I.V., Danishevs'kyy, V.V. and Kalamkarov, A.L., 2006, "Asymptotic justification of the three-phase composite model", Compos. Struct., Vol. 77, pp. 395-404.

2. Bakhvalov, N.S. and Panasenko, G.P., 1989, "Averaging Processes in Periodic Media. Mathematical Problems in Mechanics of Composite Materials", Kluwer, Dordrecht.

3. Bensoussan, A., Lions, J.L. and Papanicolaou, G., 1978, "Asymptotic Analysis for Periodic Structures", North-Holland, Amsterdam.

4. Budiansky, B., 1965, "On the elastic moduli of some heterogeneous materials", J. Mech. Phys. Solids, Vol. 13, pp. 223-227.

5. Challagulla, K.S., Georgiades, A.V. and Kalamkarov, A.L., 2007, "Asymptotic homogenization modeling of thin network structures", Compos. Struct., Vol. 3, pp. 432-444.

6. Challagulla, K.S., Georgiades, A.V., Saha, G.C. and Kalamkarov, A.L., 2008, "Micromechanical analysis of grid-reinforced thin composite generally orthotropic shells", Composites, Part B: Eng., Vol. 39, pp. 627-644.

7. Christensen, R.M., 1990, "A critical evaluation for a class of micromechanics models", J. Mech. Phys. Solids, Vol. 38, pp. 379-404.

8. Cote, F., Masson, P. and Mrad, N., 2002, "Dynamic and static assessment of piezoelectric embedded composites", Proceedings of the SPIE, Vol. 4701, pp. 316-325.

9. Drugan, W.J. and Willis, J.R., 1996, "A micromechanics-based nonlocal constitutive equation and estimates of representative volume element size for elastic composites", J. Mech. Phys. Solids, Vol. 44, pp. 497-524.

10. Hashin, Z. and Rosen, B.W., 1964, "The elastic moduli of fiber-reinforced materials", J. Appl. Mech., Vol. 31, pp. 223-232.

11. Hassan, E.M., 2011, "Modeling of 3D grid-reinforced smart anisotropic composite structures", Lambert Acad. Publ., ISBN 978-3-8433-6651-9.

12. Hassan, E.M., Georgiades, A.V., Savi, M.A. and Kalamkarov, A.L., 2011, "Analytical and numerical analysis of 3D grid-reinforced orthotropic composite structures", Int. J. Eng. Sc., Vol. 49, pp. 589-605.

13. Hassan, E.M., Kalamkarov, A.L., Georgiades, A.V. and Challagulla, K.S., 2009, "An asymptotic homogenization model for smart 3D grid-reinforced composite structures with generally orthotropic constituents", Smart Materials and Structures, Vol. 18, No. 7, 075006 (16pp).

14. Georgiades, A.V., Challagulla, K.S. and Kalamkarov, A.L., 2006, "Modeling of the thermopiezoelastic behavior of prismatic smart composite structures made of orthotropic materials", Composites, Part B: Eng., Vol. 37, pp. 569-582.

15. Kalamkarov, A.L., 1992, "Composite and Reinforced Elements of Construction", Wiley, Chichester, N.Y.

16. Kalamkarov, A.L. and Georgiades, A.V., 2002a, "Modeling of smart composites on account of actuation, thermal conductivity and hygroscopic absorption", Composites, Part B: Eng., Vol. 33, pp. 141-152.

17. Kalamkarov, A.L. and Georgiades, A.V., 2002b, "Micromechanical modeling of smart composite structures",Smart Materials Struct., Vol. 11, pp. 423-434.

18. Kalamkarov, A.L., Andrianov, I.V. and Danishevs'kyy, V.V., 2009a, "Asymptotic homogenization of composite materials and structures", Transactions of the ASME, Applied Mechanics Reviews, Vol. 62, No. 3, pp. 030802-1–030802-20.

19. Kalamkarov, A.L., Georgiades, A.V., Challagulla, K.S. and Saha, G.C., 2006, "Micromechanics of smart composite plates with periodically embedded actuators and rapidly varying thickness", J. Thermoplastic Composite Mater., Vol. 19, pp. 251-276.

20. Kalamkarov, A.L., Hassan, E.M., Georgiades, A.V., 2010, "Micromechanical modeling of 3D grid-reinforced composite structures and nanocomposites", Journal of Nanostructured Polymers and Nanocomposites, Vol. 6, Issue 1, pp. 12-20.

21. Kalamkarov, A.L., Hassan, E.M., Georgiades, A.V. and Savi, M.A., 2009b, "Asymptotic homogenization model for 3D grid-reinforced composite structures with generally orthotropic reinforcements", Compos. Struct., Vol. 89, pp. 186-196.

22. Kalamkarov, A.L. and Kolpakov, A.G., 1997, "Analysis, Design and Optimization of Composite Structures", Wiley, Chichester, N.Y.

23. Kalamkarov, A.L. and Kolpakov, A.G., 2001, "A new asymptotic model for a composite piezoelastic plate", Int. J. Solids Struct., Vol. 38, pp. 6027-6044.

24. Kalamkarov, A.L. and Liu, H.Q., 1998, "A new model for the multiphase fiber–matrix composite materials",Composites, Part B: Eng., Vol. 29, pp. 643-653.

25. Mori, T. and Tanaka, K., 1973, "Average stress in matrix and average energy of materials with misfitting inclusions", Acta Metallurgica et Materialia, Vol. 21, pp. 571-574.

26. Saha, G.C., Kalamkarov, A.L. and Georgiades, A.V., 2007a, "Effective elastic characteristics of honeycomb sandwich composite shells made of generally orthotropic materials", Composites, Part A: Appl. Sci. and Manuf., Vol. 38, pp. 1533-1546.

27. Saha, G.C., Kalamkarov, A.L. and Georgiades, A.V., 2007b, "Micromechanical analysis of effective piezoelastic properties of smart composite sandwich shells made of generally orthotropic materials", Smart Mater. Struct., Vol. 16, pp. 866-883.

28. Sanchez-Palencia, E., 1980, "Non-Homogeneous Media and Vibration Theory", Lecture Notes in Physics, Springer, Berlin.

29. Sendeckyj, G.P., 1974, "Mechanics of Composite Materials", Academic Press, New York.

30. Vinson, J.R. and Sierokowski, R.L., 1986, "The Behavior of Structures Composed of Composite Materials", Kluwer, Dordrecht.

Chapter 2

INTERFACIAL MICROMECHANICS IN FIBROUS COMPOSITES: DESIGN, EVALUATION, AND MODELS

Zhenkun Lei,[1] Xuan Li,[1] Fuyong Qin,[1] and Wei Qiu[2]

[1]State Key Laboratory of Structural Analysis for Industrial Equipment, Dalian University of Technology, Dalian 116024, China

[2]Department of Mechanics, Tianjin University, Tianjin 300072, China

ABSTRACT

Recent advances of interfacial micromechanics in fiber reinforced composites using micro-Raman spectroscopy are given. The faced mechanical problems for interface design in fibrous composites are elaborated from three optimization ways: material, interface, and computation. Some reasons are depicted that the interfacial evaluation methods are difficult to guarantee the integrity, repeatability, and consistency. Micro-Raman study on the fiber interface failure behavior and the main interface mechanical problems in fibrous composites are summarized, including interfacial stress transfer, strength criterion of interface debonding and failure, fiber bridging, frictional slip, slip transition, and friction reloading. The theoretical models of above interface mechanical problems are given.

INTRODUCTION

Polymer-matrix fibrous composites have been widely used in the aerospace and industrial locomotives fields. There exist many interfacial phenomena and all kinds of defects including inclusions, pores, and layer shrink zones in fibrous composites during design and manufacturing processes. Fiber fracture and interface debonding will appear inside the material in service and result in early fatigue, aging, damage, and failure, so it is a hidden danger to risk a major engineering accident. With development of composite material science and aerospace industry applications, the light-weight and high-tough carbon fibers have been widely applied to fibrous composites. Therefore, many researchers coming from physics, chemistry, materials, mechanics, and engineering are attracted by many basic mechanical problems in fibrous composites, such as mechanical properties characterization of high-performance fiber, fiber/matrix interface debonding, fiber bridging, fiber fracture, and matrix cracking [1].

Besides the effects of fiber surface treatments on fibrous composites have been studied; the basic problems of microscopic interfacial stress transfer and failure were focused in decades [2, 3]. Materials' microstructure configuration determines its response to external action. As a connection between reinforcing fiber and matrix, the interface is an important microstructure of fibrous composites including fiber, fiber transition region, fiber surface coating, matrix transition region, and matrix. The interface is a bridge connecting both reinforced fiber and matrix and a deliverer of mechanics information. Although the interface is much smaller than the size of composite bulk, there are many mechanical problems on the interface, such as load transfer, shear strength, interface debonding, damage, and stress singularity [4, 5]. Interfacial bonding quality directly affects the entire composites on interlaminar shear, fracture, impact, heat aging, wave propagation, and other mechanical properties. Therefore, it is necessary to study on fiber interface mechanics from a microscopic view by examining and analyzing the linkages among microstructure, interfacial mechanics properties, and macrofracture properties. The establishment and improvement of interfacial stress transfer and failure models will help to understand the composite stress transfer, debonding, and failure mechanisms from microscale experiments.

With unique advantages of nondestructive, noncontact, and high spatial resolution (1 μm), micro-Raman spectroscopy (MRS) is most likely applied to the integrity characterization of interfacial micromechanical properties in fibrous composites [6], porous silicon [7–9], and carbon nanotubes [10, 11]. In the process of interface debonding of fibrous composites, the evolution of interface mechanical parameters including the frictional shear stress, interfacial shear stress, and debonding length happened in real time, while the pulling force and displacement are also changed accordingly. At present, the main mechanical problems on fiber/matrix interface include the stress transfer, interfacial strength criterion, fiber bridging, and other aspects in fibrous composites.

Based on the outline of interface mechanics design, interface evaluation method, and fine characterization techniques of fibrous composites, the research progress on the interface mechanics by MRS is introduced in the paper, including the interfacial stress transfer, interfacial debonding and strength failure criterion, fiber bridging, interface friction, and slip transition.

At last, some theoretical models on those interface mechanics problems are summed up.

INTERFACE MECHANICS DESIGN

The mechanical properties of fibrous composites are closely related to the interface control process, material compound, material properties, and interfacial failure modes, so these are very important for the interface mechanics design and optimization of fibrous composites. However, there is no effective criterion yet to optimize the performance of interface mechanics. Based on the existing interface mechanics models and numerical analysis works, it is possible to get the interfacial mechanical parameters and material parameters having no interface failure and to give the reference of the process controls and material options for the optimization of interface mechanical properties [12, 13].

The design of interface mechanics in fibrous composites should consider the process technology, materials and environment, and other complex factors. It mainly consists of three optimization approaches: material optimization, interface optimization, and computation optimization, as shown in Figure 1. The former two approaches control the macro- and micromechanical properties of fibrous composite, respectively. Meanwhile, they are helpful to study the interfacial load transfer and failure models for different scales and to provide the theoretical and experimental basis for the interface mechanics design of fibrous composites. The third approach is utilizing the multiscale computation to associate the macro- and micromechanical models and to predict the ultimate bearing performance of the designed fibrous composites through optimizing material constituents and fiber laying configurations.

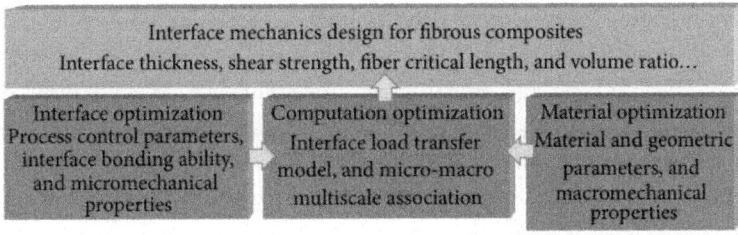

Figure 1: Interface design routes for fiber composites.

Material Optimization

The selection and optimum combination of materials is the most commonly used method for the interface design of fibrous composites. Through choosing fiber and matrix resin having specific properties, the composite laminates are formed by curing according to a certain volume ratio and fiber laying manner. Thus, the loading capacity of fibrous composites can be improved by means of the excellent mechanical properties of fibers. Typically, the macroscopic mechanical tests are used to characterize the interfacial properties of fibrous composites. There are a lot of works to get the interfacial shear strength and other interface parameters, such as the fiber critical length obtained by single fiber fragmentation test and the relationship between fiber pullout force and displacement by single fiber pullout test [17, 18].

The interface mechanical parameters obtained by the macroscopic mechanical tests are the average results for characterizing the macroscopic performance of interface bonding capability. However, it is difficult to get the fine stress distribution along the fiber/matrix interface and to observe the interfacial debonding and failure processes. Therefore, the development of microscale measurement methods is necessary [19]. In addition, there are still more researches on the mechanical testing of fiber and resin matrix itself, including tensile or compressive stress-strain behaviors of single fiber and fiber bundle, the impact of fiber surface treatment on the interface shear strength, the resin curing behavior, and wetting behavior between fiber and resin. The impact of these factors on the fiber/matrix interface physicochemical properties and geometric characteristics still needs further study.

Interface Optimization

Nowadays, it has been recognized that the way of traditional compound optimization is insufficient to improve the whole mechanical properties of fibrous composites and then the researchers turned to the interface bonding ability to improve the mechanical properties of fibrous composites. Due to different mechanical properties, compound process, and geometric conditions on the fiber/resin interface, there are thermal, mechanical, chemical, and physical coupling effects existing on the interface. Resultantly, different interfacial structures and characteristics appear and affect the fiber/resin interface bonding capacity, and then the fibrous composites exhibit different macrophysicochemical and mechanical properties [20].

As shown in Figure 2, the interface control means for fibrous composites are mostly through the fiber surface modification and compound process to obtain specific fiber/matrix interface microstructure, such as the geometric configurations, contact angle, and embedded fiber length. The specific interface microstructure will exhibit different physical and chemical properties, such as wettability, chemical bond, and van der Waals force; thus the interface bond strength is changed to improve the performance of interfacial load transfer.

At present, the fiber surface modification can be used to get appropriately bonded interface, but the physicochemical mechanisms that are how to affect the interfacial micromechanical properties as well as how to control the interfacial stress transfer have been concerned. Although a variety of interface theories in fibrous composites have been proposed, such as wettability theory, chemical bond theory, and friction theory, there is no perfect theory to explain all phenomena of interface [2].

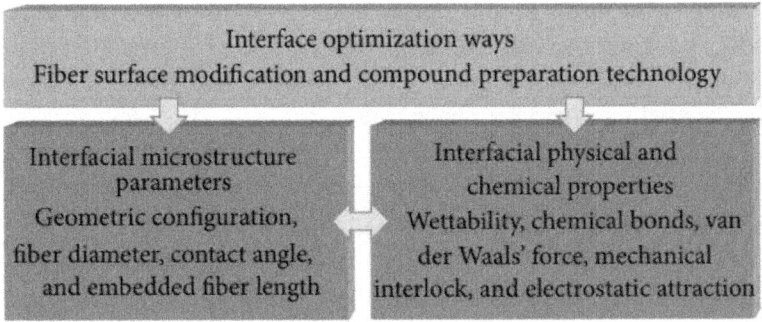

Figure 2: Interface optimization ways.

Computation Optimization

If the interface strength of fibrous composites is too low, the fiber is easy to debond, pullout, break, and fail. On the contrary, if the interface strength is very high, the stress between the fiber and matrix cannot be relaxed and the brittle fracture would occur at the interface. Therefore, the interface design can be optimized by considering the best comprehensive mechanical properties. The interfacial mechanical properties and geometrical parameters are regarded as design variables, and then certain optimization method such as genetic algorithm combines with the finite element analysis to find the best design variables. This is the fast optimization path of interface mechanical performance in fibrous composites.

However, the design variables of composite interface microstructure are not continuous so that the derivative-type optimization method will fail in the case. It is also noted that the uncertainty of initial value limits the capacity of optimization method converging to the global optimum. In addition, the existing mechanical models are imperfect to describe the micromechanical behavior of the composite interface. The mechanical properties of interface layer, residual stress, and stress singularity are the difficulties to constrain the numerical computation [12, 13]. At present, a lot of works are still to seek the appropriate computing optimization methods to solve such problems. It is inevitable and reasonable way for the computation optimization to perfectly combine with the interface evaluation tests, fine interface characterization techniques, and interface mechanical models.

INTERFACE EVALUATION AND CHARACTERIZATION

Interface Evaluation Tests

The macroscopic damage and failure criteria for fibrous composites do not consider the micromechanical properties of interface, such as fiber stress distribution, stress concentration, shear strength, and frictional shear stress on the debonding interface. In addition, there is still lack of a common understanding about the influence of interfacial microstructural parameters and physicochemical properties on the interface micromechanical properties. Currently, the research on microscale experimental mechanics characterization of the interface failure is not only the most difficult and crucial problem but also the important content of interface mechanical evaluation in fibrous composites, as shown in Figure 3.

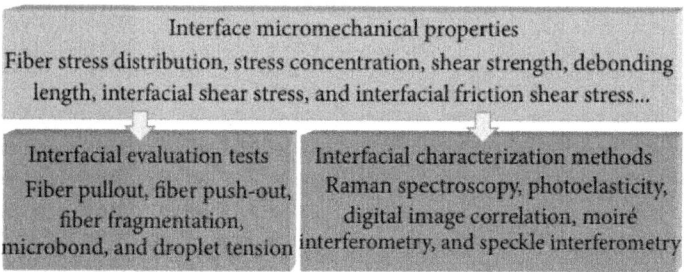

Figure 3: Evaluation tests and characterization methods for micromechanical properties of interface.

The interfacial shear strength is a commonly used parameter to evaluate interfacial bonding quality, fiber/matrix stress transfer efficiency, and the effect of fiber surface modification. The important parameter can be obtained by single fiber micromechanical testing experiments. One kind of these experiments is realized by applying the external load to single fiber, such as fiber pullout test [15], microbond test [21], microdroplet tension test [14, 22], and fiber push-out test [23]. The other is finished by applying the external load to the resin matrix, such as fiber fragmentation test [12, 13] and Broutman test [24].

During the implementation and application of these interface evaluation tests for the characterization of micromechanical properties, it is difficult to ensure the integrity, repeatability, and consistency of the interface evaluation.

The experimental results of fiber pullout test, fiber fragmentation test, and fiber push-out test vary widely at the same external conditions. Even using the same test method, the experimental results among different laboratories still have differences [25]. Further studies suggest that this difference comes from the stress singularity at the fiber end [26], so the reevaluation of these test methods and the development of new, more appropriate test methods are concerned [27].

However, the deeper reason is that the differences of many conditions (i.e., interface characteristics) in these interface evaluation tests are neglected, such as interface structure, geometric shape and dimension, and boundary and surface treatment. The testing specimens employed in the different interface evaluation methods have different geometric parameters, such as the droplet contact angle and the embedded fiber length in the microdroplets tension test. Even with the same specimen preparation procedure, it is difficult to ensure that all samples have a uniform geometry and dimension size, which affects the repeatability and consistency of the interface micromechanical parameters characterized by the interface evaluation test. The latest research of microdroplet tension test shows that the microdroplet conformations with different contact angles affect the interfacial shear stress distribution and stress transfer efficiency [28]. By optimizing the design of interface geometry to reduce or even eliminate the stress singularity, the mechanical behavior of fibrous composites can be upgraded [29]. Therefore, the further research on the interface geometry and physicochemical properties affecting the interfacial stress transfer behavior will benefit to optimize the interfacial stress distribution and reduce the stress concentrations.

Fine Characterization Techniques

Commonly, the interfacial shear strength obtained by the interface evaluation tests is used as an important characteristic parameter in the interface failure models and is an average value for characterizing the interface bonding properties. It cannot completely describe the details of the interfacial stress transfer and interfacial debonding failure processes. Therefore, more sophisticated real-time experimental data are required to quantitatively and completely characterize the micromechanical behaviors of the fiber/matrix interface [30]. The direct requirement is the use of "partial details" (i.e., stress distribution) of the interface parameters instead of the average. In addition, the respective contributions of the bonding shear stress and frictional shear stress to the interfacial shear failure mechanisms are also concerned in the fiber/matrix debonding procedure. However, most studies are lacking in the integrity of mechanical description for the procedures of the interfacial stress transfer and interfacial debonding failure. A very important reason is the lack of suitable microscale stress-strain measurement techniques and full-field observation means.

The testing methods having the ability to carry out the microscale fine characterization, including MRS and digital photoelasticity, digital image correlation, and speckle interferometry. These methods are most likely the first application to completely characterize the micromechanical properties of fiber reinforced composites. MRS measurements have unique advantages at the microscales: nondestructive, noncontact, high spatial resolution (1 μm), and the depth focus [6].

When the fiber is under deformation, it causes the movement and deformation of Raman spectrum [14, 15], as shown in Figure 4(a). Although the epoxy resin has a strong fluorescence effect, the Raman spectrum of fiber/epoxy specimen after fully curing shows a Raman spectrum overlay of fiber and epoxy resin, but this does not affect the identification of the fiber Raman peak. Raman shift has a linear relationship with the strain or stress of aramid fibers [15], as shown in Figure 4(b). Therefore, it is a potential method of microscale experimental mechanics, and it has recently been used to study the interfacial micromechanical behaviors in fibrous composites, such as fiber stress distribution, stress concentration, and interface integrity.

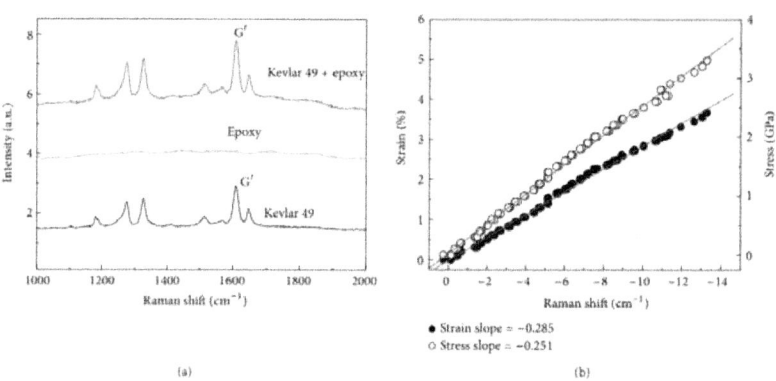

Figure 4: Raman spectra of (a) fiber/epoxy droplet specimen compared with pure epoxy and Kevlar 49 fiber [14] and (b) relationships of Raman shift with stress and strain for Kevlar 49 fiber [15].

INTERFACE MECHANICS MODELING

The interfacial stress transfer behavior between the fiber and matrix in fibrous composites is a major mechanical problem including several successive stages: the interface intact bonding, interface debonding, interface completely debonding, and fiber pullout. The elastic stress transfer in bonding area and the frictional shear stress transfer in debonded area have been widely recognized. In the process of interfacial debonding and extension, the interface mechanical parameters of bonding shear stress, debonding friction shear stress, and interface debonding length continuously evolve, and the macropulling force or stress is also changed accordingly. At present, the main interface mechanics problems in fibrous composites discussed are as follows: the elastic stress transfer, partial debonding stress transfer, interface failure criterion and fiber bridging, and so on.

Elastic Stress Transfer and Failure

One end of single fiber embeds in epoxy matrix, as shown in Figure 5, an axis tension load pulls the fiber out from the matrix. Under the assumptions of the stress uniform distribution and homogeneous isotropy, the interfacial shear strength τ_b can be calculated by the ratio of maximum pullout load F_{max} and interface bonding area; namely

$$\tau_b = \frac{F_{max}}{2\pi r l},$$

(1)

where r and l are the fiber radius and the embedded fiber length, respectively.

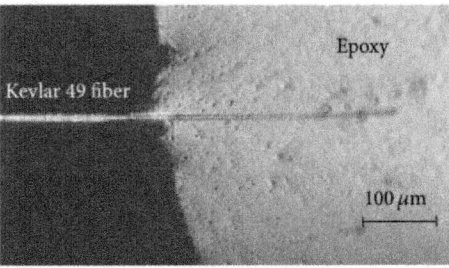

Figure 5: Single fiber pullout specimen [15].

The stress distribution along the embedded fiber cannot be obtained by the above equation, so it cannot be used to study the stress transfer between the fiber and matrix. Cox's shear-lag model [31] considers the force balance between the fiber axial stress σ and the interfacial shear stress τ along the embedded fiber. It satisfies the relationship as

$$\tau = -\frac{r}{2}\left(\frac{d\sigma}{dx}\right).$$

(2)

Piggott's model [32] is further used to describe the fiber axial stress along the embedded fiber within the elastic stress transfer, so the fiber elastic stress distribution before the interface debonding is written as

$$\sigma = \sigma_{app} \frac{\sinh\left[n\left(L - x\right)/r\right]}{\sinh\left(ns\right)},$$

(3)

where x is the distance to fiber entry, σ_{app} is the stress acting on the fiber out of the matrix, L is the fiber length that the fiber axial stress decays to zero (i.e., the effective length of stress transfer), s is the fiber aspect ratio (L/r), and n is a constant related with the geometry, material parameters of fiber, and matrix [33]. Figure 6(a) shows the fiber axial stress distribution under different strain levels in fiber pullout test. It can be seen that

the fiber axial stress increased significantly with the applied strain and the constant fiber axial stress out of the matrix ($x \leq 0$) equal to the applied load; that is, $\sigma = \sigma_{app}$. Then, the fiber axial stress along the embedded fiber was gradually reduced from the fiber entry ($x=0$) to the embedded fiber end ($x = L$) in accordance with the theoretical results (solid line) of (3). In the current 1.2% strain level, the debonding phenomenon did not occur at the embedded fiber and the entire embedded fiber was under a certain load, so the intact elastic stress transfer was presented on the fiber/matrix bonding interface.

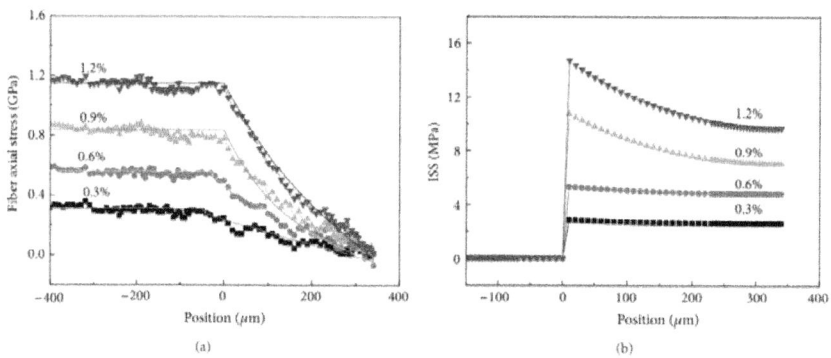

Figure 6: (a) Fiber axial stress and (b) shear stress distributions along fiber of pullout specimen under different strain levels [15].

The interfacial shear stress (ISS) along the embedded fiber is further given from (2) and (3) as

$$\tau = \sigma_{app} \frac{n \cosh\left[n\left(L - x\right)/r\right]}{2 \sinh\left(ns\right)}.$$

(4)

As shown in Figure 6(b), the ISS distribution increased with the strain levels. The ISS of fiber out of the matrix was zero and reached the maximum at the fiber entry. In the fiber pullout experiment, the aspect ratio n of the embedded fiber is large. The fiber stress and shear stress at the fiber entry (x=0) are given by the combination of (3) and (4) as

$$\sigma_m = \sigma, \qquad \tau_m = \frac{n}{2}\sigma.$$

(5)

If the applied strain continues, the fiber fracture failure occurs when the fiber stress σ on the free fiber segment is over the fiber stress strength σ_b. Similarly, the interfacial debonding failure occurs when the maximum ISS of τ_m at the fiber entry exceeds the shear strength τ_b. Then, the strength failure conditions depending on the balance of fiber strength σ_b and interfacial shear strength τ_b are written as

$$\sigma = \frac{2\tau_m}{n} \geq \sigma_b, \quad \text{Fiber fracture,}$$

$$\tau_m = \frac{n\sigma}{2} \geq \tau_b, \quad \text{Interface debond.}$$

(6)

It can be seen that the fiber/matrix interface is more likely to fail if the fiber strength σ_b increases, and the fiber tends to break if the interface shear strength τ_b increases.

Frictional Shear Stress Transfer

When the applied load further increased in the fiber pullout test (Figure 5), the interface debonding failure occurred firstly at the fiber entry and then propagated along the fiber/matrix interface. The fiber axial stress distribution in Figure 7(a) shows that the fiber has debonded from the fiber entry (Point O) to the debonding/bonding transition (Point B), and the interface frictional shear stress existed on the different stages (Figure 7(b)). This is because the debonding segments of OA and AB exhibit different interface microstructures resulting in unequal shear friction effect. The interface frictional shear stress accords with the linear distribution assumption on the debonding segments (the solid lines in Figure 7(a)). After the debonding/bonding transition (Point B), the fiber bonding interface is still intact and the fiber axial stress distribution satisfies with the Piggott's model (Segment BC).

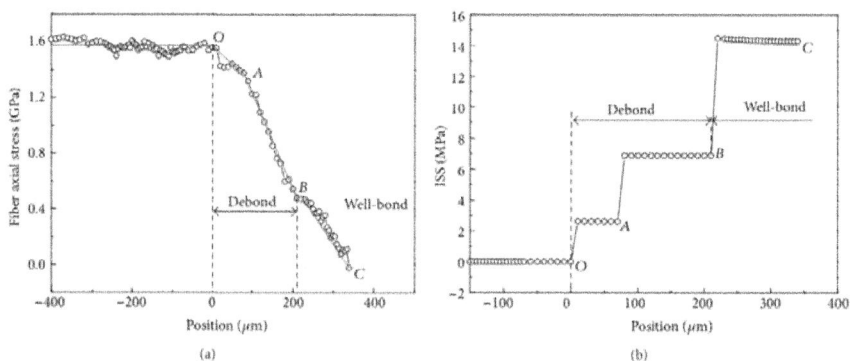

Figure 7: (a) Fiber axial stress and (b) shear stress distributions along fiber of pullout specimen under 1.6% strain [15].

Using the simple Cox's shear-lag model, the frictional stress transfer in the debonding interface can be easily analyzed. Assuming a linear distribution of the interfacial friction stress, a two-stage model of the interfacial friction shown in Figure 7(b) gives the fiber stress distributions on the debonding interface as

$$\sigma_{OA} = \sigma_{app} - 2\bar{\tau}_{OA}\frac{x}{r}, \quad 0 \le x \le L_{OA},$$

$$\sigma_{AB} = \sigma_{app} - 2\bar{\tau}_{OA}\frac{L_{OA}}{r}$$

$$- 2\bar{\tau}_{AB}\frac{(x - L_{OA})}{r}, \quad L_{OA} < x \le L_{OB},$$

(7)

where L_{OA} is the debonding fiber length on the first stage, L_{OB} is the total length of the debonding fiber, $\bar{\tau}_{OA}$ and $\bar{\tau}_{AB}$, respectively, correspond to the first and second stages of the interfacial friction shear stress constant, and x is the distance to the fiber entry (Point O). Piggott's model can be used to describe the fiber axial stress at the intact bonding interface (Segment BC in Figure 7). The fiber axial stress equals the fiber stress at the debonding/bonding transition (Point B), which can be obtained by solving (7) under the condition of $x=L_{OB}$. The interface frictional shear stresses on the debonding segments are given by the combination of (2) and (7) as

$$\tau_{OA} = \bar{\tau}_{OA}, \quad 0 \le x \le L_{OA},$$

$$\tau_{AB} = \bar{\tau}_{AB}, \quad L_{OA} < x \le L_{OB}.$$

(8)

It can be seen that the frictional shear stress plays the role of stress transfer on the debonding interface and can be described as the multistage constant distribution in this study. If the load continues to be applied, the interface debonding failure propagates forward. According to strength failure conditions (6), the fiber breakage failure occurred until the maximum fiber stress $(\sigma = \sigma_{app})$ reached the fiber strength $\sigma_{b,}$ otherwise the interfacial debonding failure will continue until the fiber is completely pulled out.

Reloading of Bridging Fiber

As shown in Figure 8, when a matrix crack vertically propagated across an embedded fiber without fiber breakage, the bridging fiber with partial debonding was across both sides of the matrix crack. The formation of bridging fiber can be regarded as two fibers pullout process. The bridging fiber contains three parts: the bonding segment, debonding segment, and bridging segment. The fiber axial stress meets Piggott's model in the bonding segment. It is affected by linear friction shear stress in the debonding segment and remains a constant in the bridging segment. In the following text, the interfacial stress transfer and failure conditions of the bridging fiber are considered to be reloading.

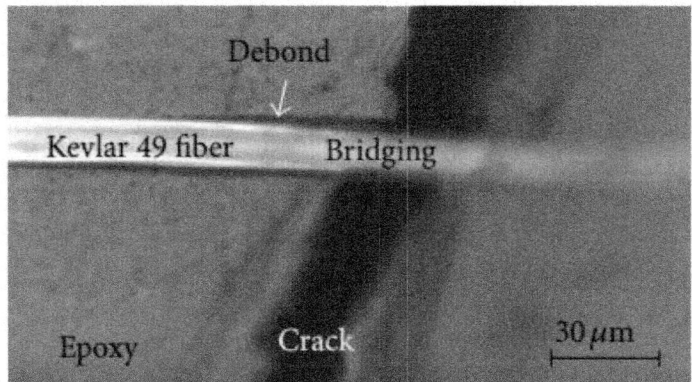

Figure 8: Bridging fiber and interfacial debonding during crack opening [16].

Slip Transform and Reloading

For the case of unloading after the formation of bridging fiber, a reverse slip will occur on the debonding segment and the fiber retraction results in residual interfacial friction stress, as shown in Figure 9(a). When the bridging fiber is reloaded, the partial slip on the debonding segment inverses its sliding direction. This will cause the different effects of interface friction force on the debonding segment, as shown in Figures 9(b) and9(c).

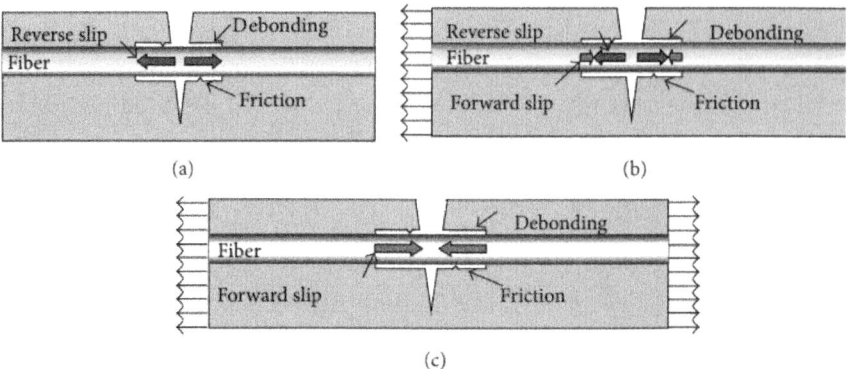

Figure 9: Slip transform for bridging fiber after reloading [16].

The reverse slip happens on the fiber debonding segment before reloading (Figure 9(a)), generating the interfacial friction in the opposite direction and resulting in compressive residual stress in the debonding segment. When the load is applied again (Figure 9(b)), the partial reverse slip on the debonding segment transforms to the forward slip resulting in the reduction of reverse slip length until all reverse slip completely converses to the forward slip (Figure 9(c)). The interfacial friction in the forward slip region makes the increase of fiber stress; on the contrary, the reverse slip results in the decrease of fiber stress. It is noted that the fiber stress remains constant in the bridging segment.

Stress Transfer Model

Raman measurements along the bridging fiber in Figure 8 gave a symmetrical axial stress distribution, as shown in Figure 10. The fiber axial stress is increased with the applied load. The stress platform is close to the bridging segment and the interfacial friction force in the slip segment should be overcome.

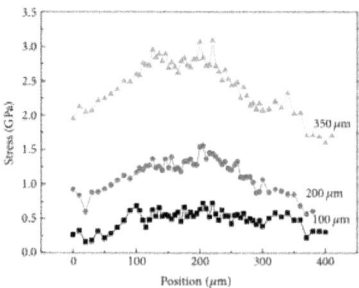

Figure 10: Stress distributions on the bridging fiber under different loads [16].

During the reloading of bridging fiber, the reverse slip in debonding segment gradually transformed into the forward slip so that the debonding fiber reloaded until the fiber stress eventually reached the maximum in the bridging segment. In fiber bridging segment, the ISS is zero due to the constant fiber stress. Setting a positive constant $\overline{\tau}$ of the interfacial friction shear stress and the fiber stress σ_m at the bonding/debonding transition point, a stress transfer model for the bridging fiber is shown in Figure 11.

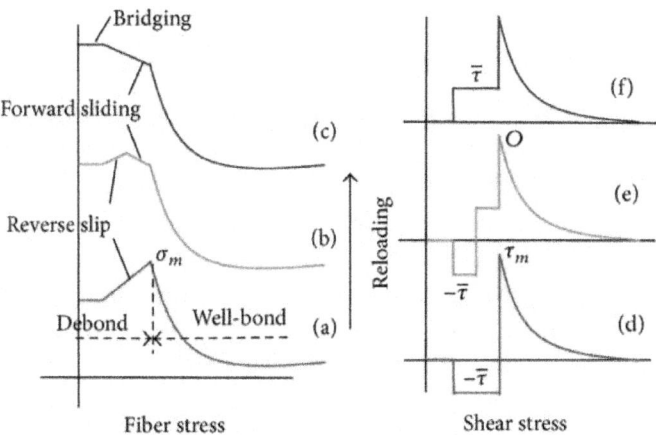

Figure 11: Stress transfer model of bridging fiber under reloading.

In the fiber debonding segment, the interfacial frictional shear stress (Figures 11(d)–11(f)) is a constant; namely.

$$\tau = \pm\overline{\tau},$$

(9)

,where the interfacial frictional shear stress takes a positive sign in the forward slip zone and a negative sign in the reverse slip zone. The fiber stress in the debonding segment meets a linear distribution of the interfacial friction shear stress as

$$\sigma = \sigma_m + 2\overline{\tau}\frac{x_1}{r} - 2\overline{\tau}\frac{x_2}{r},$$

(10)

Where r is the fiber radius, x_1 and x_2 are the forward slip length and the reverse slip length in the debonding segment, respectively.

At initial reloading stage (Figure 11(a)), the fiber stress in the debonding segment is reduced to overcome the reverse interface friction (Figure 11(d)) due to the whole debonding segment belonging to the reverse slip zone. By contrast, the fiber stressing the bridging segment keeps constant due to no interface frictional on the bridging segment.

At middle reloading stage, (Figure 11(b)), the partial reverse slip transforms into the forward slip (Figure11(e)). The fiber stress in the forward slip zone increases to overcome the positive interface friction so that the partial debonding fiber is reloaded.

At completely reloading stage (Figure 11(c)), the debonding fiber is completely reloaded due to the whole debonding segment belonging to the forward slip zone (Figure 11(f)).

It can be predicted that the debonding interface will continue to extend if the ISS at the bonding/debonding transition point reaches the interfacial shear strength. With the further increase of reloading, the fiber stress in the maximum stress plateau region will reach the fiber tensile strength so that the bridging fiber will fracture. This is the strength criteria for the bridging fiber.

For fibrous composites with stable interface, it can be seen from the above analysis that the physical and chemical nature of the interface determines the interface bonding ability, namely, the interface shear strength. The matrix crack across the fiber will cause interfacial debonding and form the bonding segment, debonding segment, and bridging segment. The fiber stress transfer among these segments has relationship with the interface bonding performance, interface friction, interfacial shear strength, and fiber strength. The balance between them determines whether the bridging fiber is stable or unstable. Once the balance is broken, the bridging fiber cannot exist stably and then transforms into the broken fiber [34]. The strength criteria for the bridging fiber can be used to explain the phenomenon that some of the debonding fibers can form a stable bridge but some of them break.

REMARKS

The interfacial mechanical design problems faced in fibrous composites elaborated from three ways of the material optimization, interface optimization, and computational optimization. The physical, chemical, geometric, and mechanical properties at microscale have a great impact on the interface behaviors. They are necessary to develop new experimental methods for reasonable evaluation on fiber/matrix interface by fine experimental testing and characterization to improve the interface micromechanical model. Micro-Raman spectroscopy was used to study main mechanical problems in fibrous composites, including the elastic stress transfer and failure criteria of well-bonding fiber, the frictional shear stress transfer behavior of partially debonded fiber, the slip transformation, and stress transfer models of bridging fiber during reloading. These works show that micro-Raman spectroscopy has ability to evaluate the stress transfer behavior of fiber/matrix interface.

CONFLICT OF INTERESTS

The authors declare that there is no conflict of interests regarding the publication of this paper.

ACKNOWLEDGMENTS

This work was supported by National Basic Research Program of China (no. 2014CB046506), National Natural Science Foundation of China (nos. 11172054, 11272232) and Fundamental Research Funds for the Central Universities (no. DUT14LK11).

REFERENCES

1. Baker, S. Dutton, and D. W. Kelly, Composite Materials for Aircraft Structures, American Institute of Aeronautics and Astronautics, Reston, Va, USA, 2 edition, 2004.

2. R. F. Gibson, Principles of Composite Material Mechanics, CRC Press, Boca Raton, Fla, USA, 3 edition, 2011.

3. R. M. Jones, Mechanics of Composite Materials, Taylor & Francis, Abingdon, UK, 2 edition, 1998.

4. Y. L. Kang, "Experimental analysis for some interfacial mechanics problems," Mechanical Engineering, vol. 21, no. 3, pp. 9–15, 1999 (Chinese).

5. W. Kim and J. A. Nairn, "Observations of fiber fracture and interfacial debonding phenomena using the fragmentation test in single fiber composites," Journal of Composite Materials, vol. 36, no. 15, pp. 1825–1858, 2002.

6. X. G. Yang and Q. L. Wu, Analysis and Application of Raman Spectroscopy, National Defense Industry Press, Beijing, China, 2008, (Chinese).

7. Y. L. Kang, Y. Qiu, Z. K. Lei, and M. Hu, "An application of Raman spectroscopy on the measurement of residual stress in porous silicon," Optics and Lasers in Engineering, vol. 43, no. 8, pp. 847–855, 2005.

8. Z. K. Lei, Y. L. Kang, H. Cen, M. Hu, and Y. Qiu, "Residual stress on surface and cross-section of porous silicon studied by micro-raman

spectroscopy," Chinese Physics Letters, vol. 22, no. 4, pp. 984–986, 2005.

9.   Z. K. Lei, Y. L. Kang, H. Cen, and M. Hu, "Variability on raman shift to stress coefficient of porous silicon," Chinese Physics Letters, vol. 23, no. 6, pp. 1623–1626, 2006.

10. W. Qiu, Q. Li, Z. K. Lei, Q. H. Qin, W. L. Deng, and Y. L. Kang, "The use of a carbon nanotube sensor for measuring strain by micro-Raman spectroscopy," Carbon, vol. 53, pp. 161–168, 2013.

11. W. Qiu, Y. L. Kang, Z. K. Lei, Q. H. Qin, Q. Li, and Q. Wang, "Experimental study of the Raman strain rosette based on the carbon nanotube strain sensor," Journal of Raman Spectroscopy, vol. 41, no. 10, pp. 1216–1220, 2010.

12. M. Nishikawa, T. Okabe, and N. Takeda, "Determination of interface properties from experiments on the fragmentation process in single-fiber composites," Materials Science and Engineering A, vol. 480, no. 1-2, pp. 549–557, 2008.

13. X. H. Wang, B. M. Zhang, S. Y. Du, Y. F. Wu, and X. Y. Sun, "Numerical simulation of the fiber fragmentation process in single-fiber composites," Materials and Design, vol. 31, no. 5, pp. 2464–2470, 2010.

14. Z. K. Lei, W. Qiu, Y. L. Kang, G. Liu, and H. Yun, "Stress transfer of single fiber/microdroplet tensile test studied by micro-Raman spectroscopy," Composites A, vol. 39, no. 1, pp. 113–118, 2008.

15. Z. K. Lei, Q. Wang, and W. Qiu, "Stress transfer of Kevlar 49 fiber pullout test studied by micro Raman spectroscopy," Applied Spectroscopy, vol. 67, no. 6, pp. 600–605, 2013.

16. Z. K. Lei, Q. Wang, and W. Qiu, "Micromechanics of fiber-crack interaction studied by micro-Raman spectroscopy: bridging fiber," Optics and Lasers in Engineering, vol. 51, no. 4, pp. 358–363, 2013.

17. S. Zhandarov and E. Mäder, "Characterization of fiber/matrix interface strength: applicability of different tests, approaches and parameters," Composites Science and Technology, vol. 65, no. 1, pp. 149–160, 2005.

18. L. A. Carlsson, D. F. Adams, and R. B. Pipes, Experimental Characterization of Advanced Composite Materials, CRC Press, Boca Raton, Fla, USA, 3 edition, 2002.

19. Galiotis, A. Paipetis, and C. Mansion, "Unification of fibre/matrix interfacial measurements with Raman microscopy," Journal of Raman Spectroscopy, vol. 30, no. 10, pp. 899–912, 1999.

20. G. Anagnostopoulos, J. Parthenios, and C. Galiotis, "Thermal stress development in fibrous composites," Materials Letters, vol. 62, no. 3, pp. 341–345, 2008.

21. R. J. Day and J. V. C. Rodrigez, "Investigation of the micromechanics of the microbond test," Composites Science and Technology, vol. 58, no. 6, pp. 907–914, 1998.

22. Z. K. Lei, Q. Wang, Y. L. Kang, W. Qiu, and X. M. Pan, "Stress transfer in microdroplet tensile test: PVC coated and uncoated Kevlar-29 single fiber," Optics and Lasers in Engineering, vol. 48, no. 11, pp. 1089–1095, 2010.

23. G. P. Tandon and N. J. Pagano, "Micromechanical analysis of the fiber push-out and re-push test,"Composites Science and Technology, vol. 58, no. 11, pp. 1709–1725, 1998.

24. R. Sinclair, R. J. Young, and R. D. S. Martin, "Determination of the axial and radial fibre stress distributions for the Broutman test," Composites Science and Technology, vol. 64, no. 2, pp. 181–189, 2004.

25. M. J. Pitkethly, J. P. Favre, U. Gaur et al., "A round-robin programme on interfacial test methods,"Composites Science and Technology, vol. 48, no. 1–4, pp. 205–214, 1993.

26. B. L. Zheng and X. Ji, "Stress singularity analyses of interface ends in micro-mechanics tests,"Composites Science and Technology, vol. 62, no. 3, pp. 355–365, 2002.

27. X. Ji, Y. Dai, B. L. Zheng, L. Ye, and Y. W. Mai, "Interface end theory and re-evaluation in interfacial strength test methods," Composite Interfaces, vol. 10, no. 6, pp. 567–580, 2003.

28. H. Cen, Y. L. Kang, Z. K. Lei, Q. H. Qin, and W. Qiu, "Micromechanics analysis of Kevlar-29 aramid fiber and epoxy resin microdroplet composite by Micro-Raman spectroscopy," Composite Structures, vol. 75, no. 1–4, pp. 532–538, 2006.

29. L. R. Xu, H. C. Kuai, and S. Sengupta, "Free-edge stress singularities and edge modifications for fiber pushout experiments," Journal of Composite Materials, vol. 39, no. 12, pp. 1103–1125, 2005.

30. S. Zhandarov and E. Mäder, "Characterization of fiber/matrix interface strength: applicability of different tests, approaches and parameters," Composites Science and Technology, vol. 65, no. 1, pp. 149–160, 2005.

31. H. L. Cox, "The elasticity and strength of paper and other fibrous materials," British Journal of Applied Physics, vol. 3, no. 3, pp. 72–79, 1952.

32. M. R. Piggott, Load Bearing Composites, Pergamon Press, Oxford, UK, 1980.

33. J. A. Nairn, "On the use of shear-lag methods for analysis of stress transfer in unidirectional composites," Mechanics of Materials, vol. 26, no. 2, pp. 63–80, 1997.

34. Z. K. Lei, Q. Wang, and W. Qiu, "Micromechanics of fiber-crack interaction studied by micro-Raman spectroscopy: broken fiber," Optics and Lasers in Engineering, vol. 51, no. 9, pp. 1085–1091, 2013.

Chapter 3

TRIBOLOGY AND MICROMECHANICS OF CHROMIUM NITRIDE BASED MULTILAYER COATINGS ON SOFT AND HARD SUBSTRATES

Juergen M. Lackner[1,], Wolfgang Waldhauser[1], Lukasz Major[2] and Marcin Kot[3]

[1]JOANNEUM RESEARCH Forschungsgesellschaft mbH, Institute of Surface Technologies and Photonics, Functional Surfaces, Leobner Strasse 94, A-8712 Niklasdorf, Austria

[2]Polish Academy of Sciences, Institute of Metallurgy and Materials Sciences, Ul. Reymonta 25, 30-059 Krakow, Poland

[3]AGH University of Science and Technology, Faculty of Mechanical Engineering and Robotics, Laboratory of Tribology and Surface Engineering, A. Mickiewicza Ave. 30, 30-059 Krakow, Poland

ABSTRACT

The tribological protection of carbon fiber reinforced epoxy composites (CFC) is essential for broadening their use from structural to functional applications, e.g., to linear bearings in mechanical engineering. However, their wear resistance in sliding and rolling contacts is low. This work focusses on the possibility of improving their tribological properties by the application of thin hard multi-layered coatings. Chromium nitride (CrN) single layer and chromium-CrN multilayer coatings of ~4 μm thickness, partly finished with a 1 μm diamond-like carbon (DLC) top layer, were deposited by magnetron sputtering at low temperatures on soft CFC and for comparison of the mechanical behavior on comparatively hard austenitic steel substrates. Structural investigations showed especially that the multilayer coatings possess a very fine grained, columnar microstructure and a very low density of intercolumnar micro-cracks, while the single layer coatings possess a coarse structure. The indentation testing and the analysis of the deformed and fractured cross-sections revealed a tougher behavior with improved plastic deformability of the multilayers in comparison to CrN single layers. However, in wear testing only coatings with DLC top layers significantly improved the tribological material properties of CFC. This is due to the reduced shear forces in sliding on low-friction DLC coatings on the soft epoxy-based CFC, decreasing the total dynamic stresses during sliding under high loads.

INTRODUCTION

High strength, fiber reinforced polymers like carbon fiber composites (CFC) are state-of-the-art ultra-high strength light-weight construction materials, which are fabricated in highly complex shapes and widely used as structural components in automotive, aerospace, medical technology, and mechanical engineering. Nevertheless, their functional performance in terms of abrasion, sliding, and impact wear resistance is low, limiting their use under higher tribological strains. Additionally, replacing steel components by light-weight CFC of similar mechanical bulk strength often requires an effective surface wear protection.

Besides nowadays in used constructions combining functionally strained metal and ceramic components with CFC, the coating of the polymer composite is seen as a mechanically and economically favorable option for the future. Thick coatings can be obtained from electrochemical and spray technologies, which guarantee high load support but lack in adhesion during overloading due to high stiffness and low plastic deformability for the following substrate deflection [1,2,3]. Mechanically, this phenomenon is described based on bending stresses [4]: During mechanical loading, brittle ceramic coatings crack under the bending stresses generated by elastic or plastic deformation of the subjacent soft substrate. The thicker such hard ceramic coating on such a deformable substrate are, the higher the bending stress is. Alternatively, thick soft polymer coatings (pure and micro-/nanoparticle strengthened lacquers) possess high elasticity following substrate deflection, but their tribological resistance is not sufficiently high [5].

Thin films of materials combining hardness and wear resistance with high compliance as well as high toughness are future candidates for the tribological protection of functionally-used polymers [6,7,8]: Single layer hard coatings like chromium nitride (CrN) on CFC substrates, presented by Kääriäinen et al. [9], do not fulfill the high demands for tribological protection: The problems, stated by the authors, were both the very inhomogeneous and highly rough CFC surface and the need for bridging high to low elasticity from polymer to hard coating. Deposition of dense, continuous coatings with high adhesion to the CFC substrate was impossible. An additional problem, described and detailed by Lackner et al. [10], is the very different thermal expansion of composites and hard coatings, leading to stresses, wrinkling and buckling phenomena and finally film delamination. Consequently, the development of coatings for CFC substrates must fulfill all these major demands: (1) Sub-millimeter sized inhomogeneous surface topography and structure require new CFC manufacturing methods for ultra-smooth CFC types. (2) Elasticity and toughness of hard coatings must be based on multilayered structures, which have generally higher resistance to cohesive and adhesive crack propagation compared to single layered coating types [11,12]. (3) Effective temperature control methods during film deposition as well as limited deposition rates are essential to prevent thermal damage (e.g., degradation) of the CFC surface as well as any thermal stress induced effects in thin film growth.

This work focusses on multilayered coating structures on soft, flexible CFC composites substrates in comparison to comparatively hard and stiff austenite steel: While the majority of commercially available state-of-the-art wear resistant coatings consists of one or a couple of single hard layers with high sensitivity to through-thickness fracture [11,12], the placement of softer layers in between hard ones may allow arresting the propagation of cracks at the internal interfaces by energy dissipation and crack deflection [13,14]. The improved toughness of such coatings is furthermore due to the nano-scale shearing of the hard layers on the softer (inter-)layers, preventing the build-up of high-bending stresses [13,14,15]. Such a material combination can simultaneously increase hardness and reduce the intrinsic growth stresses, which results in markedly improved wear resistance [16,17]. The coating materials in this work, chromium (Cr)/chromium-nitride (CrN) multilayers with diamond-like carbon (DLC, a-C:H) top layers, were manufactured by unbalanced magnetron sputtering at low temperatures to prevent damage of the substrate materials due to mechanical distortion or chemical de-polymerization. To show and discuss the improvements and their physical origins, single layer coating types as well as coating structures without DLC top layers were also investigated. High-resolution transmission electron microscopy (HR-TEM) techniques were finally used to describe the mechanisms of the multilayer deformation and for improving the tribological behavior in contact to Al_2O_3 and AISI 5210 steel counterparts.

EXPERIMENTAL SECTION

Coating Deposition

Before the deposition started, the CFC (carbon fiber composite with high-modulus carbon fibers in epoxy matrix), and polished austenitic steel substrates (DIN EN 1.4301, AISI 304) were cleaned ultrasonically by ethanol and dried. After mounting on the substrate carousel, the vacuum chamber was pumped down to the start pressure for deposition (2×10^{-3}Pa). Plasma etching by an anode layer ion source (ALS [18]) was applied to remove micrometer sized contaminations and to chemically activate the composite surface in an oxygen atmosphere. Unbalanced magnetron sputtering in an industrially-scaled, 4 rectangular cathode vacuum chamber was chosen to deposit Cr, CrN, and DLC coatings from high purity chromium (99.99%, RHP Technology GmbH, Seibersdorf, Austria) and electrographite carbon targets (99.9%, Schunk Group, Bad Goisern, Austria), respectively. The following coating architectures were developed based on former works [19,20]:

(1)   single layer Cr (3.94 µm thickness);

(2)   Cr-CrN multilayer with totally 16 Cr-CrN bilayers and a modulation ratio CrN:Cr = 2:1 (82 nm Cr + 168 nm CrN) (total thickness 3.99 µm);

(3)   Cr-CrN multilayer with totally 32 Cr-CrN bilayers and a modulation ratio CrN:Cr = 2:1 (42 nm Cr + 84 nm CrN) (total thickness 4.18 µm);

(4)   Cr-CrN multilayer similar to (3), but with a 1 µm thick DLC top coating (thickness 5.16 µm).

Cr as well as DLC deposition occurred in inert argon process gas. For CrN a mixture of Ar and N_2 was used in reactive deposition. A bias voltage of −50 V DC was applied to increase the energy density on CFC substrates. The DC magnetron power was controlled to prevent too high substrate heating exceeding the low thermal stability of the epoxy matrix, degrading above ~140 °C. To provide a homogenous film thickness of the whole coated surface and to simulate industrial batch coating deposition conditions, the substrates were rotated during ALS pre-treatment and coating.

Coating Characterization

Scanning electron microscopy (SEM, EVO 50, Zeiss, Oberkochen, Germany) was used to study the growth structure on the cross-sections and the surface morphology. Cutting with an ATM Brillant 221 system was performed up to 2/3 of the sample thickness from the backside, for the sample preparation. Fracturing of the coated CFC substrates occurred manually by tensile loading with pincers. For increasing the conductivity during microscopy imaging, the specimens were evaporated with gold.

The micro- and nanostructure of the coatings were analyzed by the application of transmission electron microscopy (TEM, TECNAI G2 F20 (200 kV FEG), Hillsboro, OR, USA) after focused ion beam based cross-section sample preparation (Quanta 200 3D DualBeam microscope equipped with in-situ OmniProbe micro manipulator, gallium ions for milling). Electron diffraction patterns in high resolution mode (HR) were used for structural analyses, while the bright field (BF) technique was applied for imaging of the coating architecture and of the fracture behavior and mechanisms below the spherical indents. Structural investigations were performed on a Bruker AXS D8 Advance diffractometer with CuKα radiation, equipped with a Sol-X detector and a Göbel mirror. The Bragg-Brentano scanning geometry (locked couple) was applied between 20° and 80° with a step width of 0.02° and 1.2 s measurement duration per step, whereby comparative measurements to the coated CFC substrates were performed with coated silicon.

The film thickness on masked steps, as well as the arithmetic roughness (R_a), were measured by stylus profilometry (Dektak 150, Veeco, Santa Barbara, CA, US). The hardness (H) and the elastic modulus (E) of the coatings were obtained from nanoindentation testing with a Vickers indenter on a Fisherscope H100C (Helmut Fischer GmbH, Sindelfingen-Machingen, Germany) device according to DIN EN ISO 14577-1:2002 [21]. The applied maximum load was 50 mN, the loading rate 20 nm·s^{-1} for all measurements, which prevented any high substrate compliance influence on the coating properties (indentation depth <10% of film thickness). The indentation results were analyzed afterwards using the Oliver and Pharr theory [22] in order to get H and E. Cone shaped diamond indenters with 20 μm tip radius were used for the indentation at 2 N normal force to achieve the desired fracture in the single and multilayer coatings by elastoplastic substrate deflection. Subsequent analysis of the deformation and coating fracture mechanisms was done by TEM and HR-TEM.

The dry friction at room temperature (21 °C) and at relative humidity of 50%–60% was evaluated on a ball-on-disc tribometer. 6 mm alumina (Al_2O_3) and hardened ferritic steel balls (AISI 5210, DIN ISO 100Cr6) were used as counterparts, sliding at wear tracks of 3.5 and 5 mm diameter, respectively. The applied load was 2 N, resulting in a maximum Hertzian contact pressure of 0.9 (Al_2O_3) and 0.8 GPa (AISI 5210). The sliding speed was set to 10 mm·s^{-1} and 10.000 laps (110 m and 157 m sliding distance, respectively) were tested. All these experiments were carried out on the as-deposited, clean surfaces. After testing, the wear tracks were inspected by optical light microscopy and profilometry to explain the wear mechanisms and analyze the wear profiles.

RESULTS AND DISCUSSION

Film Topography and Microstructure

The microstructure and surface topography of the single layer CrN and multilayer Cr-CrN coatings on CFC substrates is shown in Figure 1. The CrN single layer coating in Figure 1a reveals a columnar structure of tapered crystals and fibrous grains in the cross-section, which is assigned to the "Zone 1" structure in the Thornton structure model [23]. This structure type is generally found for PVD coatings deposited at low temperatures and characterized by micro-cracks between the columns, pinholes, transient grain boundaries and through-coating porosity. This structure type merges at higher ion bombardment during deposition and thereby activated surface diffusion to a transition "Zone T" of fibrous grains with highly decreased porosity. Such refinement of grains is also visible for the multilayer Cr-CrN coating in Figure 1b, although the ion bombardment is rather similar to the CrN single layer coating. However, the necessary renucleation at the Cr-CrN and CrN-Cr layer boundaries in the multilayer coating during growth has similar effects on structural refinement: It decreases grain size and, thus, porosity at loosely bond cone boundaries. As expected, the DLC layer on the top (Figure 1b) is amorphous without distinct columnar growth features.

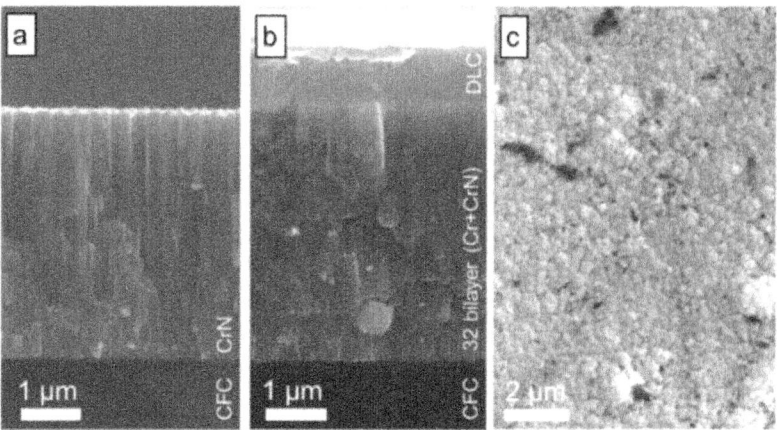

Figure 1: Growth structure and surface topography of magnetron sputtered coatings grown on fiber reinforced epoxy composites (CFC): Scanning electron microscopy (SEM) cross-sections of (**a**) ~4 μm CrN single layer coating and (**b**) ~4 μm 32 bilayer (Cr + CrN) coating with ~1 μm diamond-like carbon (DLC) top layer; (**c**) SEM top view of the coating from (b).

Compared to well-known coating topographies on polished steel substrates (e.g., shown in [24]) with fine spherical tops of the tapered crystallites, the topography of the coated CFC is much coarser and substantially different (Figure 1c): This is due to the about 30 times higher surface roughness of CFC compared to steel, which decisively influences the growth morphology (compare also to detailed descriptions in [10]) and reproduces the CFC substrate topography on the film surface. As visible with the decrease of Ra during the coating of CFC, the high roughness becomes smoother (Table 1), decreasing the roughness for the 32 (Cr-CrN) bilayer coating on CFC to a value, which is only 10 times that for the similar coating on polished austenitic steel.

Table 1: Arithmetic roughness (R_a) of uncoated substrates (carbon fiber reinforced epoxy composites (CFC), polished austenitic steel) and with deposited single layer and multilayer coatings

Material/Coating	Roughness R_a [nm]	
	CFC	Steel
Uncoated substrate	75.2	2.4
CrN single layer	59.0	6.4
16 bilayers (Cr-CrN)	64.6	8.9
32 bilayers (Cr-CrN)	68.2	6.7
32 bilayers (Cr-CrN) + 1 μm DLC	109.3	15.1

TEM imaging of the growth structures of the Cr-CrN multilayers on steel substrates reveals important effects dependent on the thickness of the individual layers of the multilayers (Figure 2): The about 50% thicker Cr layer directly on the substrate surface shows much larger facetted, columnar and coarsened Cr grains than the second deposited thinner Cr layer after the first CrN layer (Figure 2a), which was afterwards repeatedly used in the multilayer stack. In addition, the density of formed microcracks in between the grains is higher for the thicker Cr layer. As visible in Figure 2b, no substantial coarsening in the subsequently deposited Cr-CrN multilayer occurs. The microcrack-like structures in the first Cr layer do not pass the interface to CrN and are only rarely found in the second Cr layer. Generally, the Cr grains were found to be larger (15–20 nm) than the CrN grains (5–10 nm size) (Figure 3a), although the Cr layers are thinner than the CrN layers (CrN/Cr ratio = 2:1). This seems to be due to the more complex fcc CrN structure (two overlapping NaCl lattices for Cr and for N) compared to single atom bcc Cr structure [25]. Crystallographic dependence is evident at the interface: The small mismatch of lattice constants of CrN (a = 4.14 Å) and Cr (a = 4.588 Å) allows stress compensation during nucleation by introduction of step dislocations (Figure 3c) in certain crystallographic orientations (e.g., (Cr (001) ∥ CrN (001) and Cr [100] ∥ CrN [110] [26]) (Figure 3b,d).

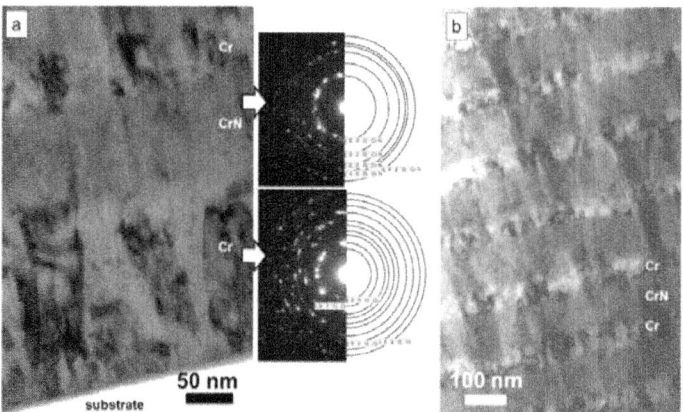

Figure 2: Transmission electron microscopy (TEM) cross-section images of a 32 bilayer (42 nm Cr + 84 nm CrN) coating on steel substrate: (**a**) multilayer film growth behavior close to the interface to the substrate; (**b**) growth close to the coating surface.

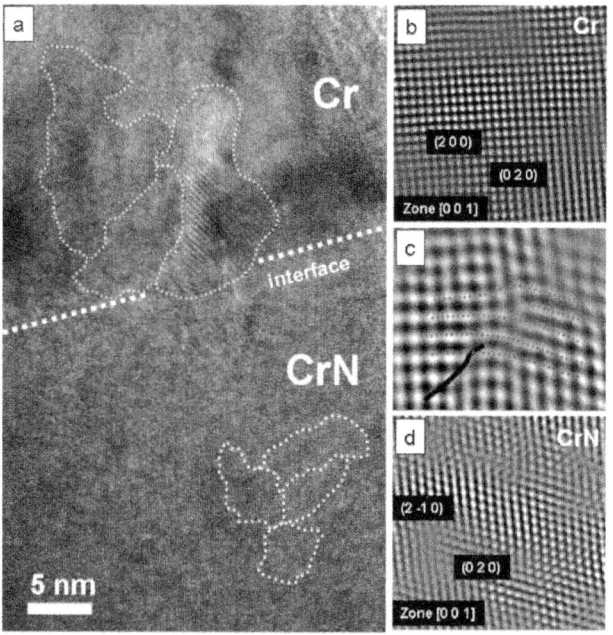

Figure 3: High resolution mode (HR)-TEM cross-section images of the interface growth mechanisms of a 32 bilayer (42 nm Cr + 84 nm CrN) coating on steel: (**a**) Overview indicating the grain structure and grain sizes; (**b**) Crystallographic analysis of the Cr layer; (**c**) Interface growth with dislocation formation; (**d**) Crystallographic analysis of the CrN layer.

The XRD phase analysis in Figure 4, comparing the coatings deposited on steel and CFC substrates, reveals high similarities in phases and orientations, although the substrate roughness is highly different: The Cr (200) peak of the coatings is more distinct on the steel than on the CFC substrate, which is connected to a higher phase content (higher crystallinity) and/or a larger Cr grain size. The peaks for all chromium nitride compounds (fcc CrN, hex β-Cr$_2$N [25]) are rather weak for the multilayer coatings, but they dominate the XRD spectra of the single layer with the (200) CrN peak. Only weak and broad (nanocrystalline to amorphous) diffraction is present for the β-Cr$_2$N phase: Nevertheless, it is stronger in (112) orientation for coated steel, but in (111) and (002) orientation for coated CFC substrates. The Cr (110) peak is only found for the films on CFC, but missing for films on steel, as also revealed by the diffraction pattern of the Cr interface layer in Figure 2a. The higher resolution in TEM diffraction patterns reveals further crystallographic phase features, not occurring in the XRD scans: strong (211) as well as weak (530) and (520) diffraction for Cr and weak (420) diffraction for CrN. The slight shift of the Cr (200) peak to lower diffraction angles could indicate tensile stresses in this phase.

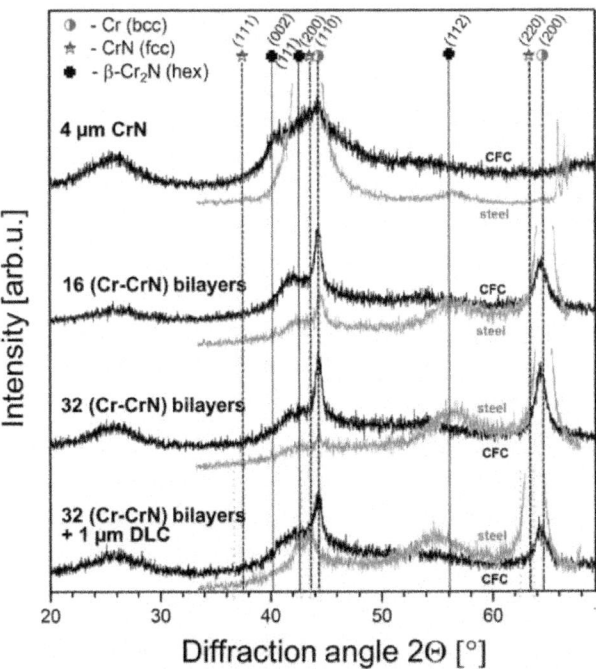

Figure 4. Locked-couple X-ray diffraction scan of single CrN layer and (Cr + CrN) multilayer coatings on CFC (black spectra) and on steel (grey spectra) including as-

signment of peaks to standard peak positions from Powder Diffraction File (ICDD) (Cr (bcc): 00-027-1402, CrN (fcc): 03-065-2899, β-Cr$_2$N: 00-035-0803).

Mechanical Properties

The multi-layered structure directly influences the hardness and the elastic modulus of the coatings. Highest hardness on austenitic steel is evident for CrN single layer coatings (Table 2), while layering with Cr results in hardness decrease towards the Cr single layer hardness of around 12 GPa. Nevertheless, the hardness as a measure of the resistance against plastic deformation increases with the nanolayered structure with 32 bilayers of (42 nm Cr and 84 nm CrN) although the total metallic Cr weight content is similar to the 16 bilayer structure (84 nm Cr and 168 nm CrN). The elastic modulus (190–200 GPa) as well as the hardness values (15–16 GPa for the Cr-CrN multilayer coatings) are quite comparable to the literature (e.g., to the work of Romero et al. [27]), if a lateral contraction coefficient of 0.25–0.30 is assumed.

The described hardness behavior follows the Hall-Petch hardening relationship, which is theoretically based on the decrease in shearing capacity by dislocation pinning at the interfaces [26]. In contrast, the elastic modulus of these multilayers roughly follows the rule of mixture between Cr (~140 GPa) and CrN (~215 GPa) [28], being quite similar for both multilayer coatings with different bilayer modulation thickness. Adding the 1 μm DLC top layer shifts hardness and elastic modulus towards the levels of single layer DLC films (H ~ 14.5 GPa, E ~ 160 GPa) due to the high enough load support for surface indentation and the low indentation depth (~0.35 μm).

Table 2: Hardness and elastic modulus of single CrN layer and multilayer (Cr-CrN) coatings, determined on steel substrates

Coating	Hardness [GPa]	Elastic modulus [GPa]
CrN single layer	17.8 ± 0.4	217 ± 4
16 bilayers (Cr-CrN)	15.0 ± 0.3	198 ± 4
32 bilayers (Cr-CrN)	16.0 ± 0.2	191 ± 3
32 bilayers (Cr-CrN) + 1 μm DLC	14.3 ± 0.3	156 ± 9

Deformation Mechanisms of the Coatings

The deformation mechanisms of coated CFC were studied by spherical indentation with high loads resulting in penetration depths of up to 7.5 μm. Such deep indents are necessary to trigger (visco-)elastoplastic flow of the epoxy matrix-rich CFC as well as deformation and fracture of the coating (top view in Figure 5a, cross-section in Figure 5b), if the yield or fracture strength is exceeded, respectively. Nevertheless, the influence of the deeper carbon fibers on surface deformation is only minor at these penetration depths in the epoxy layer.

FEM simulations of von-Mises stress distributions [7] reveal highest stressed regions below the indenter center at the coating-substrate interface, but also outside the contact, due to bending of the layer with maximum tensile stress at the coating surface. This is graphically shown in Figure 5c,d: The schematics in Figure 5d describes the loaded region in the cross-section, while the effect of elastoplastic flow of material below the indent outside the contact area, resulting in circular and radial cracks of the bulged region, is schematically shown in Figure 5c [29,30]: Ring cracks around the indent spread from the coating surface at the periphery of the contact and spread towards the substrate surface. Radial cracks in mild indentation with low local stress concentration (ball geometry, in contrast to pyramid-shaped indenters [30]) mainly occur, if alternative fracture mechanisms at lower stresses for energy dissipation are missing and the circumferential tensile stresses become too high. In the investigated series of single and multilayer coatings, such radial cracks are only observed for the 32 bilayer Cr-CrN systems, both with and without the DLC top layer. Such Cr-CrN load bearing layers possess the highest elasticity index (H/E ratio) and, thus, may also dissipate deformation energy elastically.

Ring crack formation is the main failure mechanism for all other films. Brittle crack propagation may occur at the columnar grain boundaries along existing microcracks (compare to Figure 2a) in the microstructure of the non-deformed materials due to the shear sliding of adjacent growth columns, which minimizes the shear loading [31]. The TEM cross-section image in Figure 5b shows the ring crack mode for a tough multilayer coating: Similar to preliminary investigations on the Ti-TiN multilayer coating system with soft Ti interlayers [19], the introduction of soft metal layers (Cr) in CrN for Cr-CrN multilayer structures fully changes the micromechanics of the films: While the single layered hard coatings (TiN, CrN) would fail by cohesive fracture through the whole coating thickness, such ring cracks in a multilayered film can be deflected and can change the propagation direction. The transition from tensile to shear fracture in the 32 Cr-CrN bilayer coating, is clearly visible in two instances in Figure 5b by the deflection of cracks from the direction normal to parallel to the substrate surface. After the deflection the crack propagation is arrested in the parallel shear crack.

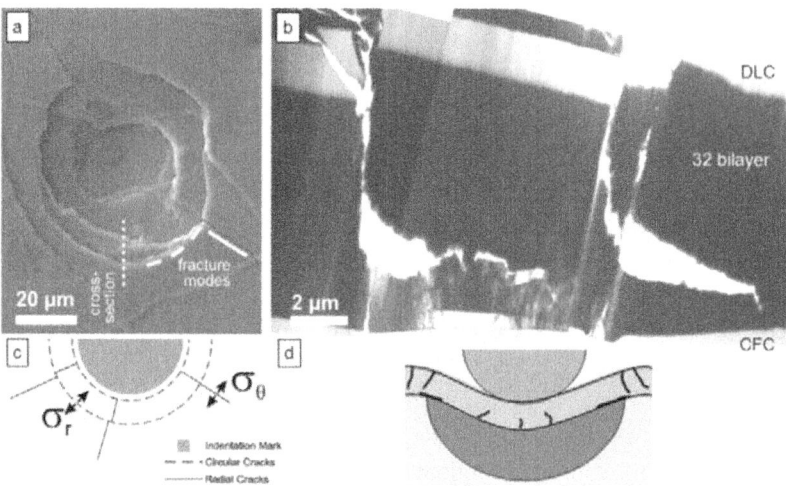

Figure 5. (a): SEM top view of an indent by 2 N loading of a spherical indenter in 32 bilayer (Cr + CrN) and 1 μm DLC top layer coated CFC; **(b)** TEM cross-section prepared by Focussed Ion Beam (FIB) from the marked position in (a) indicating deformation and fracture; **(c)** Schematics of formation of circular/ring cracks (σ_r) and radial cracks (σ_θ); **(d)** Schematics showing plastically deformed region under a ball indent in a coated compound and localization of crack initiation/formation.

In load-depth curves obtained in instrumented indentation experiments, any coating fracture is observed as kink: Our experiments showed that first cohesive coating failure occurs for the 32 bilayer multilayer coating at ~1.2 μm penetration depth (~100 mN load). For the single layer CrN coating as well as for the 16 bilayer coating, kink formation is found at much lower loads (<60 mN) and at only 0.75 μm penetration depth. In comparison, first kink formation for the 32 bilayer coating on stiffer, less compliant austenitic steel substrates is found at ~850 mN load (~2.8 μm penetration depth).

The differences of indentation micromechanics for coatings on soft and hard substrates are very apparent from the above results for very soft epoxy CFC surfaces (E: 2–3 GPa, H < 0.1 GPa) (Figure 5) and stiffer substrate materials, e.g. austenitic steel (E: 210 GPa, H: 1.5 GPa) (Figure 6): Higher load support by the steel substrate decreases total deformation and, thus, the crack density of both radial and ring cracks on the surface (Figure 6a). Nevertheless, if the load bearing capacity of the coating is exceeded, fracture occurs and leads to localization of stresses on the substrate surface, which shear locally and introduces shear steps, which is well visible at the interface (Figure 6b). As revealed by the small crack on the very left side of Figure 6b, where crack propagation is stopped by the multi-layered architecture (interfaces), fracture in this investigated region starts from the interface. A distinct shear step is missing for this very small cracking event, because the stopped crack propagation prevented sliding of the crack flanks. Nevertheless, tensile stresses on the coating surface at higher distance to the indentation center also cause ring-like cracks after high substrate deformation (Figure 6a). These ring cracks are partly deflected to shear cracks (Figure 6b, on the very right). The detail at the crack flank in Figure 6c and under high resolution in Figure 6d show in detail the mechanism of crack stopping in the multilayer: Shear deformation of the coating is concentrated in 45° planes of Cr layers, while the CrN layers show brittle fracture (fracture normal to the substrate surface).

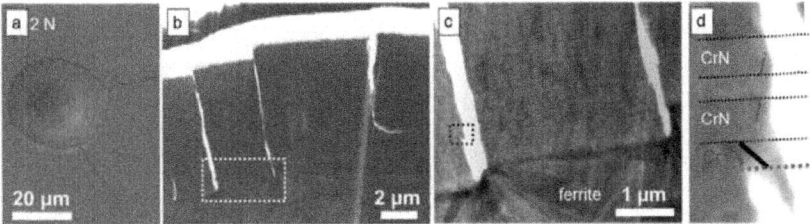

Figure 6: Deformation behaviour of Cr-CrN multilayer coatings with 32 bilayers (CrN:Cr ratio = 2:1), deposited on austenitic steel and substrates. (**a**) Surface topography after indentation with 2 N load; (**b**) TEM cross-section of the deformed/fractured indent site; (**c,d**) Details of deformation and fracture as marked in (b) and (c).

Finally, the comparison of the micromechanical results with data obtained from scratch testing reveal for all coatings high adhesion on all substrate materials, while the cohesion of the coating is mainly dependent on the used layer architecture and the substrate type (mechanical behavior): Cohesive coating failure is present for the multilayer coated CFC at critical load L_{c1} ~0.5 N and for austenitic steel at ~1 N. Adhesive failure (L_{c2}) is missing for all systems even at maximum load of the scratch tester (30 N) and for ~50 μm deformation depth.

Tribological Behavior of Coated CFC

Strains in tribological contacts are generally highly dynamical and fatigue plays an important role, even below the static yield strength. In coated systems, the compliance of the substrate decisively influences friction and wear, because bending both below and around the contact area introduces alternating tensile and compressive stresses in the substrate surface and coating. In the case of missing friction (friction coefficient $\mu = 0$), the maximum von-Mises stress is ~35% higher for the case of a stiff coating layer (E_1) on an elastic substrate ($E_2 = \frac{1}{2} E_1$) than in the non-coated case. However, friction plays a decisive role, affecting the tangential forces and, thus, the shear stresses in the coating-substrate system [31]. In the theory of pure indentation or at zero friction (μ = 0), yielding in the coated surface always initiates at the coating/substrate interface below the center of the contact (highest von-Mises stresses), and the plastic zone does not grow towards the surface below the indenter [32,33]. Applying friction load (μ > 0), the point of first yield moves from the center position backwards or forwards, depending on the friction coefficient [33]: For μ = 0.25, the highest stressed region moves closer to the surface compared to the non-coated case, being ~40% higher as non-coated and ~50% as coated surface and μ = 0. These values increase to ~65% and 125% for μ = 0.5, respectively.

Based on these theoretical simulation model results it is clear, that the superposition of large elastic deformation of the coated elastic substrate from normal loading and the friction induced shear loads easily leads to coating fracture and fatigue influenced phenomena like fragmentation and delamination, drastically influencing the tribological behavior.

Figure 7 and Figure 8 show the evolution of friction coefficients in contact to Al_2O_3 and AISI 5210 steel counterpart balls in dependency of the contact lap number. For Al_2O_3 counterparts (Figure 7), the friction coefficient of all coatings is initially very low (μ ~ 0.08). Ongoing sliding leads for the Cr-CrN multilayers to a sudden increase of friction up to 0.9, which slowly decreases to 0.6, which is the steady-state friction coefficient for uncoated CFC against Al_2O_3. In contrast, the friction for the pure CrN coating as well as the multilayer with DLC top layer steadily increases from the initial level of 0.1–0.6, which is the value for uncoated CFC. The friction for the contact to AISI 5210 steel (Figure 8) is initially lower, but after the run-in generally higher for the CrN-based coatings without a DLC top layer (μ > 0.65) than for the non-coated CFC substrate (μ ~ 0.5). Only the multilayer composition with the DLC top layer enables a lower steady-state friction (μ ~ 0.17) than is found for the non-coated contact, due to the well-known low friction coefficients of DLC-steel contacts.

A detailed analysis of the friction curves in Figure 7 and Figure 8 in combination with an analysis of the wear tracks inFigure 9 and Figure 10 is essential to understand the tribological mechanisms, which occur in sliding on the coated CFC surfaces: Based on friction stages proposed by Suh and Sin [34] the initial very low friction is mainly a result of ploughing the softer surface by asperities, starting for the investigated coatings on CFC with μ = 0.08 and 0.045 for Al_2O_3 and steel counterparts, respectively. Based on the surface hardness, this ploughing mechanism by asperities occurs in contact to Al_2O_3 on the softer coating surface, but oppositely on the steel ball surface for this tribological contact. Although the mechanisms occurring in the initial friction contact were not investigated in this work, mechanisms occurring in later stages (mainly the transfer layer formation) indicate such behavior. Nevertheless, the high waviness and roughness of the coated CFC (see Table 1) enhances this low friction mechanism.

Repeated cyclic ploughing deformation quickly results in low-cycle fatigue fracture of asperities, leading to the next stage in the frictional contact. Both plastic deformation and fracture at high stresses contribute to smoothing and polishing the wear track, whereby wear debris is trapped in grooves. Additionally, cyclic deformation occurs inside and around the contact area due to the high compliance of the epoxy matrix of the substrate. Radial and ring cracks form during overloading, shown by the static indentation experiments above. These cohesive cracks and their propagation to adhesive cracks and coating delamination contribute additionally to the specific behavior of sliding on coated CFC surfaces.

As visible in Figure 7 and Figure 8, the friction coefficient rapidly increases to >0.5 for the Cr-CrN multilayer coatings for both contact to Al_2O_3 and steel counterparts after the short run-in period, while friction coefficient for the CrN single layer hard coating in contact to Al_2O_3 as well as the coating types with DLC top layer are generally below the values for contact to uncoated CFC. The origin of this different behavior is the immediate formation of wear particles in sliding. Ploughing by wear particles results in high shear forces and, thus, high friction coefficients especially for coatings with the ability to shear like the multilayers with plastically deformable Cr layers.

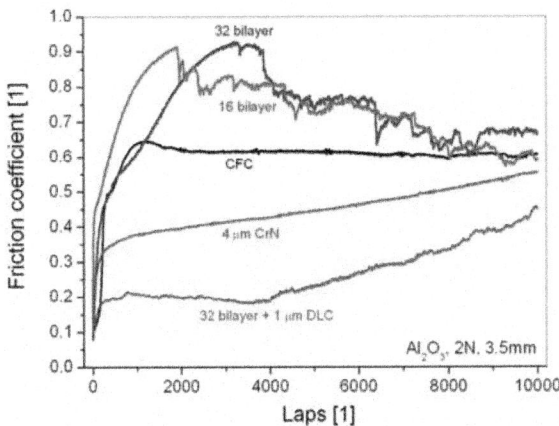

Figure 7: Friction coefficients depending on the lap number (# contact cycles) for the tribological contact of an Al_2O_3 ball (6 mm) in ball-on-disk testing with uncoated and

single and multilayer coated CFC under 2 N load and 3.5 mm wear track diameter (10,000 laps = 110 m sliding of pin).

These particles are entrapped between the sliding surfaces and may form deformable and sticky transfer layers (tribolayers), which represent the steady-state condition of friction. The friction coefficients remain constant till the substrate is exposed e.g., by ploughing by the formed wear particles or delamination of the coating from the substrate surface whereby the coating finally failed.

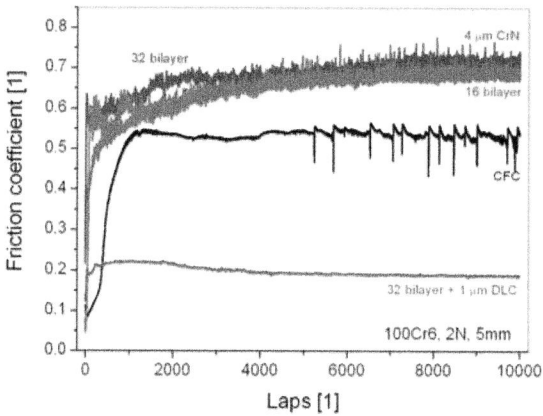

Figure 8: Friction coefficients depending on the lap number (# contact cycles) for the tribological contact of a DIN 100Cr6 / AISI 5210 steel ball (6 mm) in ball-on-disk testing with uncoated and single and multilayer coated CFC under 2 N load and 5 mm wear track diameter (10,000 laps = 157.1 m sliding of the pin).

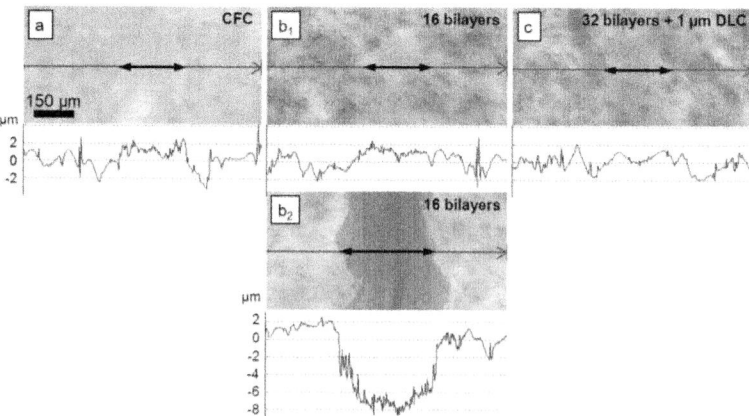

Figure 9: Light microscopy images and optical profilometry scans of the wear track after ball-on-disk testing with an Al_2O_3 ball (for experimental details see Figure 7) on

(a) uncoated CFC; (b) CFC coated with 16 bilayers (Cr-CrN) (b1: undamaged coating with transfer layer, b2: highly abraded coating and substrate, both found on the same wear track); and (c) CFC coated with 32 bilayers (Cr + CrN) and 1 μm DLC top layer. The long arrows indicate the profilometry scan lines, the short bold arrows the approximate width of the wear tracks.

Cr-CrN multilayer coating contacts to Al_2O_3 counterparts show very high friction coefficients of up to 0.94 in this steady-state regime (Figure 7). However, the shearing of metal Cr under cyclic loading, fatigue fracture, the formation of abrasive wear debris and transfer layer adhesion on the Al_2O_3 counterpart due to sticky wear debris results in quick partial failure of the coating (Figure 9b), baring the substrate surface. Transfer of epoxy polymer from these areas forms polymeric transfer layers (based on SEM EDS analysis) at the whole wear scar on ball and disc, which decreases the friction significantly. During sliding, frictional heat may chemically modify the bared polymer and the polymer transfer layers (introduction of atoms from ambient atmosphere, cross-linking and polymer chain scissoring, etc. [35]). The formation of a thick tribolayer (transfer layer) on the uncoated CFC indicates such a mechanism as the main contribution in the tribological behavior for contact to Al_2O_3 (Figure 9a). The repeated sudden changes of the friction coefficient during on-going sliding for the Cr-CrN multilayer coatings (Figure 7) indicate furthermore stepwise delamination and new formation of transfer layers.

The single layer CrN behaves differently in tribological contact to Al_2O_3 due to lack of shearing at softer Cr layers and lower adhesion of transfer layers of the worn coating on the counterpart. Furthermore, cracking in the coating around the contact area runs generally through the whole film thickness during overloading, whereby large coating fragments delaminate. This process starts very early after the run-in period, which is visible by a partly bared CFC substrate surface in the wear track. Missing adhesion of transfer layers on the Al_2O_3 keep the friction coefficient low. Nevertheless, ongoing grinding wear of CrN wear debris increases bared CFC substrate surfaces in the wear track and increases the friction coefficient towards the value measured for sliding of Al_2O_3 on uncoated CFC (Figure 7). Finally, wear of the partly bared substrate surface contributes to a transfer layer formation on both the CrN as well as on the counterpart.

In sliding of AISI 5210 counterparts on CrN and Cr-CrN coatings, a thick transfer layer is generally formed in the steady-state stage by mechanical mixing on the top of the CrN surface (Figure 10b,c). Based on SEM EDS analysis, this transfer layer consists of Fe, Cr, N, O and hydrocarbons. Its formation is mainly due to asperity fatigue mechanisms, in which the harder coating wears the softer steel counterpart. Nevertheless, the asperity fracture and coating delamination/spallation mechanisms do not stop after formation of the transfer layer, which is visible in the deep grooves down into the CFC substrate in the depth profile of the 16 bilayer coating (Figure 10b). In comparison, wear of uncoated CFC is mainly due to grinding by asperities of the much harder steel ball (Figure 10a).

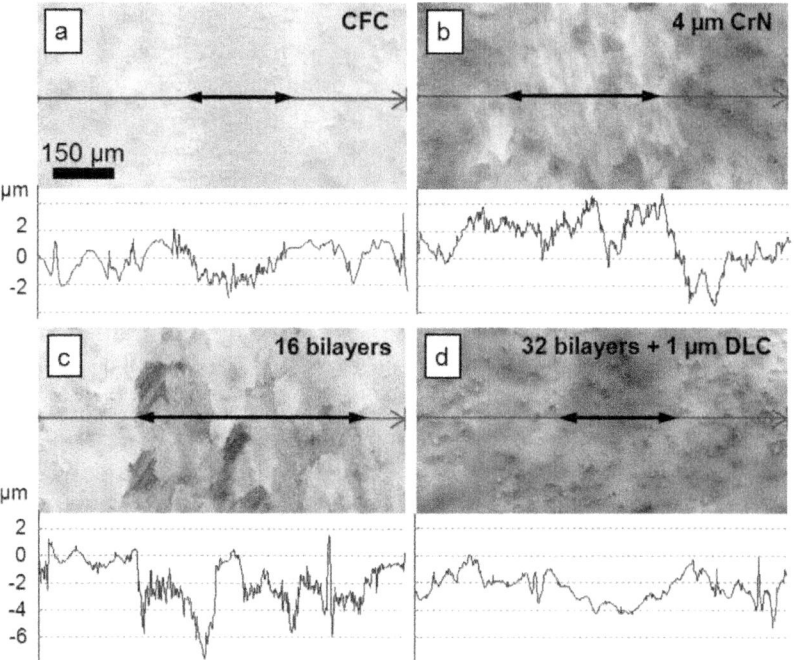

Figure 10: Light microscopy images and optical profilometry scans of the wear track after ball-on-disk testing with a DIN 100Cr6 / AISI 5210 steel ball (for experimental details see Figure 8) on (**a**) uncoated CFC; (**b**) CFC coated with ~4 μm CrN single layer coating; (**c**) CFC coated with ~4 μm 16 bilayer Cr-CrN multilayer coating; and (**d**) CFC coated with 32 (Cr + CrN) bilayers and a 1 μm DLC top layer. The long arrows indicate the profilometry scan lines, the short bold arrows the approximate width of the wear tracks.

Main tribological improvements can be achieved by low-friction DLC top layer both for the sliding of AISI 5210 steel and Al_2O_3 balls. The coated structure is not substantially damaged during the investigated sliding distance in all cases (Figure 9c and Figure 10d). This superior behavior is mainly due to the low friction coefficients and the decreased tribological stresses (low shearing component).

CONCLUSIONS

CrN single layer and Cr-CrN multilayer coatings of ~4 μm thickness, partly finished with a 1 μm DLC top layer, were deposited by magnetron sputtering at low temperatures on soft carbon-fiber strengthened composite substrates in order to increase their tribological resistance in sliding applications. All multilayer coatings possess a very fine grained, columnar microstructure and have much less severe intercolumnar microcracks than the CrN single layer coating. The coatings on CFC have dome-shaped topography and reproduce the high CFC substrate roughness. The elastic modulus decreases for the Cr-CrN multilayer coatings by the introduction of the softer Cr layers compared to CrN single layer coatings following the rule of mixture. In contrast, the hardness of the multilayers is only slightly decreased compared to CrN by the softer Cr phase.

The growth structure has significant influence on the deformation properties: The higher the tendency to crack deflection in the multilayer by the higher density of interfaces, the more pronounced is the formation of radial cracks in the surrounding of static ball indentation, as well as the crack deflection to shear cracks. However, no significant improvements by CrN single layer and Cr-CrN multilayer coatings were achieved in tribological ball-on-disc testing compared to uncoated CFC due to high friction coefficients ($\mu > 0.6$) in contact to AISI 5210 steel and Al_2O_3 counterparts. Only the application of amorphous DLC top layers significantly decreases friction coefficients to ~0.2, which drastically lowers the tangential shear forces in the coating. Minimized wear of these DLC top layers on the tough Cr-CrN multilayer films resulted in survival of these coating in tribology testing (10^4 cycles under 2 N loading) with capacity to much higher sliding cycles with low-friction behavior.

ACKNOWLEDGMENTS

The financial support of this work by the Austrian Federal Ministry of Traffic, Innovation and Technology, Austrian within the frame of the Austrian Nanoinitiative programme and the MNT-ERA.NET programme with additional support of the European Union and the Federal Country of Styria (Austria), the Polish-Austrian exchange project PL 12/2010 (funded in Austria by the Oesterrechischer Austauschdienst OeAD), and the research project of the Polish National Science Centre No. 3066/B/T02/2011/40 is highly acknowledged. The authors want to thank Harald Parizek from JOANNEUM RESEARCH for coating deposition as well as Christian Mitterer's research group at the University of Leoben for XRD analysis.

CONFLICTS OF INTEREST

The authors declare no conflict of interest.

REFERENCES

1.   Ivosevic, M.; Knight, R.; Kalidindi, S.R.; Palmese, G.R.; Sutter, J.K. Adhesive/cohesive properties of thermally sprayed functionally graded coatings for polymer matrix composites. J. Therm. Spray Technol. 2005, 14, 45–51.

2.   Palumbo, G.; Brooks, I.; Panagiotopoulos, K.; McCrea, J.; Limoges, D.; Erb, U. Strong lightweight article containing a fine-grained metallic layer. U.S. Patent 7,387,578 B2, June 2008.

3.   Friedrich, C.; Gadow, R.; Speicher, M. Protective multilayer coatings for carbon–carbon composites. Surf. Coat. Technol. 2002, 151, 405–411.

4.   He, M.Y.; Evans, A.G.; Hutchinson, J.W. Crack deflection at an interface between dissimilar elastic materials: role of residual stresses. Int. J. Solids Struct. 1994, 31, 3443–3455.

5.   Biron, M. Thermosets and Composites; Elsevier: Oxford, UK, 2004.

6.   Zhang, S.; Sun, D.; Fu, Y.; Du, H. Toughening of hard nanostructural thin films: A critical review. Surf. Coat. Technol.2005, 198, 2–8.

7.   Karimi, A.; Wang, Y.; Cselle, T.; Morstein, M. Fracture mechanisms in nanoscale layered hard thin films. Thin Solid Films 2002, 420, 275–280.

8.   Chen, Z.; Cotterell, B.; Wang, W. The fracture of brittle thin films on compliant substrates in flexible displays. Eng. Fract. Mech. 2002, 69, 597–603.

9.   Kääriäinen, T.; Rahamathunnisa, M.; Tanttari, M.; Cameron, D.C. Properties of Magnetron Sputtered Hard Coatings on Carbon and Glass Fibre Composites. In Proceedings of 49th Annual Technical Conference, Washington, DC, USA, 22–27 April 2006; pp. 548–554.

10.   Lackner, J.M.; Waldhauser, W.; Ganser, C.; Teichert, C.; Kot, M.; Major, L. Mechanisms of topography formation of magnetron-sputtered chromium-based coatings on epoxy polymer composites. Surf. Coat. Technol. 2013. in press.

11.   Voevodin, A.A.; Schneider, J.M.; Rebholz, C.; Matthews, A. Multilayer composite ceramicmetal-DLC coatings for sliding wear applications. Tribol. Int. 1996, 29, 559–570.

12.   Stueber, M.; Holleck, H.; Leiste, H.; Seemann, K.; Ulrich, S.; Ziebert, C. Concepts for the design of advanced nanoscale PVD multilayer protective thin films. J. Alloys Compd. 2009, 483, 321–333.

13. Chan, K.S.; He, M.Y.; Hutchinson, J.W. Cracking and stress redistribution in ceramic layered composites. Mater. Sci. Eng. A 1993, 167, 57–64.

14. Leyland, A.; Matthews, A. Thick Ti/TiN multilayered coatings for abrasive and erosive wear resistance. Surf. Coat. Technol. 1994, 70, 19–25.

15. Holmberg, K.; Matthews, A.; Ronkainen, H. Coatings tribology—Contact mechanisms and surface design. Tribol. Int.1998, 31, 107–120.

16. Martinez, E.; Romero, J.; Lousa, A.; Esteve, J. Nanoindentation stress–strain curves as a method for thin-film complete mechanical characterization: application to nanometric CrN/Cr multilayer coatings. Appl. Phys. A 2003, 77, 419–427.

17. Smolik, J.; Zdunek, K.; Larisch, B. Investigation of adhesion between component layers of a multi-layer coating TiC/Ti (CxN$_{1-x}$)/TiN by the scratch-test method. Vacuum 1999, 55, 45–50.

18. Lackner, J.M.; Waldhauser, W.; Schwarz, M.; Mahoney, L.; Major, L.; Major, B. Polymer pre-treatment by linear anode layer source plasma for adhesion improvement of sputtered TiN coatings. Vacuum 2008, 83, 302–307.

19. Kot, M.; Rakowski, W.; Major, Ł.; Lackner, J.M. Load-bearing capacity of coating-substrate systems obtained from spherical indentation tests. Mater. Des. 2012, 43, 99–111.

20. Lackner, J.M.; Major, L.; Kot, M. Microscale interpretation of tribological phenomena in Ti/TiN soft-hard multilayer coatings on soft austenite steel substrates. Bull. Pol. Acad. Sci. Tech. Sci. 2011, 59, 343–355.

21. DIN EN ISO 14577-1:2002 Metallic Materials—Instrumented Indentation Test for Hardness and Materials Parameters. Part 1: Test Method; International Organization for Standardization: Geneva, Switzerland, 2002.

22. Oliver, W.C.; Pharr, G.M. Improved technique for determining hardness and elastic modulus using load and displacement sensing indentation experiments. J. Mater. Res. 1992, 7, 1564–1583.

23. Thornton, J.A. Influence of apparatus geometry and deposition conditions on the structure and topography of thick sputtered coatings. J. Vac. Sci. Technol. 1974, 11, 666–670.

24. Jung, M.J.; Nam, K.H.; Jung, Y.M.; Han, J.G. Nucleation and growth behavior of chromium nitride film deposited on various substrates by magnetron sputtering. Surf. Coat. Technol. 2002, 171, 59–64.

25. JCPDS-International Centre for Diffraction Data. Cards No. 00-027-1402 (Si), 00-035-0803 (-Cr2N), 03-06899 (CrN), 00-027-1402 (Cr).

26. Hall, E.O. The deformation and ageing of mild steel: III discussion of results. Proc. Phys. Soc. B 1951, 64.

27. Romero, J.; Esteve, J.; Lousa, A. Period dependence of hardness and microstructure on nanometric Cr/CrN multilayers. Surf. Coat. Technol. 2004, 188, 338–343.

28. Chen, H.Y.; Tsai, C.J.; Lu, F.H. The Young's modulus of chromium nitride films. Surf. Coat. Technol. 2004, 184, 69–73.

29. Hutchinson, J.W.; Thouless, M.D.; Liniger, E.G. Growth and configurational stability of circular, buckling-driven film delaminations. Acta Metall. Mater. 1992, 40, 295–308.

30. Tabor, D. Indentation Hardness and Its Measurement: Some Cautionary Comments. In Microindentation Techniques in Material Science and Engineering; Blau, P.J., Lawn, B.R., Eds.; American Society of Testing of Materials: Ann Arbor, MI, USA, 1986; p. 129.

31. Holmberg, K.; Matthews, A. Coatings Tribology, Properties, Techniques and Applications in Surface Engineering; Elsevier: Amsterdam, The Netherlands, 1996.

32. Komvopoulos, K. Elastic-plastic finite element analysis of indented layered media. Trans. ASME J. Tribol. 1989, 111, 430–439.

33. O'Sullivan, T.C.; King, R.B. Sliding contact stress field due to a spherical indenter on a layered elastic half-space. J. Tribol. 1998, 110, 235–240.

34. Suh, N.P.; Sin, H.C. The genesis of friction. Wear 1981, 69, 91–114.

35. Stachowiak, G. Wear: Materials, Mechanisms and Practice; Wiley: Chichester, UK, 2005.

Chapter 4

A MICROMECHANICS-BASED INTERFACE MESOMODEL FOR VIRTUAL TESTING OF LAMINATED COMPOSITES

Pierre Ladevèze[1]; Federica Daghia[2]; Emmanuelle Abisse[3]; and Camille Le Mauff[4]

[1]LMT-Cachan (ENS Cachan, CNRS-UMR8535, UPMC, PRES UniverSud Paris)

[2]LMT-Cachan (ENS Cachan, CNRS-UMR8535, UPMC, PRES UniverSud Paris)

[3]LMT-Cachan (ENS Cachan, CNRS-UMR8535, UPMC, PRES UniverSud Paris)

[4]LMT-Cachan (ENS Cachan, CNRS-UMR8535, UPMC, PRES UniverSud Paris)

ABSTRACT

Background

The prediction of the behavior of laminated composite structures up to final fracture continues to be a challenge today. Indeed, failure may occur due to the interaction of small-scale degradations, such as transverse intraply cracks and interface delamination, which are difficult to account for in calculations on the structure's scale.

Methods

Here, in order to model the interaction of intralaminar and interlaminar degradations, we develop a new and relatively simple micromechanics-based interface mesomodel which differs from classical cohesive interface models, since it includes the coupling between transverse intraply cracks and interface delamination.

Results

The new interface model was implemented in a finite element code and used in the simulation of tensile tests on unnotched and holed specimens. Simulations with a classical cohesive interface model (not including coupling) were also carried out.

Conclusions

The simulations highlight the need for introducing intra-/interlaminar's behavior coupling in order to accurately predict the damage evolution and failure stress and mode.

BACKGROUND

The last quarter-century has witnessed considerable research efforts in the mechanics of composites in order to understand and predict the behavior of these materials, the ultimate goal being the design of the materials/structures/ manufacturing processes. Even in the case of laminated composites, the prediction of the evolution of damage up to and including final fracture remains a major challenge which is at the heart of today's 'virtual structural testing' revolution engaged in by the aeronautical industry. Virtual structural testing consists, whenever possible, in replacing the numerous experimental tests used today by virtual tests.

An answer to the virtual structural testing challenge is what is called the 'damage mesomodel for laminated composites', developed at LMT-Cachan since the 1980s [1, 2]. The main assumption is that the behavior of any laminate under any loading up to final fracture can be described using two elementary entities: the ply and the interface. The ply is described as a full three-dimensional orthotropic and damageable continuum. In particular, transverse macrocracks running parallel to the fibers (such as splits) are modeled as completely damaged zones; these may appear thicker numerically than the cracks observed experimentally. The interface is a surface entity, i.e. a cohesive interface [3]. An enhanced ply model based on micromechanics has been introduced in [4, 5]. Today, several similar mesoscopic approaches are being developed [6].

The starting point of this paper was the need to improve the predictions of the standard mesomodel in terms of delamination. Even though it led to realistic calculated responses for complex engineering problems [7–10], it was shown to underpredict the delaminated areas in some industrially significant test cases, such as low-velocity impact [10]. This means that a standard cohesive interface model, even combined with a ply mesomodel, may not be capable of producing realistic responses in terms of delamination. A heuristic remedy was proposed in [10] and more elaborate corrections were introduced in [11, 12].

The description of the interaction between delamination and transverse microcracking is a rather ancient question in micromechanics [13–22]. In all the referred works, two-dimensional discrete models are used. Both transverse intraply cracks and delamination cracks are described in detail; thus, the competition between the two mechanisms can be modeled directly. Indeed, the physics of the problem is very well-known (see the review papers [1, 21–23]). Today, the difficulty lies elsewhere, namely in the fact that the discrete modeling of every single discontinuity becomes unfeasible for complex engineering problems involving several thousands of cracks. On the one hand, even with high-performance computational tools [24], the computational micromechanical model introduced in [1, 23, 25] still leads to prohibitive computational efforts and, thus, is far from meeting the virtual structural testing requirements. On the other hand, when a mesoscale damage approach is used, some of the information regarding the detailed microscopic stress/strain state is lost. Therefore, the ply/interface coupling proposed in this article is necessary in order to restore the correct physical description in terms of transverse microcracking-induced delamination.

Apart from purely microscopic and mesoscopic approaches, intermediate approaches have recently been proposed in the literature in order to account for the interaction between transverse cracking and delamination. For example, in works such as [26, 27], classical cohesive interfaces are used for both transverse cracks and delaminations; in this case, however, a priori information about the cracking pattern (e.g.the position of the splits) needs to be introduced in order to carry out the simulations. Another approach consists in introducing discrete cracks thanks to techniques such as the X-FEM [28]; once again, the interaction between transverse cracks and delamination occurs naturally, but the local stress/strain field is still poorly represented compared to a purely microscopic approach, and a minimum crack spacing (which is generally much larger than in reality) related to the element size chosen needs to be introduced. These intermediate approaches are helpful for one's understanding of the degradation mechanisms. Unfortunately, because of the approximations introduced in the physics and the a priori information which they require, they cannot be considered to be predictive models.

In this paper, we present a new and relatively simple micromechanics-based interface model which takes into account the interaction between delamination and microcracking. We consider an ($\alpha/-\alpha$) interface between two plies with different microcracking densities; both in-plane and out-of-plane mesostresses are taken into account. In the first Section, the classical micromechanical description of the damage mechanisms and the main features of the bridge between micro- and mesomechanics [4, 5, 29] are reviewed.

Out-of-plane mesostresses are discussed in the second Section, in which the homogenized interface stiffness is derived using what is known as the basic interface problem, which is part of the micro-meso bridge [29]. This problem, defined over a 3D cell, is solved numerically for realistic situations involving out-of-plane mesostresses. Classical interface damage evolution laws are retained because their identification relies on standard delamination tests. In-plane mesostresses are discussed in the third Section using, once again, the basic interface problem. In-plane mesostresses can induce local delamination at the tips of transverse microcracks after saturation of the microcracking mechanism. It is shown that these local delaminations are generally unstable and, therefore, a criterion for the delamination of an interface, associated to the mesostress state of each adjacent ply, is proposed. In order to illustrate the predictive capabilities of the enhanced interface mesomodel and the importance to introduce it to ensure sufficient predictive capabilities to the model, we use the example of a simple tensile test, namely the $[0_m/90_n]_s$, and a more structural one namely an open-hole tensile test (fourth Section). No further information concerning the cracking pattern is introduced in the model.

METHODS

The damage mechanisms on the microscale

Four scenarios can be distinguished. The first two mechanisms have been studied for laminated composites by the micromechanics community (see the reviews [1, 21–23]). Matrix microcracking (Scenario 1) is driven by the ply's microstructure: usually, matrix microcracks originate perpendicular to the plane of the ply, then run throughout the ply's thickness, and finally grow parallel to the fibers' direction. Moreover, the microcracking pattern can be considered to be locally periodic: thus the amount of microcracking can be quantified by the microcracking rate $\rho=H/L$, where H is the ply thickness and L the distance between two cracks (see Figure 1). Local delamination (Scenario 2) generally occurs after the saturation of matrix microcracking: it is caused by the stress concentrations at the tips of the intraply cracks. This mechanism is quantified by the local delamination ratio τ =e/H, where e is the length of the delaminated zone (see Figure 1). Diffuse intra- and interply damage mechanisms (Scenarios 3 and 4) were introduced into the damage mechanics of laminates a long time ago, but they are usually not taken into account in micromechanics. As they occur at the fiber's scale, they can be homogenized and they are introduced directly through damage variables and evolution laws at the ply's scale.

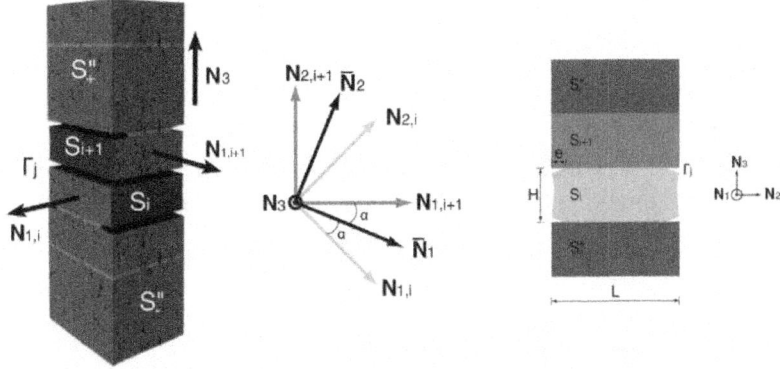

a) Three-dimensional basic interface problem   b) Two-dimensional reduction

Figure 1: The basic interface problem (interface Γ_j).

In order to handle these mechanisms, a computational micromodel was introduced in [1, 23] and developed in [24, 29]. This micromodel reproduces the key points observed in the micromechanics of laminates [1, 23] quite well.

The bridge between micromechanics and mesomechanics

The enhanced damage mesomodel [5, 11, 12] is the homogenized version of the micromechanical model introduced in [1]. The details of the bridge we derived between micromechanics and mesomechanics are given in [4, 5, 29]. The idea is to impose that the potential energy stored in the plies and in the interfaces must be the same on the microscale and on the mesoscale, which leads to the following relation between the microquantities and mesoquantities:

$$\pi \varepsilon_{meso} \pi = \frac{1}{\text{mes}(\Gamma)} \int_\Gamma \pi \varepsilon_{micro} \pi \, dS, \quad \sigma_{meso} N_3 = \frac{1}{\text{mes}(\Gamma)} \int_\Gamma \sigma_{micro} N_3 dS \tag{1}$$

where π is the projection operator onto the plane and Γ is an arbitrary section of the unit cell perpendicular to vector N_3 (see Figure 1). Thus, there are two basic problems, one associated with in-plane loading and the other associated with out-of-plane loading.

The problem associated with out-of-plane loading, which defines the mesodescription of the interface, is summarized in Figure 1. Considering an interface Γ_j (in this case, a 3D matrix layer of thickness $\frac{H_g}{20}$, where H_a is the thickness of the elementary ply) between two cracked plies S_i and S_{i+1}, the upper part S_+'' and the lower part S_-'' of the laminate are homogenized. Periodic boundary conditions are defined. Uniform elementary loadings are introduced on the cracked surfaces: this residual problem can be superposed to an uncracked problem in order to obtain the full solution of the cracked cell under elementary loadings. More details on the definition of the interface problem can be found in [29].

Using the finite element method, the 3D reference problem on the microscale was solved for different sets of parameters (thickness, stiffness, $\rho \in [0, 0.7], \tau \in [0, 0.4])$ which are likely to be encountered in practice, leading to a set of mesodamage indicators associated to the preferential directions of the interface $(\bar{N}_1, \bar{N}_2)$ defined in Figure 1. It was shown that the mesodamage of the interface depends only on the interface itself and on the microcracking rates of the adjacent plies [29].

The interface's damage mesomodel - the concept of interface stiffness

First, let us study the change in the stiffness of the interface mesomodel due to microcracking in the adjacent plies in the general case of different microcracking rates. To obtain these stiffness changes, the basic interface problem must be solved under out-of-plane loading. With only a limited loss of accuracy, one can consider the solution to be the superposition of the solutions of two 2D problems (one of which is depicted in Figure 1), which are associated with the fiber directions of Ply S_i and Ply S_{i+1} [29].

Properties of the basic 2D interface problem

The basic 2D interface problem is defined in Figure 1. Since the results are quasi-independent of the stacking sequence, a sequence of [90/0/902] with x≡N1 was chosen. h denotes the interface's thickness; the main parameters are the microcracking rate ρ and the delamination ratio τ The typical properties of carbon/epoxy unidirectional plies are considered:

$$E_1 = 148 \text{ GPa}, \quad E_{2,3} = 10.8 \text{ GPa}, \quad \nu_{12,13} = 0.3, \quad \nu_{23} = 0.4,$$

$$G_{12,13} = 5.8 \text{ GPa}, \quad G_{23} = \frac{E_2}{2(1 + \nu_{23})}, \quad H = H_e = 0.125 \cdot 10^{-3} \text{ m}.$$

For the interfaces, which are considered to be thin 3D matrix layers made of isotropic material, the material properties are $E = 2.4$ GPa, $\nu = 0.33$, $h = H_e/20$.
.

The problem to be solved is elastic and follows the generalized plane strain assumption (i.e. the displacement in direction N_1 is constant). It has been proven that the mesobehavior of interface Γ_j depends only on interface Γ_j and ply S_i, i.e. on parameters $\lambda=2 \tau \rho,\rho$ and on the ply thickness [29].

The cell was analyzed for unit values of stresses $\sigma_{33}, \sigma_{23}, \sigma_{13}, \sigma_{22}$ and σ_{12} using a relatively refined FE mesh, leading to a residual energy expressed as a surface energy:

$$
\begin{aligned}
\Delta e = {} & c_{33}\,(\sigma_{33})^2 + c_{23}\,(\sigma_{23})^2 + c_{13}\,(\sigma_{13})^2 + c_{22}\,(\sigma_{22})^2 + c_{12}\,(\sigma_{12})^2 \\
& + c_{3313}\sigma_{33}\sigma_{13} + c_{3323}\sigma_{33}\sigma_{23} + c_{1323}\sigma_{13}\sigma_{23} + c_{2212}\sigma_{22}\sigma_{12} + c_{2233}\sigma_{22}\sigma_{33} \\
& + c_{2213}\sigma_{22}\sigma_{13} + c_{2223}\sigma_{22}\sigma_{23} + c_{1233}\sigma_{12}\sigma_{33} + c_{1213}\sigma_{12}\sigma_{13} + c_{1223}\sigma_{12}\sigma_{23}
\end{aligned}
$$

$$(2)$$

The values of the coupling coefficients have been computed

$$
\alpha_{ijkl} = \frac{c_{ijkl}}{\left(c_{ij}c_{kl}\right)^{\frac{1}{2}}}, \quad \bar{\alpha}_{ijkl} = \max\big|_{\text{calculated points}} \left|\alpha_{ijkl}\right|
$$

$$(3)$$

and the calculated points were τ =(0.1, 0.2) and ρ=(0.2, 0.4, 0.6, 0.8).

Except for $c_{,,,,}$, these coupling coefficients are negligible, the maximum being around $6.1 \cdot 10^{-13}$. Thus, Δ e can be taken as:

$$
\Delta e = c_{33}\,(\sigma_{33})^2 + c_{23}\,(\sigma_{23})^2 + c_{13}\,(\sigma_{13})^2 + c_{22}\,(\sigma_{22})^2 + c_{12}\,(\sigma_{12})^2 + c_{2233}\sigma_{22}\sigma_{33}
$$

$$(4)$$

Moreover, the last three terms, which are proportional to h, are small compared to the ply's residual energy, which is proportional to H, so they, too, are negligible. Consequently, the interface's residual energy can be taken as

$$
\Delta e = c_{33}\,(\sigma_{33})^2 + c_{23}\,(\sigma_{23})^2 + c_{13}\,(\sigma_{13})^2
$$

$$(5)$$

Now, let us introduce approximations for coefficients c_{33}, c_{13} and c_{23}, which depend on λ=2 τ ρ and ρ. These approximations are derived from the analysis of the extreme cases: small ρ, large ρ and λ equal to 0 or 1.

Let us introduce the damage parameters $d_{33,i}$, $d_{13,i}$ and $d_{23,i}$ associated to the 2D basic interface problem involving Ply S_i:

$$
\frac{d_{33,i}}{1 - d_{33,i}} = c_{33}\frac{2E}{h}, \quad \frac{d_{13,i}}{1 - d_{13,i}} = c_{13}\frac{2G}{h}, \quad \frac{d_{23,i}}{1 - d_{23,i}} = c_{23}\frac{2G}{h}
$$

$$(6)$$

As shown in Figure 2, the following approximations work quite well:

$$
\bar{d}_{33,i} = \lambda, \quad \bar{d}_{13,i} = \lambda, \quad \frac{\bar{d}_{23,i}}{1 - \bar{d}_{23,i}} = \frac{\lambda}{1 - \lambda} + A(\rho), \quad A(\rho) = \frac{a(\rho)}{1 - a(\rho)}
$$

$$(7)$$

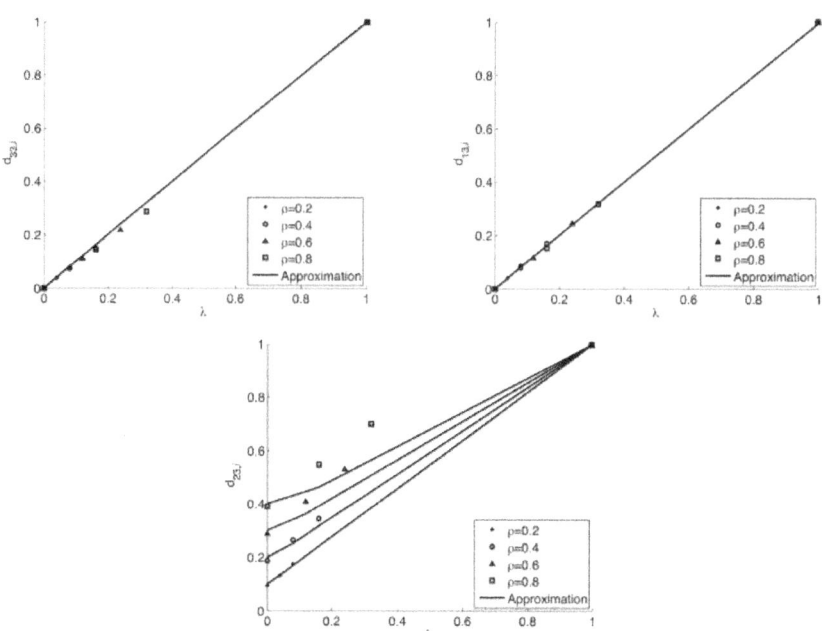

Figure 2: The interface's mesodamage parameters.

with the material function a(ρ) assumed to be linear (a(ρ)=0.5ρ for the material studied).

Taking into account the delaminated area in the other direction $\mathbf{N'}_1$, one gets:

$$\Delta e \sim \frac{h}{2}\left[\frac{\lambda}{1-\bar{\lambda}}\left(\frac{\sigma_{13}^2}{G}+\frac{\sigma_{33}^2}{E}\right)+\frac{\sigma_{23}^2}{G}\left(\frac{\lambda}{1-\bar{\lambda}}+A(\rho)\right)\right]$$

(8)

where

$\lambda = 2\tau\rho = \frac{e}{L}, \lambda' = 2\tau'\rho'$ and $\left(1-\bar{\lambda}\right) = \left(1-\lambda'\right)\left(1-\lambda\right)$.

The microcracking/stiffness interaction of the interface meso-model

One must add up the residual energies of the two basic 2D interface problems:

$$\Delta e = \frac{h}{2}\left[\frac{\lambda+\lambda'}{1-\bar{\lambda}}\frac{\sigma_{33}^2}{E}+\frac{\sigma_{13}^2+\sigma_{23}^2}{G}\frac{\lambda}{1-\bar{\lambda}}+\frac{\sigma_{1'3}^2+\sigma_{2'3}^2}{G}\frac{\lambda'}{1-\bar{\lambda}}+\frac{\sigma_{23}^2}{G}A(\rho)+\frac{\sigma_{2'3}^2}{G}A(\rho')\right]$$

(9)

Let $\bar{\sigma}_{33}$, $\bar{\sigma}_{13}$ and $\bar{\sigma}_{23}$ be the mesostress components written in the interface's basis $(\bar{\mathbf{N}}_1, \bar{\mathbf{N}}_2)$ and let 2α be the angle between the fiber directions of the adjacent plies. One has:

$$\sigma_{13} = \cos\alpha\,\bar{\sigma}_{13} - \sin\alpha\,\bar{\sigma}_{23}, \quad \sigma_{23} = \sin\alpha\,\bar{\sigma}_{13} + \cos\alpha\,\bar{\sigma}_{23}$$

$$\sigma_{1'3} = \cos\alpha\,\bar{\sigma}_{13} + \sin\alpha\,\bar{\sigma}_{23}, \quad \sigma_{2'3} = -\sin\alpha\,\bar{\sigma}_{13} + \cos\alpha\,\bar{\sigma}_{23} \tag{10}$$

Neglecting the term $\lambda\,\lambda$' which is very small compared to 1, one easily obtains:

$$\Delta e = \frac{h}{2}\left\{ \frac{\bar{\lambda}}{1-\bar{\lambda}}\frac{\bar{\sigma}_{33}^2}{E} + \frac{\bar{\sigma}_{13}^2}{G}\left[\frac{\bar{\lambda}}{1-\bar{\lambda}} + (A+A')\sin^2\alpha\right] + \frac{\bar{\sigma}_{23}^2}{G}\left[\frac{\bar{\lambda}}{1-\bar{\lambda}} + (A+A')\cos^2\alpha\right] \right.$$

$$\left. + \frac{2\bar{\sigma}_{13}\bar{\sigma}_{23}\sin\alpha\cos\alpha}{G}(A-A') \right\} \tag{11}$$

The new interface mesomodel - stiffness and damage

Let us note that the interface mesomodel is described as a cohesive interface with a very small 'thickness' compared to the cell's dimensions. The contributions due to microcracking should be viewed as relatively long-wavelength contributions. Thus, the energy of the interface mesomodel is

$$\bar{e} = \frac{h}{2}\left[\frac{\langle -\sigma_{33}\rangle^2}{E} + \frac{\langle\sigma_{33}\rangle^2}{E(1-d_{33})} + \frac{\bar{\sigma}_{13}^2}{G(1-d_{13})} + \frac{\bar{\sigma}_{23}^2}{G(1-d_{23})} + \frac{\omega}{G}\bar{\sigma}_{13}\bar{\sigma}_{23} \right] \tag{12}$$

where the purpose of the positive part <·> is to account for crack opening and crack closure. The usual damage variables, deduced from the micro-meso energy equivalence, are

$$d_{33}, \quad d_{13} = \frac{d_{33}+(1-d_{33})(A+A')\sin^2\alpha}{1+(1-d_{33})(A+A')\sin^2\alpha}, \quad d_{23} = \frac{d_{33}+(1-d_{33})(A+A')\cos^2\alpha}{1+(1-d_{33})(A+A')\cos^2\alpha} \tag{13}$$

with the coupling term ω written as $\omega = 2\sin\alpha\cos\alpha\,(A-A')$.

In previous papers [11, 12], a simplified expression was considered, based on $\bar{\rho} = \frac{\rho+\rho'}{2}$. This expression is equivalent to $\rho = \rho'$, τ small (i.e. $d_{33} \to 0$) and $\alpha \sim 45°$.

It is remarkable that this energy depends only on ρ, ρ' and $\bar{\lambda}$. As mentioned previously, $a(\rho)$ is a material function which can be identified from the basic 2D interface problem. In the present work, we used a linear law.

Computation of the dissipation

The dissipation work associated with the new interface model is:

$$\dot{D} = \Delta \dot{e} = Y_I \dot{d}_{33} + Y_{II} \dot{d}_{13} + Y_{III} \dot{d}_{23} + \frac{h}{2G} \omega \bar{\sigma}_{13} \bar{\sigma}_{23} \tag{14}$$

where $\omega \dot{\omega}$ depends on $\dot{\rho}$ and $\dot{\rho}'$.

One can easily see that $\dot{D} \geq 0$, as $\dot{d}_{33}$, $\dot{\rho}$ and $\dot{\rho}'$ are positive or equal to zero. Using (11) and (7):

$$\dot{D} = \frac{h}{2} \left[\frac{\dot{d}_{33}}{(1 - d_{33})^2} \left(\frac{\bar{\sigma}_{33}^2}{E} + \frac{\bar{\sigma}_{13}^2 + \bar{\sigma}_{23}^2}{G} \right) + (\bar{\sigma}_{13} \sin \alpha + \bar{\sigma}_{23} \cos \alpha)^2 \frac{\dot{A}}{G} \right.$$
$$\left. + (\bar{\sigma}_{13} \sin \alpha - \bar{\sigma}_{23} \cos \alpha)^2 \frac{\dot{A}'}{G} \right] \tag{15}$$

Since $\dot{A} = \frac{dA}{d\rho} \dot{\rho}$ is positive, it follows that $D \geq 0$ thus, the interface mesomodel is compatible with the principles of thermodynamics.

The damage mesomodel - delamination criteria

Two different fracture mechanisms should be considered for out-of-plane loading and in-plane loading. The first fracture mechanism, associated with out-of-plane loading, is described through classical interface damage laws involving the normal stress vector.

The second fracture mechanism is due to in-plane stresses leading to microdelamination cracks at the tips of the transverse microcracks in plies. This is shown to be an unstable mechanism with a characteristic length of the same order of magnitude as the cell's dimensions.

Delamination criterion for out-of-plane loading

The standard interface model is extended as follows. The elementary damage forces are Y_I, Y_{II} and Y_{III} defined as:

$$Y_I = \frac{1}{2} \frac{h}{E} \frac{\langle \bar{\sigma}_{33} \rangle^2}{(1 - d_{33})^2}, \quad Y_{II} = \frac{1}{2} \frac{h}{G} \frac{\bar{\sigma}_{13}^2}{(1 - d_{13})^2}, \quad Y_{III} = \frac{1}{2} \frac{h}{G} \frac{\bar{\sigma}_{23}^2}{(1 - d_{23})^2} \tag{16}$$

The effective damage force, which is responsible for the increase in the interface's damage, is:

$$Y = \left[(Y_I)^r + (\gamma_{II} Y_{II})^r + (\gamma_{III} Y_{III})^r \right]^{1/r} \tag{17}$$

where γ_{II} and γ_{III} are two equal material coupling coefficients and the exponent r, which is also a material constant, is generally taken as 1. One has:

$$d_{33} = \left(\frac{n}{n+1} \frac{\langle \bar{Y} - Y_0 \rangle_+}{Y_c - Y_0} \right)^n \text{ if } d_{33} < 1, \quad d_{33} = d_{13} = d_{23} = 1 \quad \text{otherwise}$$

$$(18)$$

where $\bar{Y}|_t = \sup_{\tau \le t} Y|_\tau$ and k, n, Y_c, Y_0 are material constants which can be identified using standard delamination tests. Let us note that the interface mesomodel is independent of the angle 2α between the fiber directions of the adjacent plies.

Delamination criteria for in-plane loading

In order to analyze the microdelamination due to in-plane loading, let us review the modeling of transverse microcracking going back to the basic 2D interface problem.

The modeling of microcracking

Using finite fracture mechanics [14, 21, 30], the fracture criterion is classically written as:

$$l(\rho, \sigma) \equiv \left(\frac{G_I^u \langle \tilde{\sigma}_{22} \rangle^2}{G_I^c} + \frac{G_{II}^u \tilde{\sigma}_{12}^2}{G_{II}^c} + \frac{G_{III}^u \tilde{\sigma}_{23}^2}{G_{III}^c} \right) \eta,$$

$$(19)$$

$$\eta = \frac{H}{H_e} \text{ for } \frac{H}{\bar{H}} \le 1, \quad \eta = \frac{\bar{H}}{H_e} \text{ for } \frac{H}{\bar{H}} \ge 1$$

$$(20)$$

where the unit finite energy release rates G_I^u, G_{II}^u and G_{III}^u are calculated using the 'derivative'

$$\check{f}(\rho) = \frac{f\left(\frac{2\rho}{q}\right) - f\left(\frac{\rho}{q}\right)}{\frac{\rho}{q}}$$

$$(21)$$

q being a parameter (equal to about 1.5) associated with the stochastic behavior of microcracking [22]. The effective stress $\sigma \tilde{\sigma}$ is considered and H $\bar{H}$ is the transition thickness between thick ply and thin ply behavior.

The fracture model is relatively simple:

$$\begin{cases} \dot{\rho} \ge 0 & l(\rho, \tilde{\sigma}) \le 1 \\ \dot{\rho} \left[l(\rho, \tilde{\sigma}) - 1 \right] = 0 \end{cases}$$

$$(22)$$

Remark

The transverse damage d_{22} associated with $\tilde{\sigma}_{22}$ is a function of ρ which tends to $d_{22} = 1$ for large values of ρ.

The solving of the 2D generic basic interface problem leads also to a residual energy of the layer adjacent to the interface in term of out of plane stresses. However, for the ply this contribution is not as important as the contribution over the interface which explains why it is not introduced in the present version of the enhanced mesomodel [5, 11, 12]. However, it will be considered in a companion paper.

The part of the plies in contact with a completely delaminated interface should behave, regarding microcracking, as half a ply [21, 22].

The modeling of microdelamination

With ρ constant, the energy release rates related to microdelamination can be calculated as λ-derivatives. For τ =0, they are equal to zero. Let us use finite fracture mechanics again and consider the τ values:

$$\tau = (0.05, \ 0.1, \ 0.15, \ 0.2); \quad \Delta\tau = 0.05$$

The curves giving the unit energy release rates are shown in Figure 3. It follows that the initiation criterion can be defined as:

$$g(\tilde{\sigma}, \rho) = \left[\frac{Q_{22}^{u} \, (\tilde{\sigma}_{22})^2}{Q^c} + \frac{\gamma_{12} Q_{12}^{u} \tilde{\sigma}_{12}^2}{Q^c} \right] \frac{H}{H_e} \tag{23}$$

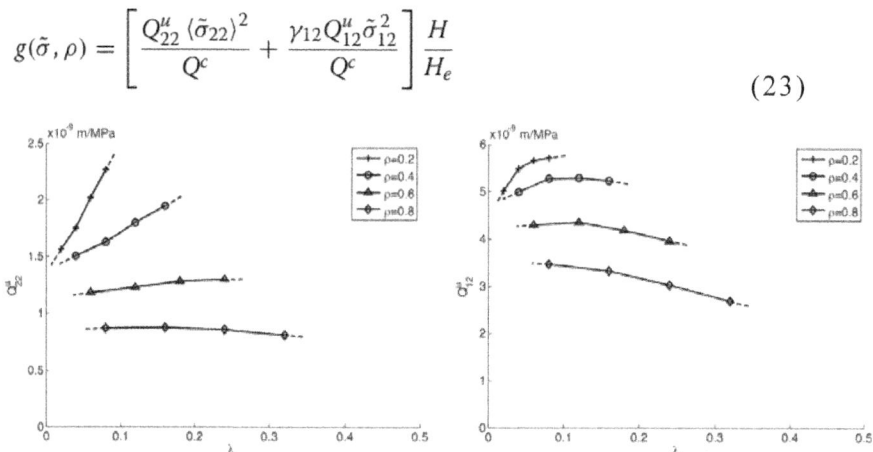

Figure 3: The unit energy release rates Q_{22}^{u} and Q_{12}^{u} related to microdelamination (parameter: ρ).

where Q_{22}^{u} and Q_{12}^{u} are the unit energy release rates associated to ρ, τ =0.05, Δ τ =0.05. Q^c and γ_{12} are critical material values. One has:

$$\begin{cases} \dot{\tau} \geq 0 \quad g(\tilde{\sigma}, \rho) \leq 1 \\ \dot{\tau}\left[g(\tilde{\sigma}, \rho) - 1\right] = 0 \end{cases}$$

$$(24)$$

Here, the out-of-plane effective stress $\tilde{\sigma}_{23}$ is not considered. Indeed, it is negligible except in high-gradient zones (e.g. because of edge effects), in which case it is taken into account by the interface model.

The curves of Figure 3 are either increasing or flat and show that in most cases the microdelamination mechanism is unstable. When it is activated, one can consider that the interface has been completely fractured; thus, Equations (23)-(24) can be viewed as a mesodelamination criterion.

A remark on the identification of the mesodelamination criterion

The criterion given in Equations (23)-(24) depends on two material constants Q^c and γ_{12} which can be identified by taking advantage of available experimental results related to microcracking saturation.

Let us consider the case $\tilde{\sigma}_{22} \neq 0, \tilde{\sigma}_{12} = 0$. From $[0_m/90_n]_s$ tensile tests, one can identify the material constant ρ_s which represents the microcracking density at saturation [5]. From an energy point of view, this saturation is associated to the decrease in the microcracking strain energy release rate and the corresponding nearly constant strain energy release rate for microdelamination (see Figure 4). This quantity is associated with the onset of microdelamination. It follows that Q^c can be identified as:

$$Q^c = \frac{Q_{22}^u(\rho_s)}{G_I^u(\rho_s)} G_I^c(\text{ply})$$

$$(25)$$

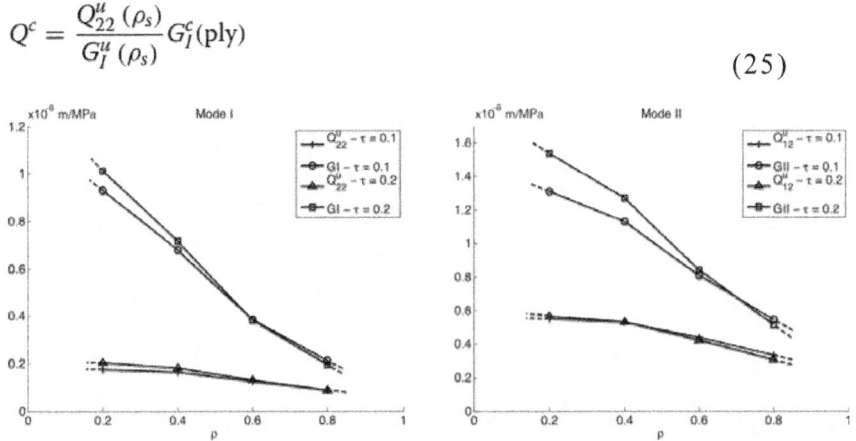

Figure 4: The unit energy release rates G_I^u and Q_{22}^u and G_{II}^u and Q_{12}^u as functions of the microcracking rate ρ for different values of τ.

where Q^μ and G_I^μ are evaluated for $\rho = \rho_s$. In the case $\tilde{\sigma}_{12} \neq 0, \tilde{\sigma}_{22} = 0$, a saturation value seems to exist, but it may be different from that observed in mode I.

REMARK

The constant γ_{12} can be identified from a tensile test of a $[+45/-45]_{ns}$ stacking sequence or a tensile test of a holed specimen, in which shear plays an important role. Otherwise, one can take the value related to the interface model.

The new interface mesomodel - in-plane loading

The following criterion is added to the interface mesomodel:

*   if $g(\tilde{\sigma}, \rho) < 1$ and $g'(\tilde{\sigma}', \rho) < 1$, then no extra condition; otherwise, $d_{33}=1$.

$g(\tilde{\sigma}, \rho)$ and $g'(\tilde{\sigma}', \rho')$ are associated with the adjacent plies of the interface interface being considered.

Results and discussion

The objective of this section is to illustrate the improvement brought by the new interface model described in this paper. One should note that this is not a complete experimental validation, but an example to demonstrate the need for the in-plane mesodelamination criterion in some classical test cases.

To do this, two different interface models are used and compared: the enhanced model described in this paper and a more classical cohesive interface model which does not include the coupling between the ply and interface behavior.

In a first time, the enhanced model is tested on a classical tension test in order to demonstrate its capability to mirror simple tests and to predict damage evolutions.

In a second time, a more complete comparison is performed with the two models, based on a structural test case: an open-hole tensile test on a quasi-isotropic laminate. This example allows then, on one hand, to highlight the need of introducing the intra-interlaminar coupling to mirror correctly the damage evolution, and, on the other hand, to illustrate the improvement brought by the enhanced model in the accuracy of the damage state prediction.

Tension test on $[0/90_4]_s$: a first validation of the proposed interface model

Experimental results are well-known and analysed in many papers such as [21]. The top part of Figure 5 gives an overview of the sequence of damage mechanisms. Three dimensional finite element calculations are performed with a very refined mesh, a small initial defect being introduced at the center of the plate. The elastic material properties used in the simulation are the same as the ones given in Section Properties of the basic 2D interface problem. As for the parameters associated to fiber breaking and diffuse damage, they are taken as typical values for carbon/epoxy composites. The energy release rates associated with transverse cracking are: $G_f^c = 200$ J/m^2 and $G_{II}^c = 800$ J/m^2. The ones related to the interfaces are assumed to be the same. Finally, the values of the parameters introduced are taken from the curves shown in the previous section. The enhanced interface model is used combined with the ply mesomodel [5, 11, 12].

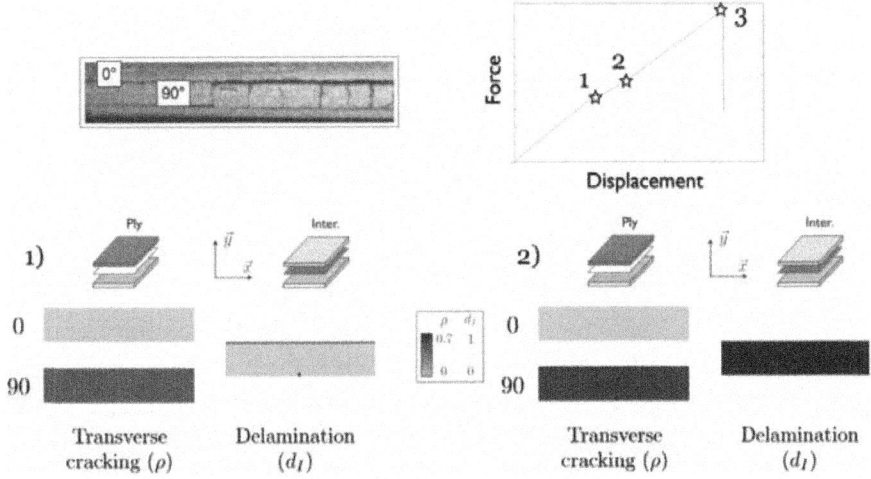

Figure 5: Experimental damage mechanisms, stress/strain curve and damage prediction in a cross-ply tensile test $[0/90_4]_s$ with the new interface model.

Figure 5 shows that the simulation reproduces correctly the damage physics. Until (1), transverse microcracking development is observed. Diffuse damage remains weak and is not shown in the damage charts. From (1) to (2), delamination develops very quickly and, in the end, the specimen fails by fiber failure.

For this test case, a finite element calculation carried out with a classical cohesive interface, would not reproduce correctly the interface damage physics. Indeed, in this type of model, the delamination is activated by out-of-plane stresses which are really small in these cases and would not be sufficient to activate the damage mechanism.

Moreover, the enhanced interface model proposed in this paper bring a real improvement in the damage prediction compared to the former model used previously as in [12]. Indeed, this former model uses the mean value of the microcracking densities in the two adjacent plies of the interface to trigger delamination. Then, in this particular case where only one adjacent ply of the interface is damaged, the former model fails in predicting the interface breaking.

Open-hole tensile test: need of the coupling introduction

The test case used hereafter is part of Wisnom and Hallet's experimental campaign on open-hole tensile tests [31]. Series of tests were carried out on quasi-isotropic IM7/8552

carbon/epoxy specimens with a $[45_m/90_m/-45_m/0_m]_{ns}$ lay-up and the geometry described in Figure 6.

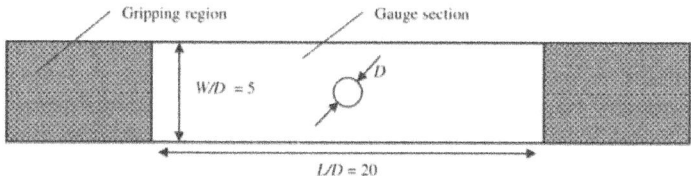

Figure 6: Geometry of the specimens.

The lay-up of the specimen chosen for this illustration is $[45_4/90_4/-45_4/0_4]_s$ with a ply thickness h=0.5 mm, the hole diameter is D=6.35 mm and the ratio W/D=5. Experimental results reported in [31] show that this specimen experiences a delamination-dominated failure: the spread of transverse cracking in the plies, and the important amount of delamination associated lead to the coupon failure. Hence, the failure relies on the interaction between the transverse cracking in the plies and the delamination of the interface.

Concerning the damage evolution, the experiments show that the transverse cracking first develops in the upper 45° ply, resulting in damage in the 45/90 interface. Then, transverse cracking reaches the 90° plies. Damage goes through plies and interfaces until the degradation of the −45/0 interface on the whole width of the coupon, which corresponds to the failure.

Because a large amount of subcritical damage occurs, the stress-strain curve experiences a slope change before the final breakdown.

In order to highlight the influence of the interface models on the damage evolution prediction, the test case is simulated using the enhanced interface model and a more classical one which does not include the intra-interlaminar coupling.

REMARK

Details concerning the material properties and finite element simulation features are presented in the paper [12].

Simulation results: global behavior

The stress-strain curves given by the two simulations are presented in Figure 7.

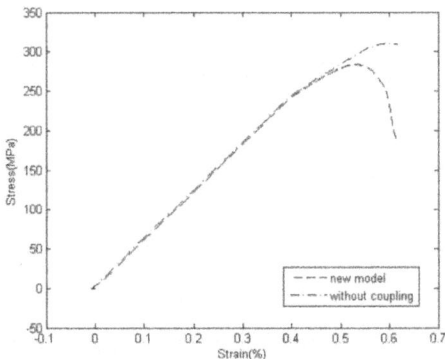

Figure 7: Stress-strain curves issued from the two simulations.

The two simulations show a slope change for a imposed strain $\varepsilon=0.38\%$. This corresponds to the development of subcritical damage in the coupon which matchs the experimental observations.

The model including coupling predicts a failure stress close to the experimental one: σ_{max} =280 MPa for the simulation versus σ_{max} =285 MPa for the experimental value. The second one, that does not include coupling, predict a failure stress higher than the experimental one $\sigma_{max} = 280$ MPa for the simulation versus $\sigma_{max} = 285$ MPa.

In the following, the damage evolution predicted by the two models are compared. The study focuses on transverse cracking in the plies (represented in the damage charts by the variable ρ) and delamination in the interfaces (represented by the variable d_I) as they are the main mechanisms concerned by the interface model.

The models are compared at four strain level:

1. $\varepsilon=0.38\%$: transverse cracking appears in the plies
2. $\varepsilon=0.42\%$: all plies experience transverse cracking
3. $\varepsilon=0.52\%$: transverse cracking has spread all over the width of the upper 45° ply $\varepsilon=0.58\%$: specimen has failed

Damage prediction comparison: need of the intra-interlaminar coupling

For ε=0.38% (Figure 8), the two models give similar results in terms of transverse cracking. It appears in the upper ply and goes through plies and interfaces as described in [31].

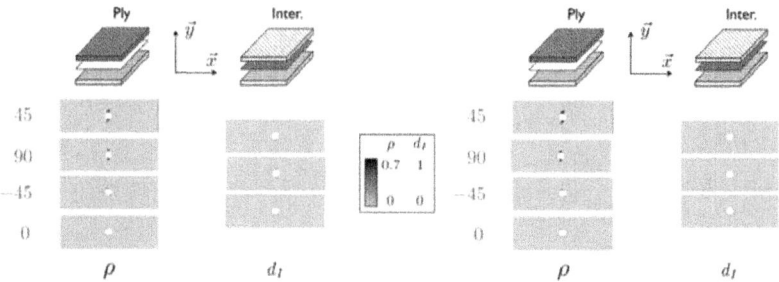

Figure 8: Damage charts yielded by models with (left) and without (right) coupling for a strain ε= 0.38%.

For ε=0.42% (Figure 9) and ε=0.52% (Figure 10), the two models go on predicting similar behavior in terms of transverse cracking. However, whereas the model including coupling predicts a spread of delamination in the interfaces, the model without it does not predict any degradation of the interfaces.

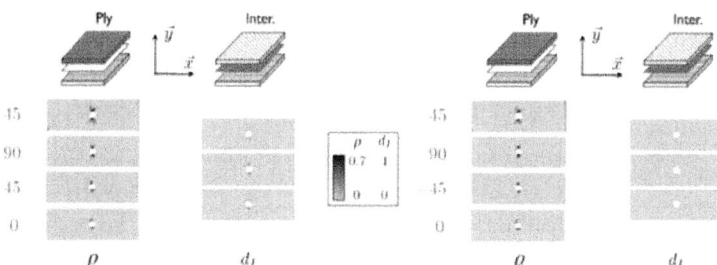

Figure 9: Damage charts yielded by models with (left) and without (right) coupling for a strain ε= 0.42%.

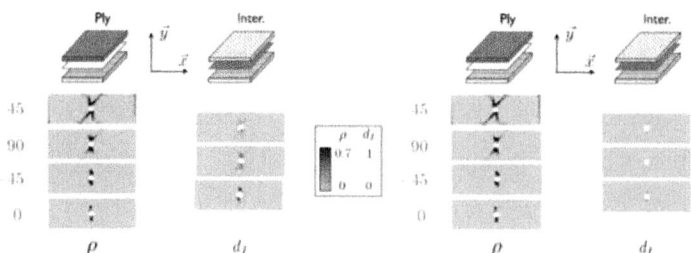

Figure 10: Damage charts yielded by models with (left) and without (right) coupling for a strain ε= 0.52%.

For ε=0.58% (Figure 11), the model including coupling leads to a delamination-dominated failure, as reported in [31], whereas the second model yields a fiber breaking dominated failure.

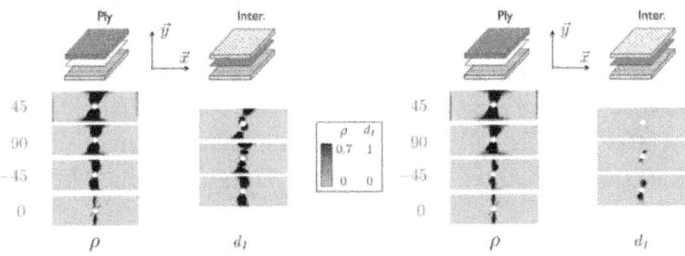

Figure 11: Damage charts yielded by models with (left) and without (right) coupling for a strain ε= 0.58%.

To resume, the two simulations predict similar behaviors for the transverse cracking, which match experimental observations. However, the enhanced model predicts a spread of delamination in the different interfaces almost as soon as transverse cracking appears, whereas the second model do not predict any delamination until an equivalent strain of ε=0.5%. This difference of behavior leads to different failure mode: the new model predicts a delamination dominated failure matching the experimental observations, the second model predicts a delayed failure due to fiber breaking.

These results highlight the need for introducing intra-/interlaminar's behavior coupling in order to accurately predict the damage evolution and failure stress and mode. More, the comparison with the experimental results illustrates the good capabilities of the enhanced interface model to predict the damage evolution and the failure pattern in the case of structural test cases such as open-hole tensile tests. Let us note that for this case the former version of our interface model gives similar results to the enhanced present one [12].

CONCLUSION

A new and relatively simple interface mesomodel taking into account the coupling with microcracking in the adjacent plies has been derived from the description of the damage scenarios on the microscale. This is a general model in which the damage states of the adjacent plies can be very different. Classical tests suffice to enable the identification of the material constants. The resulting enhanced mesomodel (ply and interface) is a computational model which is suitable for virtual testing. Indeed, it includes a physically sound description of situations involving intra/interlaminar coupling, thus it goes beyond the domain of validity of the standard mesomodel. Let us note also that the micro-meso bridge developed in this paper could be extended to the study of carbon/epoxy laminates interfaces interleaved with thermoplastic particles [32].

In this paper, the simulation of $[0_m/90_4]_s$ and open-hole tensile tests showed that this model reproduces experimental observations quite well. A more complete validation, for plates with holes and low-velocity impact tests, even for ultra-thick laminates [33], will be addressed in companion papers. In these more complex cases, there remain some issues related to the numerical treatment of isolated transverse macrocracks, which tend to be wider in the simulations than in reality. This is a general question which is currently being addressed.

Moreover, computational cost being prohibitive for designers, another challenge has to be tackled using the laminates model presented here: the building of virtual charts i.e, reduced models including the description of uncertainties [34].

COMPETING INTERESTS

The authors declare that they have no competing interests.

AUTHORS' CONTRIBUTIONS

All authors participated in the modeling and simulation work and in the redaction of the paper. All authors read and approved the final manuscript.

REFERENCES

1.   Ladevèze P: Multiscale computational damage modelling of laminated composites, No. 474 in series, CISM Courses and Lectures.. New York: SpringerWien; 2005a.

2.   Herakovich CT: Mechanics of fibrous composites.. New York: Wiley; 1998.

3.   de Borst R, Schipperen JHA: Continuum damage mechanics of materials and structures; chap. Computational Methods for delamination and fracture in composites.. Amsterdam: Elsevier; 2002. pp 325–352 pp 325–352

4.   Ladevèze P, Lubineau G: On a damage mesomodel for laminates: micro-meso relationships, possibilities and limits. Composite Sci Technol 2001,61(15):2149–2158. 10.1016/S0266-3538(01)00109-9

5.   Ladevèze P, Lubineau G: An enhanced mesomodel for laminates based on micromechanics. Composite Sci Technol2002,62(4):533–541. 10.1016/S0266-3538(01)00145-2

6. Lopes CS, Camanho PP, Gürdal Z, Maimí P, Gonzáles EV: Low-velocity impact damage on dispersed stacking sequence laminates. Part II: Numerical simulation. Composite Sci Technol 2009,69(7–8):937–947.

7. Flesher ND, Herakovich CT: Predicting delamination in composite structures. Composite Sci Technol 2006,66(6):745–754. 10.1016/j. compscitech.2004.12.039

8. Greve L, Pickett AK: Delamination testing and modelling for composite crash simulation. Composite Sci Technol 2006,66(6):816–826. 10.1016/j. compscitech.2004.12.042

9. Johnson AF, Holzapfel M: Modelling soft body impact on composite structures. Composite Struct 2003,61(1–2):103–113.

10. Guinard S, Allix O, Guédra-Degeorges D, Vinet A: A 3D damage analysis of low-velocity impacts on laminated composites.Composite Sci Technol 2002,62(4):585–589. 10.1016/S0266-3538(01)00153-1

11. Daghia F, Ladevèze P: Identification and validation of an enhanced mesomodel for laminated composites within the WWFE-III. J Composite Mater 2013,47(20–21):2675–2693. 10.1177/0021998313494095

12. Abisset E, Daghia F, Ladevèze P: On the validation of a damage mesomodel for laminated composites by means of open-hole tensile tests on quasi-isotropic laminates. Composite Part A 2011, 42: 1515–1524. 10.1016/j.compositesa.2011.07.004

13. Harris CE, Morris DH: Delamination and Debonding of Materials; chap. In Role of delamination and damage development on the strength of thick notched laminates STP 876. Philadelphia: ASTM; 1985:424–447.

14. Nairn JA, Hu S: The initiation and growth of delaminations induced by matrix microcracks in laminated composites. Int J Fracture1992,57(1):1–24. 10.1007/BF00013005

15. Finn SR, He YF, Springer GS: Delaminations in composite plates under transverse impact loads - experimental results. Composite Structures 1993,23(3):191–204. 10.1016/0263-8223(93)90222-C

16. Eggers H, Goetting HC, Bäuml H: Synergism between layer cracking and delaminations in multidirectional laminates of carbon-fibre-reinforced epoxy. Composite Sci Technol 1994,50(3):343–354. 10.1016/0266-3538(94)90022-1

17. Chen WH, Yang SH: Multilayer hybrid-stress finite element analysis of composite laminates with delamination cracks originating from transverse cracking. Eng Fracture Mech 1996,54(5):713–729. 10.1016/0013-7944(95)00118-2

18. Johnson P, Chang FK: Characterisation of matrix crack-induced laminate failure - Part II: Analysis and verifications. J Composite Mater 2001b,35(22):2037–2074.

19. Li S, Reid SR, Zou Z: Modelling damage of multiple delaminations and transverse matrix cracking in laminated composites due to low velocity lateral impact. Composite Sci Technol 2006,66(6):827–836. 10.1016/j.compscitech.2004.12.019

20. Zhang H, Minnetyan L: Variational analysis of transverse cracking and local delamination in $[\theta_m/90_n]_s$ laminates. Int J Solid Struct 2006, 43: 7061–7081. 10.1016/j.ijsolstr.2006.03.004

21. Nairn JA, Hu S: Damage Mechanics of Composite Materials; chap. Matrix Microcracking. Elsevier Science; 1994. pp 187–243 pp 187–243

22. Nairn JA: Comprehensive Composite Materials: Polymer Matrix Composites; chap. Matrix Microcracking in Composites.. Oxford: Pergamon Press; 2000.

23. Ladevèze P: Mechanics of the 21st Century; chap. A bridge between the micro- and mesomechanics of laminates fantasy or reality?. Dordrecht,: Springer; 2005b. pp 187–201 pp 187–201

24. Violeau D, Ladevèze P, Lubineau G: Micromodel-based simulations for laminated composites. Composite Sci Technol2009,69(9):1364–1371. 10.1016/j.compscitech.2008.09.041

25. Ladevèze P, Lubineau G, Violeau D: A computational damage micromodel of laminated composites. Int J Fracture 2006a,137(1–4):139–150. 10.1007/s10704-005-3077-x

26. Hallett SR, Jiang WG, Khan B, Wisnom MR: Modelling the interaction between matrix cracks and delamination damage in scaled quasi-isotropic specimens. Composite Sci Technol 2008,68(1):80–89. 10.1016/j.compscitech.2007.05.038

27. Bouvet C, Castanié B, Bizeul M, Barrau J-J: Low velocity impact modelling in laminate composite panels with discrete interface elements. Int J Solids Struct 2009,46(14–15):2809–2821.

28. Van der Meer FP, Sluys LJ: Mesh-independent modeling of both distributed and discrete matrix cracking in interaction with delamination in composites. Eng Fracture Mech 2010,77(4):719–735. 10.1016/j.engfracmech.2009.11.010

29. Ladevèze P, Lubineau G, Marsal D: Towards a bridge between the micro- and mesomechanics of delamination for laminated composites. Composite Sci Technol 2006b, 66: 698–712. 10.1016/j.compscitech.2004.12.026

30. Hashin Z: Finite thermoelastic fracture criterion with application to laminate cracking analysis. J Mech Phys Solids 1996, 7: 1129–1145.

31. Wisnom MR, Hallett SR: The role of delamination in strength, failure mechanism and hole size effect in open hole tensile tests on quasi-isotropic laminates. Composites 2009,40(1):335–342.

32. Gao F, Jiao G, Lu Z, Ning R: Mode II delamination and damage resistance of carbon/epoxy laminates interleaved with thermoplastic particles. J Composite Mater 2007, 41: 111–123.

33. Czichon S, Zimmermann K, Middendorf P, Vogler M, Rolfes R: Three-dimensional stress and progressive failure analysis of ultra thick laminates and experimental validation. Composite Structures 2011, 93: 1394–1403. 10.1016/j.compstruct.2010.11.009

34. Chinesta F, Ladeveze P, Cueto E: A short review on model order reduction based on Proper Generalized Decomposition. Arch Comput Methods Eng 2011, 18: 395–404. 10.1007/s11831-011-9064-7

Chapter 5

MICROMECHANICAL ANALYSIS OF HYBRID COMPOSITES REINFORCED WITH UNIDIRECTIONAL NATURAL FIBRES, SILICA MICROPARTICLES AND MALEIC ANHYDRIDE

Leandro José da Silva[I]; Túlio Hallak Panzera, [I] André Luis Christoforo[I]; Juan Carlos Campos Rubio[I]; Fabrizio Scarpa[III]

[I]Department of Mechanical Engineering, Federal University of São João Del Rei - UFSJ, Brazil, Praça Frei Orlando, 170, São João Del Rei, MG, Brazil

[II]Department of Mechanical Engineering, Federal University of Minas Gerais - UFMG, Belo Horizonte, MG, Brazil

[III]Advanced Composites Centre for Innovation and Science, University of Bristol, Bristol, UK

ABSTRACT

The work describes the analytical and experimental characterisation of a class of polymeric composites made from epoxy matrix reinforced with unidirectional natural sisal and banana fibres with silica microparticles and maleic anhydride fabricated by manual moulding. The analytical models, ROM rule of mixtures and Halpin-Tsai approach, have been used in conjunction with a Design of Experiments (DOE) analysis from tensile tests carried out on 24 different composites architectures. The following experimental factors were analyzed in this work: type of fibres (sisal and banana fibres), volume fraction of fibres (30% and 50%) and modified matrix phase by adding silica microparticles (0%wt, 20%wt and 33%wt) and maleic anhydride (0%wt and 2%wt). The ROM approach has shown a general good agreement with the experimental data for composites manufactured with 30%vol of natural fibres, which can be attributed to the strong adhesion found between the phases. On the opposite, the semi empirical model proposed by Halpin and Tsai has shown greater fidelity with composites manufactured from 50%vol of natural fibres, which exhibit a weak interfacial bonding. The addition of microsilica and maleic anhydride in the system did not enhance the adhesion between the phases as expected.

INTRODUCTION

Biocomposites made from polymeric matrices and natural fibres appeared during the last decade as a sustainable alternative material for many applications related to aerospace, automotive and structural engineering. Interest towards an increased use of biocomposite structural materials has grown from the surging demand for low cost materials from environmental-friendly renewable sources, and the possibility of finding an alternative to traditional composites made of synthetic fibres[1]. The research and development activity on the use and disposal/recycling of synthetic fibres and resins derived from petroleum has also been motivated by increasingly stringent requirements by legal authorities, as well as by the high cost of the use of synthetic fibres for some non high-end engineering applications[1].

A variety of natural fibres have been evaluated as reinforcement phase in polymeric composites, such as the bagasse from sugar cane, sisal, jute, curauá, flax, piassava and banana plant[2]. Among them, the sisal fibre constitutes perhaps the most promising due to its low cost, high mechanical properties and market availability. Direct extraction of banana fibres is not common practice[3], however this particular type of biofibre can be considered as a side waste product from the cultivation of banana plants. The fibres extracted from the pseudo-stem of banana plant exhibit interesting mechanical properties for polymeric composites reinforcements[1].

One of the main difficulties when dealing with natural composites is the adhesion between fibres and matrices[4,5], mainly due to the hydrophilic and hydrophobic characteristics showed by the fibres and the polymers, respectively. However, the chemical affinity between the cellulose and the polymeric matrix can be improved by the modification of the fibre surface[6,7] or the polymer[8-11], using chemical additives like maleic anhydride.

Mishra et al.[6], Naik and Mishra[7], evaluated the effect of adding maleic anhydride on the sisal and banana fibres surface adhesion, observing a significant reduction of the water absorption and an increase of the modulus of elasticity, hardness and impact strength. An alternative method to improve the mechanical performance of biocomposites is by adding a second reinforcement phase. In that sense, several studies involving the fabrication of hybrid composites of polymeric matrices reinforced with fibres and nano or microparticles of ceramic minerals have been reported in open literature[12-17]. During a biocomposite failure, the crack initiates in the matrix phase and increases the debonding between fibre and the matrix itself[18]. When the crack propagation reaches the ceramic particles along the fibre-matrix interface, the crack is impeded to penetrate through the locations where the particles are concentrated, because of the high strength provided by the ceramics. Hence, additional effort is required for the crack to propagate through the fibre-particle interface or the particle-matrix interfaces (whichever is longer). This additional effort not only reduces the crack propagation velocity, but also increases the mechanical strength of the composite. One of the major difficulties in developing hybrid composites reinforced with fibres and particles is the prediction of the effective mechanical and physical properties[19]. The problem is more accentuated for structural biocomposites, because their natural fibres exhibit large variations in properties, due to the uncertainties associated to the environmental conditions (moisture, soil, temperature) in which these natural materials are produced[20]. The absence of robust micromechanical models predicting the mechanical properties of these hybrid biocomposites is a major obstacle towards the design of structural components using these novel types of composite[19].

Several models have been used in open literature to predict the effective properties of composite materials reinforced with long and short fibres, such as Rule of mixture (ROM)[19,20], Halpin-Tsai[20,21], shear-lag analysis[20,22] and Hashin-Strickman[23]. The rule of mixture (ROM) has shown its effectiveness on predicting the tensile strength of different natural fibres reinforced HDPE (high-density polyethylene)[24]. Halpin-Tsai model is also found to be the most effective equation in predicting the Young's modulus of composites containing different types of natural fibres[21.]

The Rule of mixture (ROM) is the simplest available micromechanical analysis model that can be used to predict the elastic properties of a composite material[19,20]. As an application of the ROM approach, Equation 1 shows how to estimate the effective modulus of elasticity (E*) of the composite, as a function of the properties of the fibres and matrix materials, considering the direction of fibre alignment. E_F, E_M, V_F and V_M are the modulus and volume fractions of the fibre and matrix materials respectively.

$$E^* = E_f V_f + E_m V_m$$

(1)

The rule of mixture assumes a perfect bonded interface between matrix and reinforcement(s). This assumption may be unrealistic for the majority of real manufactured composites, and therefore it is useful to adopt semi-empirical models like the Halpin-Tsai one[25], which compensates for non-perfect interface conditions. The calculation of the uniaxial tensile Young's modulus for a unidirectional composite according to Halpin-Tsai's approach is illustrated in Equation 2:

$$E^* = \frac{E_m(1+\xi\eta V_f)}{1-\eta V_f}$$

$$(2)$$

Where η is given as:

$$\eta = \frac{E_f - E_m}{E_f + \xi E_m}$$

$$(3)$$

The parameter ξ in Equations 2 and 3 is a shape fitting variable to fit the Halpin-Tsai equation to the experimental data, which describes also the packing arrangement and the geometry of the reinforcing fibres[20]:

$$\xi = \frac{E_f(E^* - E_m) - V_f E^*(E_f - E_m)}{E_m(E_f - E^*) - V_m(E_f - E_m)}$$

$$(4)$$

In this work it was evaluated the mechanical behaviour of a polymeric composite reinforced with unidirectional natural fibres, such as sisal and banana fibres, by the use of micromechanical models and experimental tests. The maleic anhydride was also investigated as a coupling agent between the phases. The experimental Young's modulus was compared with the uniaxial one predicted by rule of mixture and Halpin-Tsai equations. From the micromechanical analyses presented, it is possible to observe the effect of the hybridization and chemical additive on the interfacial adhesion between the constitutive phases.

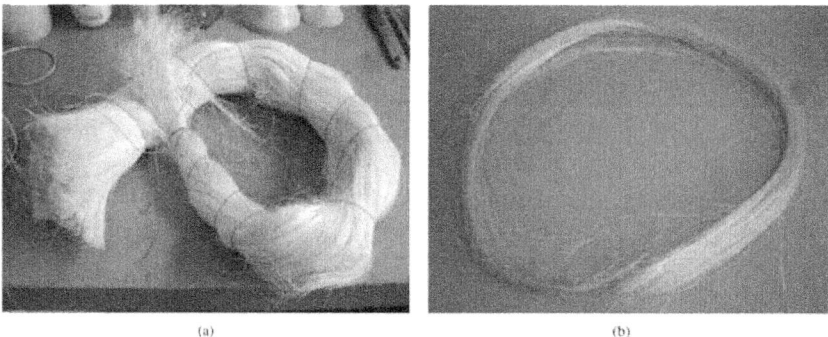

(a)   (b)

Figure 1: Sisal fibres (a) and banana fibres (b).

Materials and Methods

The polymeric composites were fabricated from modified and non-modified epoxy matrix, supplied by Resiqualy Company (São Paulo - Brazil), and from dispersive phase of unidirectional sisal (Figure 1a) and banana fibres (Figure 1b) supplied by Sisalsul Company (São Paulo - Brazil). The fibres were extracted, washed and combed by the supplier, with no chemical treatment. The matrix phase was modified by the addition of silica microparticles and maleic anhydride. The silica microparticles were supplied by Moinhos Gerais Company (Minas Gerais - Brazil), and classified by sieving process in monomodal range of 400-500 US-Tyler (0.037-0.025 mm).Table 1 exhibits the physical and mechanical properties of the silica powder were provided by Moinhos Gerais Company. The apparent density was determined using a gas pycnometer by Micromeritics model AccuPyc 1330 and the mechanical properties were estimated via dynamic ultra micro hardness tester by Shimadzu model DUH-211. The resin and the hardener were combined; afterwards the silica microparticles were added and hand-mixed by 5 minutes in room temperature around 22 °C.

Table 1: Properties of silica particles supplied by Moinhos Gerais Company

Properties	Unity	Lower limit	Higher limit
Apparent density	kg/m³	2170	2220
Young's modulus	GPa	56	74
Tensile strength	MPa	45	155
Compressive strength	MPa	1100	1600

Table 2: Experimental conditions

Conditions	Type of fibres	Volume fraction (%)	Maleic anhydride (%wt)	Silica addition (%wt)
C1	Sisal	30	0	0
C2	Sisal	30	2	0
C3	Sisal	30	0	20
C4	Sisal	30	2	20
C5	Sisal	30	0	33
C6	Sisal	30	2	33
C7	Sisal	50	0	0
C8	Sisal	50	2	0
C9	Sisal	50	0	20
C10	Sisal	50	2	20
C11	Sisal	50	0	33
C12	Sisal	50	2	33
C13	Banana	30	0	0
C14	Banana	30	2	0
C15	Banana	30	0	20
C16	Banana	30	2	20
C17	Banana	30	0	33
C18	Banana	30	2	33
C19	Banana	50	0	0
C20	Banana	50	2	0
C21	Banana	50	0	20
C22	Banana	50	2	20
C23	Banana	50	0	33
C24	Banana	50	2	33

Tensile tests were carried out according to ASTM D3822-07[26] and ASTM D638-03[27] standards to determine the tensile strength and modulus of elasticity of the fibres and matrix phase, respectively. The test speeds were set as 3 mm/min for sisal and banana fibres and 2 mm/min for polymeric matrices.

Figure 2 shows the samples for the non-modified and modified matrices which were used to evaluate the physical properties such as apparent density, apparent porosity and water absorption based on BS 10545-3[28]standard. The samples were fabricated manually by hand-mixed of epoxy resin, silica particles and maleic anhydride, for 5 minutes in room temperature around 22 °C. As a part of the overall mechanical characterisation of the composites, the tensile strength of the pure epoxy matrices was also determined experimentally.

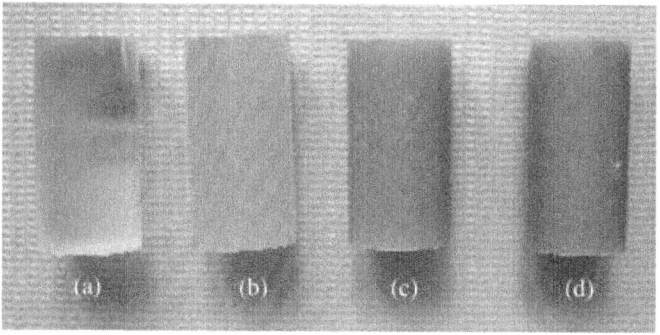

Figure 2: Non-modified matrix (a) and modified matrices by addition of 2%wt of maleic anhydride (b), 20%wt of silica microparticles (c) and 33%wt of silica microparticles (d).

The Design of Experiment (DOE) activity was carried out considering as experimental factors the type of natural fibres (sisal and banana), volume fraction of fibres (30% and 50%), maleic anhydride (0%wt and 2%wt) and silica microparticles (0%wt, 20%wt and 33%wt). The combination of these factors leads to investigate a total of 24 experimental conditions (see Table 2).

Preliminary tests were conducted in order to set the upper volume fraction of fibres (50%) and silica particles (33%wt) in the system, to obtain a suitable surface finishing and lower porosity. A large percent of natural fibres contributes to an overall lower cost of the composite (i.e. for a 50/50%vol laminate, the cost of epoxy resin corresponds nearly to ten times higher than the sisal phase) and also a more sustainable composite material in terms of recycling and sourcing. Mixture time (5 minutes), cure time (7 days), room temperature (~22 °C) and the epoxy resin matrix were kept constants during the DOE process.

The biocomposite laminates were fabricated aligning manually the fibres by the aid of a metal frame. The manual moulding process was carried out over a glass plate covered by a cloth parting (Armalon), providing good surface finishing to the lamina. The polymeric matrix (modified and non-modified) was spread on the fibres by the use of spatula and roller aerator. A glass fibre composite was used to protect the specimen ends at the clamping area, avoiding premature crack during the tensile testing (see Figure 3).

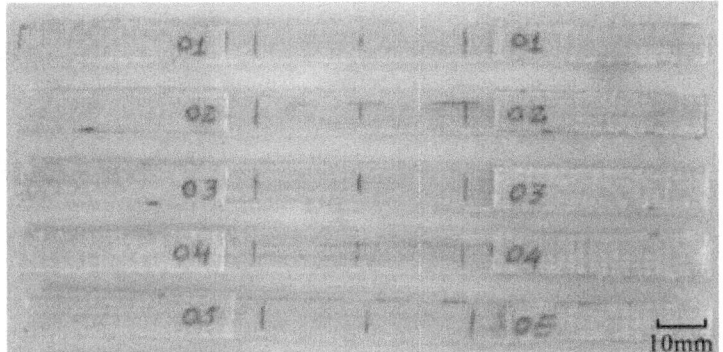

Figure 3: Tensile test specimens.

A scanning electron microscope (SEM - Hitachi T-3000) was used to observe the cross section of the composites. The tensile and flexural testing was carried out following the BSI standard 2747[29] using an Autograph machine monitored by a Topazium software with load cell maximum capacity of 20 kN. The test speed of the tensile tests was 2 mm/min.

A randomization procedure was adopted during the sample fabrication and experimental tests. This randomization let an arbitrary ordering of the experimental conditions, avoiding that non controlled factors affect the responses[30].

The effective modulus of the composites was estimated using the ROM and Halpin-Tsai model. The particulate phase was not directly considered in the micromechanical analysis, using instead non-modified and modified matrices mechanical properties in the models. The comparison between experimental and predicted results allows verifying whether the particles and/or the chemical additions contribute to the fibre-matrix adhesion.

Results

Table 3 shows the physical and mechanical properties of sisal and banana fibres evaluated within this work[31].Table 3 shows the mean values of the properties with the respective standard deviations. The banana fibres exhibited in general a lower density and higher porosity than sisal fibres. The tensile strength between the different types of fibres is quite similar; however the banana fibres appear stiffer than the sisal ones, showing a modulus of elasticity of 31.6GPa ± 2.8. The critical constituent responsible for natural fibre strength and stiffness are cellulose microfibrils. These microfibrils have a width ranging from 5 to 30 nm, are highly crystalline materials formed by the aggregation of long thread like bundles of molecules stabilized laterally by hydrogen bonds between hydroxyl groups and oxygens of adjacent molecules[32]. According to Joseph et al.[33] the percent of cellulose in sisal and banana fibres is nearly 70 and 83%, respectively.

Table 3: Properties of sisal and banana fibres (mean values and standard deviation)

Properties	Sisal fibre	Banana fibre
Diameter (μm)	192.5 (±26.3)	131.1 (±17.7)
Apparent density (g/cm³)	1.41 (±0.12)	1.35 (±0.09)
Apparent porosity (%)	76.21 (±2.01)	86.69 (±1.76)
Tensile strength (MPa)	887 (±143)	1063 (±259.5)
Modulus of elasticity (GPa)	16.4 (±2.5)	31.56 (±2.8)

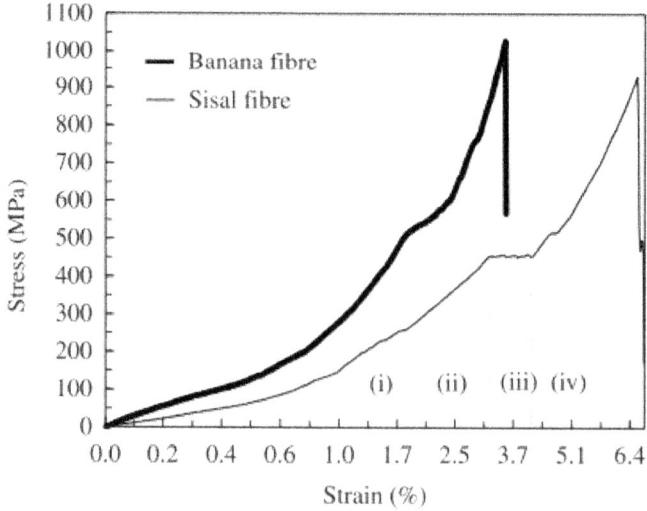

Figure 4. Tipical stress/strain curves for sisal and banana fibres.

Figure 4 shows the stress-strain behaviour of the sisal and banana fibres under tensile loading. The stress-strain curve of the sisal (Figure 4a) and banana (Figure 4b) fibres can be approximately divided in four stages. During stage (i), the stiffness reaches a maximum of 200 MPa, increasing to a value of 450 MPa during stage (ii). The 450 MPa value was used to calculate the Young›s modulus to be inserted in the Equations 1-4 for micromechanics analysis. Stage (iii) features a large elongation of the fibres, which can be attributed to the initial fraying effect. During stage (iv) one can observe a significant increase in tensile stress, achieving a maximum value close to 1000 MPa for sisal fibres and 1200 MPa for banana ones.

Table 4 shows the physical and mechanical properties related to the modified and non-modified termoset matrices. Table 4 shows the mean values of the properties with the respective standard deviations. It is apparent that, although, the addition of maleic anhydride did not affect the physical properties of the matrices, the addition of silica microparticles increased the material›s density, which can be attributed to the higher density of the silica particles ($\sim$2.2 g.cm^{-3}). However, the inclusion of silica in the composites did not affect the apparent porosity and water absorption of the matrices

Table 4: Properties of polymeric matrices (mean values and standard deviation)

Setup	Apparent density (g/cm³)	Apparent porosity (%)	Water absorption (%)	Tensile strength (MPa)	Modulus of elasticity (GPa)
Epoxy resin	1.16 (±0.00)	0.30 (±0.07)	0.26 (±0.06)	31.99 (±2.72)	0.83 (±0.05)
2% of MA	1.15 (±0.00)	0.30 (±0.06)	0.26 (±0.06)	35.73 (±0.87)	0.81 (±0.03)
20% of silica	1.28 (±0.01)	0.30 (±0.06)	0.24 (±0.05)	26.26 (±1.36)	0.95 (±0.03)
33% of silica	1.34 (±0.02)	0.29 (±0.07)	0.22 (±0.05)	22.54 (±2.64)	1.10 (±0.07)

Figure 5 shows the mechanical behaviour of the modified matrices for 0%wt and 2%wt of maleic anhydride dispersions, revealing an increase of the tensile strength and tenacity when the chemical agent is added. However, the chemical additive did not affect significantly the value of the modulus of elasticity. From Figure 5 it is possible to observe that adding silica microparticles leads to a decrease of the tensile strength of the matrices. As expected the Young›s modulus of the epoxies increased by 14% and 32% when 20%wt and 33%wt of silica were added, respectively. This behaviour can be attributed to the high stiffness of the particulate phase, contributing to increase the modulus of elasticity of the matrices. These results are in accordance with previous works published in the open literature[13-14], where the Young's modulus of the composites increases by the addition of particles into polymeric matrix.

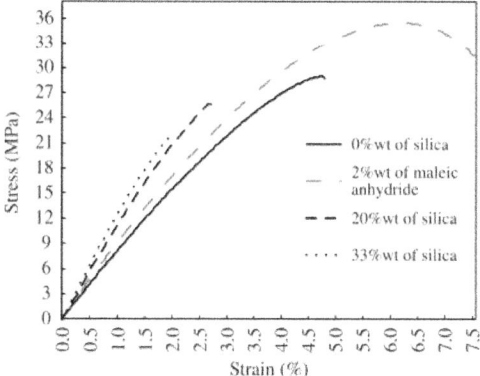

Figure 5: Stress/strain curves of the non-modified and modified matrices with maleic anhydride and silica microparticles.

Table 5 contains the results related to the experimental (E) and numerical (N) unidirectional Young's modulus of the fibres. The analytical values from the micromechanical models, were calculated based on the individual mean properties of the fibres and matrix phases from Tables 3 and 4. Table 5 shows the relation between the experimental (E) and numerical (N) modulus of elasticity.

Table 5: Results obtained from tensile testing and micromechanical analysis

Experimental conditions	Experimental modulus of elasticity (MPa)	Rule of mixture (MPa)	E/N*	Halpin-Tsai (MPa)	E/N**
C1	5722	5502	1.04	2400	2.38
C2	5559	5487	1.01	2376	2.34
C3	5632	5583	1.01	2526	2.23
C4	6158	5423	1.14	2275	2.71
C5	5440	5690	0.96	2693	2.02
C6	5217	5766	0.90	2811	1.86
C7	5912	8616	0.69	4923	1.20
C8	5030	8605	0.58	4902	1.03
C9	4978	8673	0.57	5035	0.99
C10	5298	8559	0.62	4813	1.10
C11	4866	8750	0.56	5183	0.94
C12	4762	8804	0.54	5287	0.90
C13	9105	10050	0.91	3770	2.42
C14	8650	10035	0.86	3745	2.31
C15	8045	10131	0.79	3897	2.06
C16	7685	9971	0.77	3644	2.11
C17	7270	10238	0.71	4066	1.79
C18	7850	10314	0.76	4185	1.88
C19	8255	16196	0.51	8721	0.95
C20	7920	16185	0.49	8699	0.91
C21	6450	16253	0.40	8834	0.73
C22	6940	16139	0.43	8609	0.81
C23	7260	16330	0.44	8984	0.81
C24	6565	16384	0.40	9089	0.72

*Relation between experimental and predicted modulus of elasticity by rule of mixture model. **Relation between experimental and predicted modulus of elasticity by Halpin-Tsai equations.

If the value E/N is higher than 1.0, indicates that the experimental value is higher than the predicted one.

Composites fabricated with sisal fibres

Figure 6 shows the comparison between ROM, Halpin-Tsai and experimental results for the C1 to C12 sisal fibres composites.

Figure 6a shows the modulus of elasticity for the C1 and C7 composites; i.e. those composites manufactured with 30%vol and 50%vol of sisal fibres, respectively, with no silica and maleic anhydride addition.

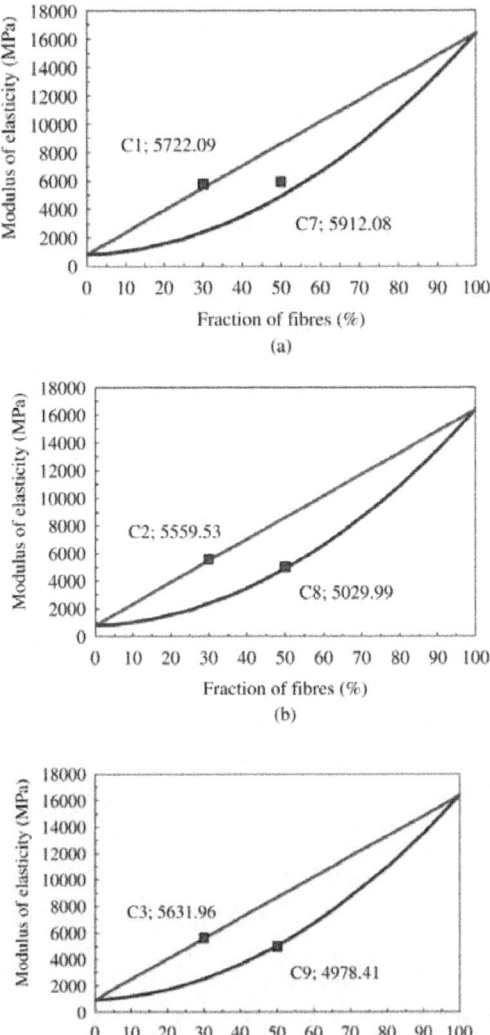

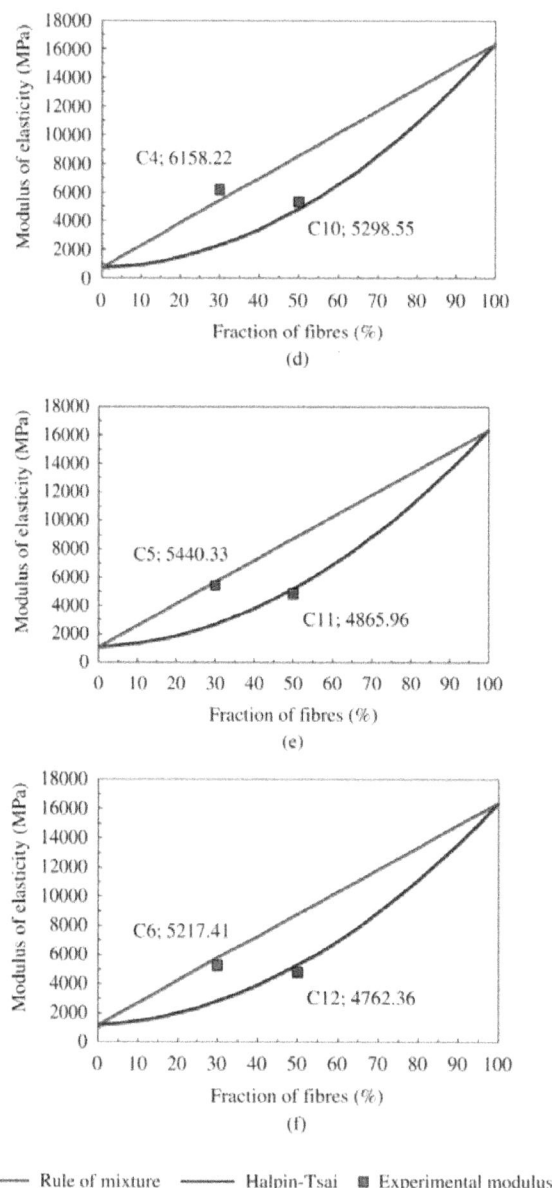

Figure 6. Micromechanical analysis and experimental results for C1 to C12 composite

It can be observed that the composite with 30%vol of fibres follows the ROM model ($E/N^* = 1.04$), while the Halpin-Tsai approach underestimates the effective modulus of elasticity ($E/N^{**} = 2.38$). This result indicates a strong interfacial adhesion for C1 composite.

On the opposite, the experimental results for C7 composite showed higher agreement with the Halpin-Tsai equation (see Table 5), which suggests the presence of an non-perfect bonding condition at the interface between fibres and matrix. This behaviour can be attributed to the small amount of matrix phase in the system (50%), which affects the matrix wetting capacity around the fibres, and consequently, the increase in porosity of the composites.

Figure 6b features the Young's modulus of the composites fabricated with maleic anydride addition, corresponding to C2 and C8 composites (30%vol and 50%vol of sisal fibres, respectively). Similarly to the composites with non-modified epoxy resin shown in Figure 6a, the interface condition can be considered perfect for the low level of sisal fibres (30%vol), exhibiting a E/N* ratio of 1.01. The improved agreement provided by the Halpin-Tsai approach (E/N** = 1.06) suggests also in this case the existence of an imperfect interfacial condition was achieved for the composites with high level of sisal fibres addition (50%vol).

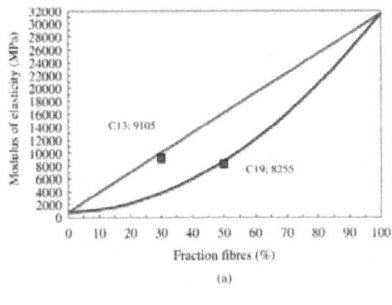

(a)

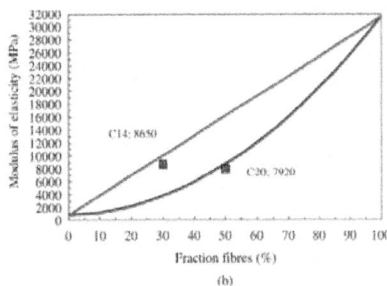

(b)

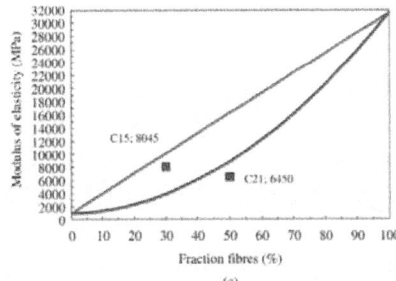

(c)

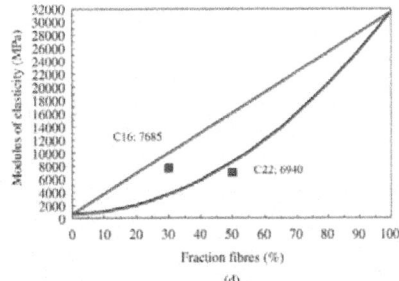

(d)

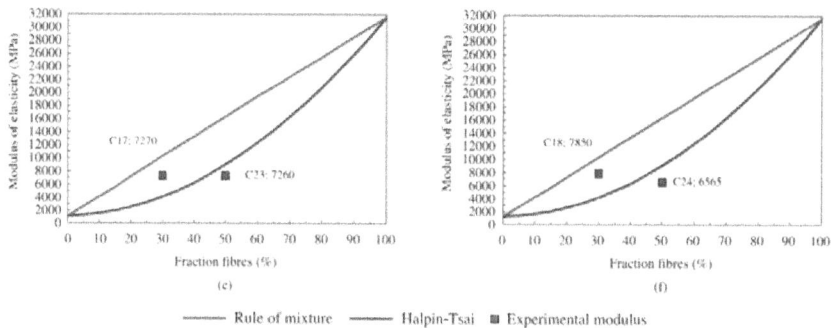

Figure 7: Micromechanical analysis and experimental results for C13 to C24 composites.

The micromechanical analyses of the composites made with 20%wt of microsilica addition (C3 and C9 composites) is shown in Figure 6c. It can be observed a particularly good agreement (E/N* = 1.01) between the experimental and rule of mixture modulus predicted for the composite manufactured with 30%vol of sisal fibres (C3 composite). On the opposite, the composite with high level of sisal fibres (50%vol) revealed an imperfect interface condition, exhibiting a large deviation of rule of mixture prediction (E/N* = 0.57); while the Halpin-Tsai model provided a higher fidelity prediction (E/N** = 0.99). Figure 6d shows the estimated modulus for the composites made with 20%wt of silica and 2%wt of maleic anhydride addition (C4 and C10 composites). The results indicate that the maleic anhydride affected the rheology of the system, increasing the elastic modulus of the composites. The increase of the fibre fraction reduced the interfacial adhesion between the phases, a fact also confirmed by the low fidelity of the ROM prediction in this case (E/N* = 0.62). Figure 6e presents the behaviour of the C5 and C11 composites, i.e. those composites manufactured with 30%vol and 50%vol of sisal fibres and 33%wt of silica addition. The composites made with 30%vol of fibres featured a higher agreement between the rule of mixture model and experimental Young›s modulus (E/N* = 0.96). However, the composite with 50% of fibres was better described by the Halpin-Tsai approach (E/N** = 0.94).

Although the addition of 33%wt of silica microparticles increased the stiffness of the matrix (see Table 4), it was not able to increase the modulus of elasticity of the composites. This behaviour can be attributed to the increase of surface area due to the silica particles addition, affecting not only the rheology of the system but also the matrix wetting capacity on the fibre surface. The addition of silica microparticles also increases the porosity of the composites, therefore contributing to the reduction of the mechanical properties.

Divergence between the experimental results and predicted stiffness by ROM and Halpin-Tsai models is shown inFigure 6f for the C6 and C12 composites, which were manufactured with 30%vol and 50%vol of sisal fibres, respectively, and 33%wt of silica microparticles and 2%wt of maleic anhydride added into the polymeric matrix phase. The modulus of elasticity of the matrix was increased by silica and maleic anhydride addition (Figure 5). However, the stiffer matrices did not originate stiffer composites. This behaviour confirms the hypothesis of fibre-matrix interface adhesion reduction provided by the addition of high content of silica micro particles.

Composites manufactured with banana fibres

Figure 7 shows the analytical predicted and experimental results related to the composites reinforced with banana fibres (C13 to C24, see Table 2). The same discussions performed for the composites reinforced with sisal fibres (see section 3.1) can be extended for the composites reinforced with banana fibres.

The composites manufactured with low level of banana fibres (30%vol) showed a better agreement with the rule of mixture model, while the composites with high level of fibres (50%vol) were better described using the Halpin-Tsai approach, indicating therefore a poor interface condition for those composites compared to the ones of the previous case. However, based on the results shown in Table 5, it is possible to verify that the elastic moduli of banana fibre composites are in general lower than the elastic moduli estimated by the micromechanical analysis. Higher divergence between experimental moduli and predicted moduli were observed for the banana fibre composites in comparison to sisal fibre composites, for both micromechanical models. This result implies that the composites fabricated with sisal fibres show in general a sounder fibre-matrix adhesion than the banana fibre composites.

Figure 8 shows the backscatter mode SEM images at 100× of magnification featuring the failure surface of the sisal (Figure 8a) and banana (Figure 8b) fibre composites after tensile testing. The sisal fibre composites shows a fracture mode more brittle than the banana fibre composites.

Based on the investigations of Facca et al.[20] and Ku et al.[24], the Halpin-Tsai model well described the experimental modulus of HDPE (High-density polyethylene) composites reinforced with different types of short natural fibres and volume fractions varying from 0%wt up to 40%wt. In contrast, the rule of mixture was not able to predict the experimental data. In the present work, the composites were fabricated with unidirectional banana fibres. The Young's modulus was better predicted by Halpin-Tsai model when 50%vol of fibres were added. However, when the composites were fabricated with 30%vol of natural fibres, the rule of mixture presented a better prediction, especially for the sisal fibres. This result reveals the Rule of mixture (ROM) can be acceptably applied to estimate the tensile modulus of biocomposites with good interfacial adhesion.

CONCLUSIONS

The experimental and analytical Young's moduli of structural biocomposites based on sisal and banana fibres were evaluated in this work. The experimental results were generated through a Design of Experiments approach. The main conclusions from this work are the following:

- The banana fibres have shown a general higher stiffness than sisal fibres, however the sisal fibres exhibited a superior tensile strength than the banana ones;

- The mechanical behaviour under tensile loading is very similar for both natural fibres, featuring four different stages in their stress-strain behaviour. A fraying effect of fibres was observed when the stress is around 450 MPa, subsequently the stiffness is increased, achieving a maximum tensile stress close to 1000 MPa for sisal fibres and 1200 MPa for banana fibres;

- The addition of 2%wt of maleic anhydride into matrix phase provided not only the increase of tensile strength, but also the tenacity of the polymer itself;

- The addition of silica microparticles into the matrix phase led not only to the reduction of tensile strength, but also to the increase of the Young's modulus of the polymer;

- The micromechanical analysis provided some indications about the interfacial conditions between fibres and matrices within the natural composites. The rule of mixture showed higher fidelity when low levels of fibres (30%vol) enhancing the wetting capacity of the matrix were added. The Halpin-Tsai results were providing a higher correlation with those composites fabricated with 50%vol of fibres, revealing the presence of a poor interfacial adhesion;

- The sisal fibres are less porous than banana fibres, therefore absorbing less matrix phase. Low porosity indicates better interfacial condition, and higher fidelity of the predictions provided by both ROM and Halpin-Tsai models;

- The addition of silica microparticles increases the stiffness of the matrices, but does not seem sufficient to improve the elastic moduli of the composites, with a decrease of the level of adhesion between the composites phases;

- The addition of maleic anhydride did not show a relevant effect on the interfacial adhesion, featuring instead a small increase of the Young's modulus for the composites C4 and C8 (sisal fibres - 20%wt of silica addition) and C18 and C24 (banana fibres - 33%wt of silica addition). Further investigation need to be performed to assess the effect of this material as an efficient coupling agent.

ACKNOWLEDGEMENTS

The authors would like to thank the financial support of CAPES and the material suppliers: Resiqualy Company (epoxy resin), Sisalsul Company (sisal fibres) and Moinhos Gerais Company (silica particles).

REFERENCES

1. Silva RV. Compósito de resina poliuretana derivada de *óleo* de mamona e fibras vegetais. [Tese]. São Carlos: Escola de Engenharia de São Carlos, Universidade de São Paulo; 2003.

2. Bledzki AK, Mamun AA and Faruk O. Abaca fiber reinforced PP composites and comparison with jute and flax PP composite. Express Polymer Letters. 2007; 1:755-762.http://dx.doi.org/10.3144/expresspolymlett.2007.104.

3. Mukhopadhyay S, Fangueiro R and Shivankar V. Variability of Tensile Properties of Fibers from Pseudostem of Banana Plant. Textile Research Journal. 2009; 79:387-393.http://dx.doi.org/10.1177/0040517508090479 .

4. Maldas D and Kokta BV. Influence of polar monomers on the performance of wood fiber reinforced polystyrene composites. I. Evaluation of critical conditions. International Journal of Polymeric Materials. 1990; 14:165-189. http://dx.doi.org/10.1080/00914039008031512.

5. Lyons JS and Ahmed MR. Factors Affecting the Bond Between Polymer Composites and Wood. Journal of Reinforced Plastics and Composites. 2005; 24:404-405.http://dx.doi.org/10.1177/0731684405044898.

6. Mishra S, Naik JB and Patil YP. The compatibilising effect of maleic anhydride on swelling and mechanical properties of plant-fiber-reinforced novolac composites. Composites Science and Technology. 2000; 60:1729-1735. http://dx.doi.org/10.1016/S0266-3538(00)00056-7.

7. Naik JB and Mishra S. Esterification Effect of Maleic Anhydride on Swelling Properties of Natural Fiber/High Density Polyethylene Composites. Journal of Applied Polymer Science. 2007; 106:2571-2574. http://dx.doi.org/10.1002/app.25329.

8. López Manchado MA, Arroyo M, Biagiotti J and Kenny JM. Enhancement of mechanical properties and interfacial adhesion of PP/EPDM/Flax fiber composites using maleic anhydride as a compatibilizer. Journal of Applied Polymer Science. 2003; 90:2170-2178. http://dx.doi. org/10.1002/app.12866.

9. Sombatsompop N, Yotinwattanakumtorn C and Thongpin C. Influence of type and concentration of maleic anhydride grafted polypropylene and impact modifiers on mechanical properties of PP/Wood sawdust composites. Journal of Applied Polymer Science. 2005; 97:475-484. http://dx.doi.org/10.1002/app.21765.

10. Kim S, Moon J, Kim G and Ha S. Mechanical properties of polypropylene/ natural fiber composites: Comparison of wood fiber and cotton fiber. Polymer Testing. 2008; 27:801-806.http://dx.doi.org/10.1016/j. polymertesting.2008.06.002.

11. Soleimani M, Tabil L, Panigrahi S and Opoku A. The effect of fiber pretreatment and compatibilizer on mechanical and physical properties of flax fiber-polypropylene composites. Journal of Polymers and the Environment. 2008; 16:74-82. http://dx.doi.org/10.1007/s10924-008-0102-y.

12. Rosso P, Ye L, Friedrich K and Sprenger S. A Toughened Epoxy Resin by Silica Nanoparticle Reinforcement.Journal of Applied Polymer Science. 2006; 100:1849-1855. http://dx.doi.org/10.1002/app.22805.

13. Isik I, Yilmazer U and Bayram G. Impact modified epoxy/montmorillonite nanocomposites: synthesis and characterization. Polymer. 2003; 44:6371-6377. http://dx.doi.org/10.1016/S0032-3861(03)00634-7.

14. Yasmin A, Abot JL and Daniel IM. Processing of clay/epoxy nanocomposites by shear mixing. Scripta Materialia. 2003; 49:81-86. http://dx.doi.org/10.1016/S1359-6462(03)00173-8.

15. Haque A, Shamsuzzoha M, Hussain F and Dean D. S2-Glass/Epoxy Polymer Nanocomposites: Manufacturing, Structures, Thermal and Mechanical Properties. Journal of Composite Materials. 2003; 20:1821-1837.http://dx.doi.org/10.1177/002199803035186.

16. Subramaniyan AK and Sun CT. Enhancing compressive strength of unidirectional polymeric composites using nanoclay. Composites Part A: Applied Science and Manufacturing. 2006; 37:2257-2268.http://dx.doi. org/10.1016/j.compositesa.2005.12.027.

17. Tsai JL and Cheng YL. Investigating Silica Nanoparticle Effect on Dynamic and Quasi-static Compressive Strengths of Glass Fiber/Epoxy

Nanocomposites. Journal of Composite Materials. 2009; 43:3143-3155. http://dx.doi.org/10.1177/0021998309345317.

18. Cao Y and Cameron J. Impact properties of silica particle modified glass fiber reinforced epoxy composite.Journal of Reinforced Plastics and Composites. 2006; 25:761-769.http://dx.doi. org/10.1177/0731684406063536.

19. Casaril A, Gomes ER, Soares MR, Fredel MC and Al-Qureshi HA. Análise micromecânica dos compósitos com fibras curtas e partículas. Matéria. 2007; 12:408-419.

20. Facca AG, Kortschot MT and Yan N. Predicting the elastic modulus of natural fibre reinforced thermoplastics.Composites Part A: Applied Science and Manufacturing. 2006; 37:1660-1671.http://dx.doi. org/10.1016/j.compositesa.2005.10.006.

21. Biagiotti J, Fiori S, Torre L, López-Manchado MA and Kenny JM. Mechanical Properties of Polypropylene Matrix Composites Reinforced With Natural Fibers: A Statistical Approach. *Polymer Composites.* 2004; 25:26-36.http://dx.doi.org/10.1002/pc.20002.

22. Nairn JA. On the Use of Shear-Lag Methods for Analysis of Stress Transfer in Unidirectional Composites.Mechanics of Materials. 1997; 26:63-80. http://dx.doi.org/10.1016/S0167-6636(97)00023-9.

23. Sejnoha M and Zeman J. On adequacy of the Hashin-Shtrikman variational principles applied to polymer matrix based random fibrous composites. In: Proceedings of the VIII International Conference on Computational Plasticity - COMPLAS VIII; 2005; Barcelona. Barcelona; 2005.

24. Ku H, Wang H, Pattarachaiyakoop N and Trada M. A review on the tensile properties of natural fiber reinforced polymer composites. Composites Part B: Engineering. 2011; 42:856-873.http://dx.doi.org/10.1016/j. compositesb.2011.01.010.

25. Halpin JC and Kardos JL. The Halpin-Tsai Equations: A Review. Polymer Engineering & Science. 1976; 16:344-352. http://dx.doi.org/10.1002/ pen.760160512.

26. American Society for Testing and Materials - ASTM. ASTM D 3822: Standard Test Method for Tensile Properties of Single Textile Fibers. ASTM; 2007.

27. American Society for Testing and Materials - ASTM. ASTM D 638: Standard Test Method for Tensile Properties of Plastics. ASTM; 2003.

28. British Standard. BS 10545-3: Ceramic Tiles - Part 3: Determination of water absorption, apparent porosity, apparent relative density and bulk density. BS; 1997.

29. British Standard. BS 2747: Glass fibre reinforced plastics, tensile test. BS; 1998.

30. Werkema MCC and Aguiar S. Planejamento e análise de experimentos: como identificar e avaliar as principais variáveis influentes em um processo. Belo Horizonte: Fundação Christiano Ottoni, Escola de Engenharia da UFMG; 1996.

31. Silva LJ. Estudo Experimental e Numérico das propriedades mecânicas de compósitos poliméricos laminados com fibras vegetais. [Dissertação]. São João Del Rei: Universidade Federal de São João Del Rei; 2011. 151 p.

32. Deepa B, Abraham E, Cherian BM, Bismarck A. Blaker JJ, Pothan LA et al. Structure, morphology and thermal characteristics of banana nano fibers obtained by steam explosion. Bioresource Technology. 2011; 102:1988-1997. PMid:20926289. http://dx.doi.org/10.1016/j.biortech.2010.09.030.

33. Joseph K, Tolêdo Filho RD, James B, Thomas S and Carvalho LH. A Review on Sisal Fiber Reinforced Polymer Composites. Revista Brasileira de Engenharia Agrícola e Ambiental. 1999; 3:367-379.

Chapter 6

STRAIN RATE DEPENDENT DEFORMATION OF A POLYMER MATRIX COMPOSITE WITH DIFFERENT MICROSTRUCTURES SUBJECTED TO OFF-AXIS LOADING

Xiaojun Zhu, Xuefeng Chen, Zhi Zhai, Zhibo Yang, Xiang Li, and Zhengjia He

State Key Lab for Manufacturing Systems Engineering, Xi'an Jiaotong University, Xi'an 710049, China

ABSTRACT

This paper aims to investigate the comprehensive influence of three microstructure parameters (fiber cross-section shape, fiber volume fraction, and fiber off-axis orientation) and strain rate on the macroscopic property of a polymer matrix composite. During the analysis, AS4 fibers are considered as elastic solids, while the surrounding PEEK resin matrix exhibiting rate sensitivities are described using the modified Ramaswamy-Stouffer viscoplastic state variable model. The micromechanical method based on generalized model of cells has been used to analyze the representative volume element of composites. An acceptable agreement is observed between the model predictions and experimental results found in the literature. The research results show that the stress-strain curves are sensitive to the strain rate and the microstructure parameters play an important role in the behavior of polymer matrix.

INTRODUCTION

In the last few decades, polymer matrix composite materials (PMCs) have been developed rapidly to meet the demands for better materials with higher standards of performance and reliability in structures and machines [1, 2]. In some of these applications such as marine structures, aerospace, and lightweight armor, the PMCs are often subjected to complex loadings under extreme circumstances [3, 4] in which the properties of the PMCs exhibit highly nonlinear and rate dependence, so it is necessary for structural design and analysis to characterize and model the nonlinearity and strain rate dependence of the composite.

Polymers are known to have a strain rate dependent deformation response that is nonlinear above 1 or 2% strain [5]. Many experimental studies have been made to determine the effects of strain rate on the PMCs [6]. Weeks [7] conducted experiments using an MTS machine and the split Hopkinson pressure bar for AS4/PEEK composite and produced strain rates ranging from 0.00001/s to 1000/s. Uniaxial tension tests were conducted on various off-axis coupon specimens to obtain stress/strain curves for various strain rates [8]. Haque and Ali [9] adopted a systematic experimental approach to identify the damage progression at various stress levels and the strain rate effects on composites. Shokrieh and Omidi [10] studied tensile failure properties unidirectional glass/epoxy composites at various strain rates from 0.001/s to 100/s using a high-speed servohydraulic testing apparatus. Experimental results showed a significant increase of the tensile strength by increasing the strain rate.

On the other hand, there are also many macromechanical and micromechanical models to predict the behavior of composite materials subjected to different strain rates [11, 12]. Weeks and Sun [13] developed a macromechanical, rate dependent constitutive model to analyze the inelastic response of carbon reinforced composites. Thiruppukuzhi and Sun [14] later directly incorporated the rate dependence of the material response into the constitutive model. Espinosa et al. [15] presented a 3D finite deformation anisotropic viscoplasticity model to analyze the effects of strain rate and temperature on a woven composite made of S-2 glass fibers. A 3D model based on finite elastoplasticity was applied to study the effect of temperature and strain rate on the tensile behaviour on a series of polymeric matrix unidirectional glass-fibre composites [16]. Recently, a Johnson-Cook based modeling approach was used to represent the apparent strain rate dependency of textile reinforced composites in laminate through-thickness direction [17]. A phenomenological-based approach was proposed by Raimondo et al. [18] for the three-dimensional modeling of strain rate in unidirectional polymer composites. A nonlinear constitutive model for large deformation loading at different strain rate condition was developed to represent tensile progressive damage of the nonlinear large deformation rate dependent behavior of polymer-based composite materials [19].

Compared with macromechanical model, which considered composites as anisotropic medium with homogeneous distribution, the micromechanical model only needs to test the ingredient properties of composites, while macromechanical model needs to do repetitive experiments for composites [20].

Therefore, many scholars have done a lot of research on the micromechanical model for years. A 3D micromechanical formulation was proposed [21] for fiber composites with viscoplastic matrix properties. The nonlinear responses of composites under various cyclic loading conditions were predicted accurately by their analysis. Goldberg and Stouffer [22] adopted a four-region micromechanics method, in which the composite unit cell is divided into a number of slices to analyze polymer matrix composites subject to different strain rates. Later, the micromechanical model was implemented in the nonlinear finite element software LS-DYNA [23]. By combining the bridging micromechanics model [24] with classical lamination, a general constitutive relationship was established for the inelastic and failure analysis of laminate structures [25]. Paley and Aboudi [26] proposed the generalized method of cells (GMC) to deal with the representative volume element (RVE) with complex microstructures. Ogihara et al. [27] adopted the GMC to study the nonlinear behavior of unidirectional carbon-epoxy laminates subjected to off-axis loading. The epoxy matrix was predicted using the one-parameter plasticity model. Tsai and Chen [28] employed the GMC to characterize the nonlinear rate-dependent behaviors of graphite/epoxy composites. The epoxy matrix is described by a three-parameter viscoplasticity model. However, the comprehensive effect of three microstructure parameters (fiber cross-section shape, fiber volume fraction, and fiber off-axis orientation) and the strain rate on the macroscopic property of composites has been seldom reported in the above studies. In this paper, by combining the GMC with the modified Ramaswamy-Stouffer viscoplastic state variable model, and with no need to judge whether the material is in elastic or plastic stage and is more convenience and effective to predict the matrix behavior [29], a new general constitutive relationship was established for the inelastic analysis of the comprehensive influence of three microstructure parameters and strain rate on the stress-strain behavior of the polymer matrix composite.

In this paper, the rest outline is as follows. Section 2 introduces the micromechanical model based GMC. In Section 3, the modified Ramaswamy-Stouffer viscoplastic state variable model is incorporated into GMC. Composites with three microstructure parameters are considered to analyze the rate dependent stress-strain response in Section 4. Conclusions are given in Section 5.

MICROMECHANICAL MODEL BASED GENERALIZED MODEL OF CELLS

Generalized Model of Cells

The two-dimensional generalized method of cells is a micromechanical model developed originally by Paley and Aboudi [26] for predicting the response of unidirectional matrix composites with periodic microstructures. The GMC was then reformulated in terms of the interfacial subcell tractions substituting the subcell strains as the basic unknowns by Pindera and Bednarcyk [30], which can significantly increase the calculation efficiency when the number of subcells became larger.

When a micromechanical approach is used to model the mechanical response of fiber reinforced composites with periodic microstructures, a proper RVE is required to represent the microstructures of the materials such that the overall composites responses can be predicted directly from the representative volume element. In this study, three kinds of fiber cross-section shapes, such as square, circular, and elliptical, were considered as shown in Figure 1. In this figure, the fiber and matrix are indicated by the black and white, respectively.

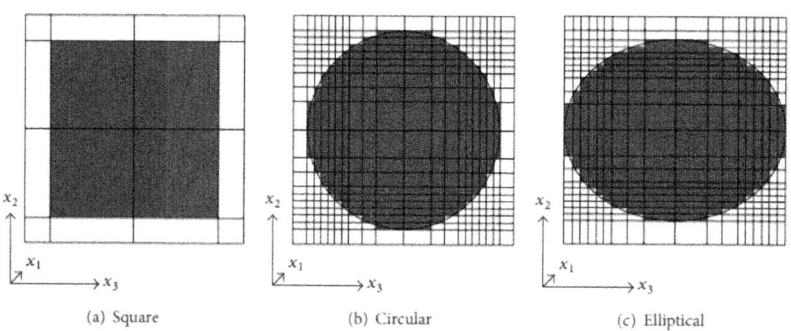

(a) Square   (b) Circular   (c) Elliptical

Figure 1: Three kinds of fiber shapes.

In the GMC analysis, the representative volume element is usually divided into $N_\beta \times N_\gamma$ subcells as shown in Figure 2.

In general, each of these subcells is assumed to be occupied by a material that exhibits inelastic behavior. The subcell material's inelastic behavior can be modeled by a lot of constitutive theories, such as classical incremental plasticity, linear viscoelasticity, or unified viscoplasticity theories. Therefore, the representative volume element, which consists of $N_\beta \times N_\gamma$ different inelastic materials, can represent a multiphased, inelastic composite.

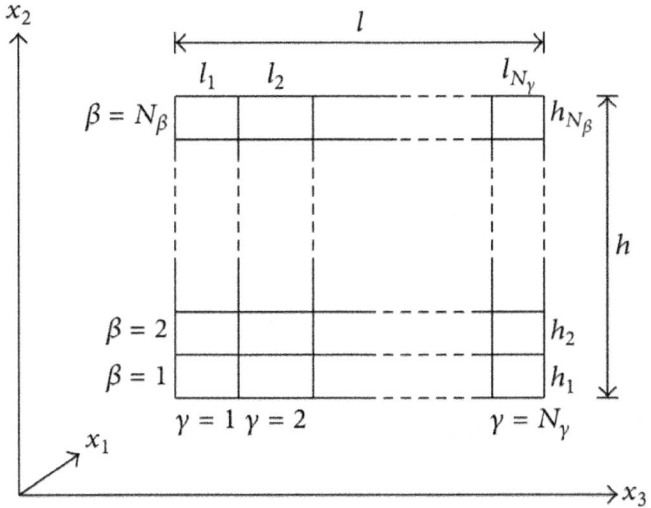

Figure 2: A typical RVE divided into $N_\beta \times N_\gamma$ subcells in GMC analysis.

Based on the displacement continuity on the interface of the adjacent subcells in conjunction with the periodicity condition of the RVE, the relation between overall strain and the subcell strain is expressed as

$$\bar{\varepsilon}^{(\beta\gamma)} = A^{(\beta\gamma)}\bar{\varepsilon},$$

(1)

where (β_γ) indicates the subcell whose location in RVE is at the βth row and γ^{th} column, and A is the matrix linking the micro- and macrostrain. At the same time, the interfacial traction continuity conditions can be expressed as

$$\bar{\sigma}_{22}^{(1\gamma)} = \bar{\sigma}_{22}^{(2\gamma)} = \cdots = \bar{\sigma}_{22}^{(N_\beta\gamma)} = T_{22}^{(\gamma)} \quad \left(\gamma = 1,\ldots,N_\gamma\right),$$

$$\bar{\sigma}_{33}^{(\beta1)} = \bar{\sigma}_{33}^{(\beta2)} = \cdots = \bar{\sigma}_{33}^{(\beta N_\gamma)} = T_{33}^{(\beta)} \quad \left(\beta = 1,\ldots,N_\beta\right),$$

$$\bar{\sigma}_{21}^{(1\gamma)} = \bar{\sigma}_{21}^{(2\gamma)} = \cdots = \bar{\sigma}_{21}^{(N_\beta\gamma)} = T_{21}^{(\gamma)} = T_{12}^{(\gamma)}$$
$$\left(\gamma = 1,\ldots,N_\gamma\right),$$

$$\bar{\sigma}_{31}^{(\beta1)} = \bar{\sigma}_{31}^{(\beta2)} = \cdots = \bar{\sigma}_{31}^{(\beta N_\gamma)} = T_{31}^{(\beta)} = T_{13}^{(\beta)}$$
$$\left(\beta = 1,\ldots,N_\beta\right),$$

$$\bar{\sigma}_{23}^{(1\gamma)} = \bar{\sigma}_{23}^{(2\gamma)} = \cdots = \bar{\sigma}_{23}^{(N_\beta\gamma)} = T_{23}^{(\gamma)} = T_{23}$$
$$\left(\gamma = 1,\ldots,N_\gamma\right),$$

$$\bar{\sigma}_{32}^{(\beta1)} = \bar{\sigma}_{32}^{(\beta2)} = \cdots = \bar{\sigma}_{32}^{(\beta N_\gamma)} = T_{32}^{(\beta)} = T_{23}$$
$$\left(\beta = 1,\ldots,N_\beta\right).$$

(2)

For each subcell of composites, the constitutive relationship of each subcell can be written as

$$\bar{\varepsilon}^{(\beta\gamma)} = \mathbf{S}^{(\beta\gamma)}\bar{\sigma}^{(\beta\gamma)} + \bar{\varepsilon}^{p(\beta\gamma)} + \boldsymbol{\alpha}^{(\beta\gamma)}\Delta T. \tag{3}$$

Substituting (3) into (1) and then combining (2), the relations between subcell tractions and overall strains can be obtained as

$$\bar{\sigma}_{ij}^{(\beta\gamma)} = \mathbf{C}_{ijkl}^{(\beta\gamma)} \mathbf{A}^{(\beta\gamma)}\bar{\varepsilon}. \tag{4}$$

Based on the homogenization theory, the overall stress of the RVE can be written as

$$\bar{\sigma} = \frac{1}{hl}\sum_{\beta=1}^{N_\beta}\sum_{\gamma=1}^{N_\gamma}h_\beta l_\gamma \bar{\sigma}^{(\beta\gamma)}. \tag{5}$$

Substituting (4) into (5), the overall stress and strain relation of the RVE are established as

$$\bar{\sigma} = \mathbf{C}^* \left(\bar{\varepsilon} - \bar{\varepsilon}^p - \boldsymbol{\alpha}^*\Delta T\right), \tag{6}$$

where $\mathbf{C}^*$ indicates the overall elastic stiffness matrix $\bar{\varepsilon}^p = [\bar{\varepsilon}_{11}^p, \bar{\varepsilon}_{22}^p, \bar{\varepsilon}_{33}^p, \bar{\varepsilon}_{23}^p, \bar{\varepsilon}_{13}^p, \bar{\varepsilon}_{12}^p]^T$ T indicates the overall plastic strain, and $\boldsymbol{\alpha}^* = [\alpha_{11}^*, \alpha_{22}^*, \alpha_{33}^*]^T$ represents the overall thermal expansion coefficient vector. It should be noted that the elements of matrixes $\mathbf{C}^*$, $\bar{\varepsilon}^p$, and α^* can be explicitly obtained in terms of the subcell material and geometric parameters and subcell plastic strains, so when the subcell ingredient properties and the RVE geometry are known, (6) can be used to model the responses of fiber composites.

VISCOPLASTIC CONSTITUTIVE MODEL

The matrix viscoplastic constitutive model is based on the modified Ramaswamy-Stouffer viscoplastic state variable model. The Ramaswamy-Stouffer viscoplastic state variable model [31] was originally developed to simulate the rate dependent inelastic response of metals. However, the relationship between load and deformation in resins is more complicated than that in metals since the hydrostatic component of the stress has a significant effect even at low level of stress [32]. The effect of the hydrostatic stresses was considered by modifying the effective stress term in the flow law of Ramaswamy-Stouffer model [22]. In the modified Ramaswamy-Stouffer model, the total strain rate, $\dot{\varepsilon}_{ij}$, is composed of elastic strain rate $\dot{\varepsilon}^e_{ij}$, rate, $\dot{\varepsilon}^I_{ij}$; ; that is,

$$\dot{\varepsilon}_{ij} = \dot{\varepsilon}^e_{ij} + \dot{\varepsilon}^I_{ij}. \tag{7}$$

The elastic strain rate can be obtained according to the time derivative of Hook's law. The inelastic strain rate is defined in the following form:

$$\dot{\varepsilon}^I_{ij} = D_0 \exp \left[-\frac{1}{2} \left(\frac{Z_0^2}{3K_2} \right)^n \right] \times \frac{s_{ij} - \Omega_{ij}}{\sqrt{K_2}}, \tag{8}$$

where D_0, Z_0, and n are all material constants. D_0 denotes the maximum inelastic strain rate, Z_0 indicates the initial, isotropic "hardness" of the material before any load is applied, n represents the rate dependence of deformation response, S_{ij} is the deviatoric stress component, and Ω_{ij} is the internal stress state variable. The relation between the internal stress rate, $\dot{\Omega}_{ij}$ and Ω_{ij}, is defined as follows:

$$\dot{\Omega}_{ij} = \frac{2}{3} q \Omega_m \dot{\varepsilon}_{ij} - q \Omega_{ij} \dot{\varepsilon}^I_e, \tag{9}$$

where q and Ω_m are both material constants. q represents the "hardening" rate, Ω_m represents the maximum value of the internal stress, and $\dot{\varepsilon}^I_e$ is the effective inelastic strain rate, which is defined as follow

$$\dot{\varepsilon}^I_{ij} = \sqrt{\frac{2}{3} \dot{\varepsilon}_{ij} \dot{\varepsilon}_{ij}}. \tag{10}$$

The term K_2, which represents the effective stress, is defined in the original Ramaswamy-Stouffer model in the following form:

$$K_2 = \frac{1}{2} \left(S_{ij} - \Omega_{ij} \right) \left(S_{ij} - \Omega_{ij} \right). \tag{11}$$

In the modified Ramaswamy-Stouffer model, in order to consider the effect of hydrostatic stresses, (11) is rewritten as follows:

$$K_2 = \frac{1}{2} \left[K_{11} + K_{22} + K_{33} + 2 \left(K_{12} + K_{13} + K_{23} \right) \right]. \tag{12}$$

The normal terms in the above expression are the same as the original definition while the shear terms are modified and can be written as

$$K_{12} = \alpha \left(S_{12} - \Omega_{12} \right) \left(S_{12} - \Omega_{12} \right),$$

$$K_{13} = \alpha \left(S_{13} - \Omega_{13} \right) \left(S_{13} - \Omega_{13} \right),$$

$$K_{23} = \alpha \left(S_{23} - \Omega_{23} \right) \left(S_{23} - \Omega_{23} \right),$$

(13)

Where

$$\alpha = \left(\frac{\sigma_m}{\sqrt{J_2}} \right)^{\beta}.$$

(14)

In (14), σ_m is the mean stress, J_2 is the second invariant of the deviatoric stress tensor, and β is a rate independent material constant which is determined empirically by fitting data from uniaxial composites with shear dominated fiber orientation angles, such as [15∘]. The other material constants, such as D_0, Z_0, and n, are determined through the method discussed in the article written by Goldberg and Stouffer [22]. Through the above introduction of the modified Ramaswamy-Stouffer model, it can be seen that the model does not depend on the yield rule and the inelastic strains are assumed to be present at all values of stress. Therefore, there is no need to judge whether the material is in elastic or plastic stage.

RESULTS AND DISCUSSION

Model Validation

To verify the ability of the micromechanics model and the viscoplastic constitutive model in the prediction of rate effects of composites several examples are considered and discussed in this section. The material considered here is a composite composed of carbon AS4 fibers in a PEEK thermoplastic matrix. For the AS4 fibers, the longitudinal elastic modulus is 214 GPa, the transverse and in-plane shear modulus is 14 GPa, the longitudinal Poisson's ratio is 0.2, and the transverse Poisson's ratio is 0.25 [22]. The material properties of PEEK resin can be seen in Table 1. The fiber volume fraction (V_f) used here is 0.62 and the fiber cross-section shape is square (seen in Figure 1). For comparison purposes, the experimental data obtained by Weeks [7] is shown as well. Two different strain rates, 0.1/sec and 10^{-5}/sec (which is written as 1E-05 in the figures for convenience), are considered. From Figures 3, 4, and 5, it can be seen that the results predicted by the presented micromechanics model and viscoplastic constitutive model exhibit good agreement with the experimental results.

Table 1: Material properties of PEEK resin

E (GPa)	v	D_0 (1/sec)	n	Z_0 (MPa)	q	Ω_m (MPa)	β
4.0	0.4	10^4	0.7	630	310	52	0.40

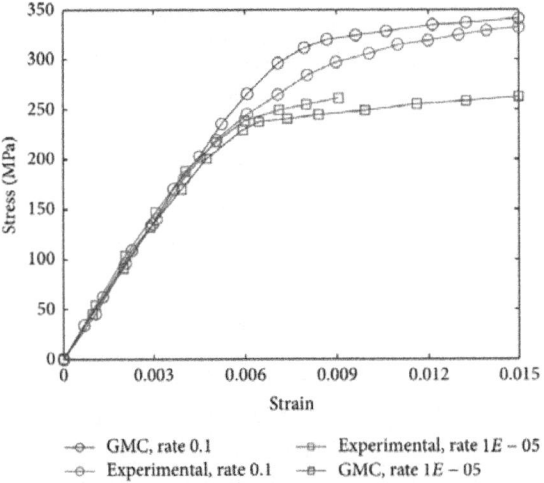

Figure 3: Stress-strain response of AS4/PEEK [15°] laminate at strain rate of 0.1/sec and 10^{-5}/sec

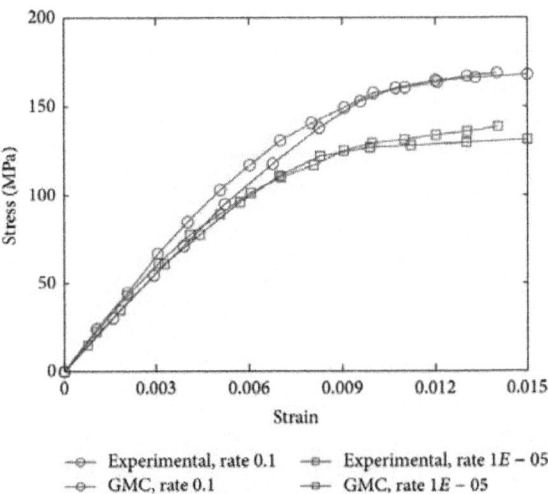

Figure 4: Stress-strain response of AS4/PEEK [30°] laminate at strain rate of 0.1/sec and 10^{-5}/sec.

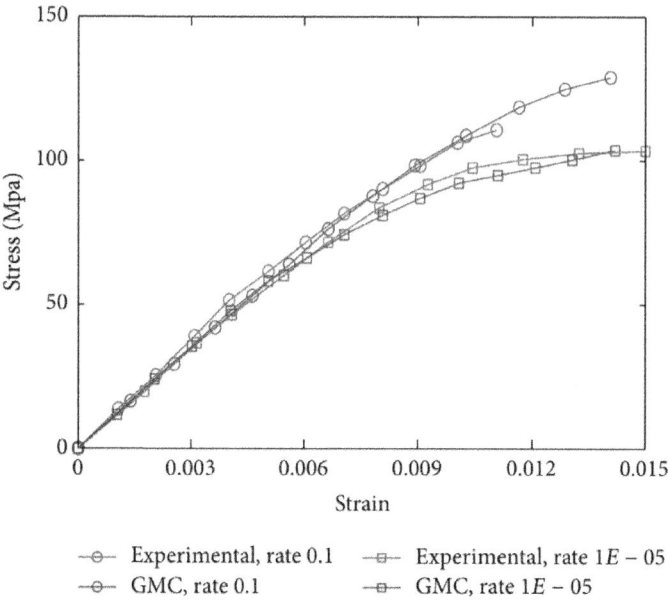

Figure 5: Stress-strain response of AS4/PEEK [45°] laminate at strain rate of 0.1/sec and 10^{-5}/sec.

Stress-Strain

Response of Composites with the Same Fiber Volume Fraction but Different Fiber Shapes and Different Strain Rates in Different Fiber Off-Axis Orientations. Figure 6 presents the responses for 15° , 45° , 60° , and 75° off-axis orientations in the case that composites contain 0.15 fiber volume fraction with different fiber shapes and strain rates. In the case of the elliptical fibers, the transverse loading is applied in the principal material directions of the long axis. In this kind of composites with very low fiber volume fraction,the effect of fiber on the composites behavior is small, so it can be seen that the composites response is hardly affected by the fiber cross-section shape, but it could be affected by the off-axis orientation and the stain rate. Among the four kinds of off-axis orientations, the one with 15° off-axis orientation exhibits the stiffest response while the one with 60° off-axis orientation exhibits the most compliant response. For all the off-axis orientations, when the strain rate changes from 10^{-5}/sec to 0.1/sec, the composites provide an effective increase in the flow stress while the elastic behavior almost remain unchanged. This is because the fact that when the strain rate is smaller, the composites have more time to occur plastic flow and unload.

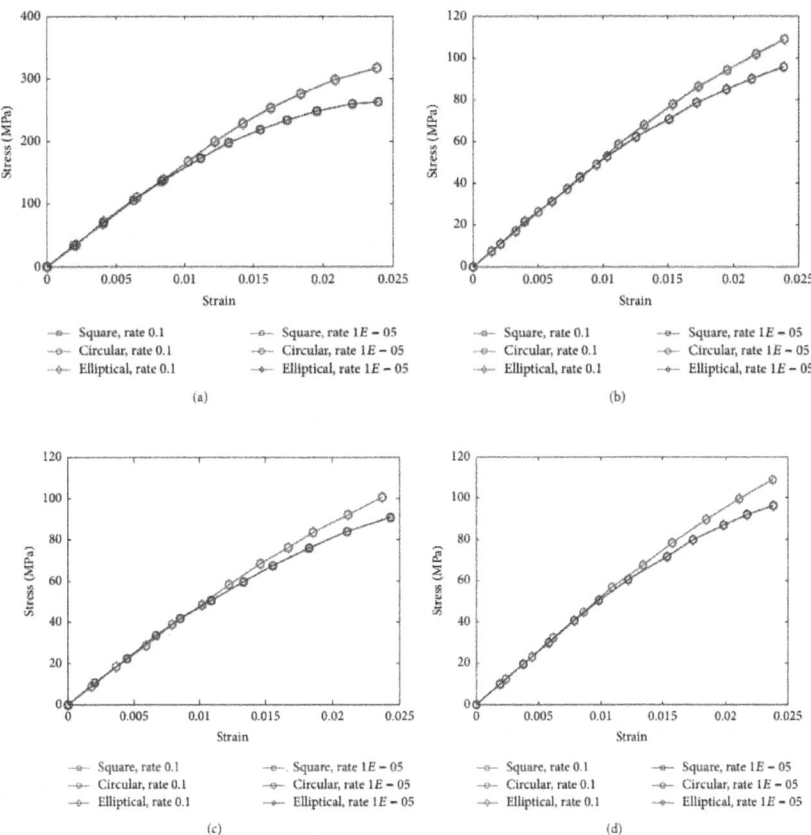

Figure 6: Off-axis responses of AS4/PEEK laminate ($V_f = 0.15$) at strain rate of 0.1/sec and 10^{-5}/sec: (a) 15° , (b) 45° , (c) 60° , and (d) 75°

Increasing the fiber volume fraction further accentuates the differences in the composite's transverse response due to the fiber's cross-sectional shape. Figure 7 shows the stressstrain responses of composites when the fiber volume fraction is increased to 0.30. It can be seen that when the off-axis angle is smaller than 75° , the composites response is hardly affected by the fiber cross-section shape. However, when the off-axis angle is increased to 75° , the effect of the fiber cross-sectional shape on the transverse response in the plastic region becomes discernible, with the square fibers being the most effective in increasing the flow stress of the composite. Figure 7(d) shows that the responses of composites with circular fibers and elliptical fibers with an aspect ratio of 4/3 are almost the same, which are lower than the responses of composites with square fibers. With the increasing of the offaxis orientations of composites, the response of composites decreases first and then increases.

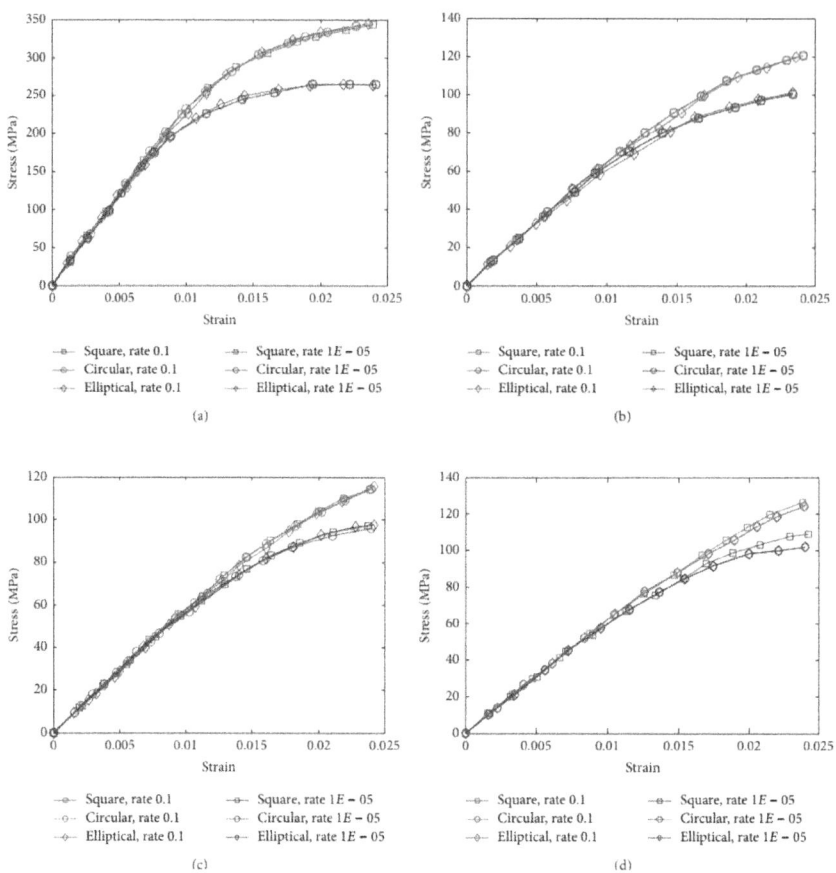

Figure 7: Off-axis responses of AS4/PEEK laminate (V_f = 0.30) at strain rate of 0.1/sec and 10^{-5}/sec: (a) 15°, (b) 45°, (c) 60°, and (d) 75°.

Figure 8 presents the results that correspond to those shown in the preceding two figures when the fiber volume fraction is further increased to 0.45. In this case, compared with the stress-strain curves when the fiber volume fraction is 0.30, it can be seen that when the off-axis angle is 60°, the composites response has already been affected by the fiber cross-section shape although the difference is small. But when the off-axis angle is increased to 75°, a substantial difference between the unit cell with the square fiber and the remaining unit cells is now apparent in the plastic region. In Figure 8(d), the square fiber provides a 20% increase in the transverse flow stress of the composite relative to that of the elliptical and circular fibers when the strain rate is 10^{-5}/sec, while the square fiber provides a 10% increase when the strain rate is 0.1/sec. This is because the square fiber can provide a higher magnitude of hydrostatic stress in the matrix phase relative to the circular fiber, which can delay localized yielding and provide constraint on the expansion of the plastic zone throughout the matrix phase.When the strain rate is smaller, the composites have more time to occur plastic strain and unload. Therefore, it can be noted that when the strain rate is 10^{-5}/sec, the difference of composites response between the square fiber and the circular fiber is larger than the case when the strain rate is 0.1/sec.

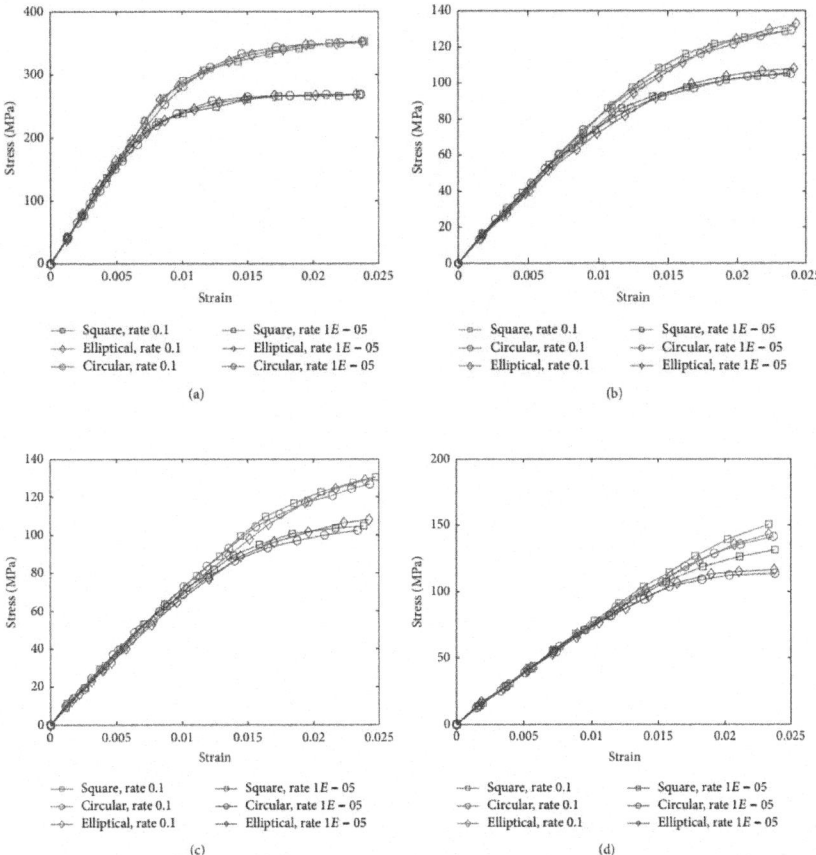

Figure 8: Off-axis responses of AS4/PEEK laminate (V_f = 0.45) at strain rate of 0.1/ sec and 10^{-5}/sec: (a) 15° , (b) 45° , (c) 60° , and (d) 75° .

Figure 9 shows the stress-strain curves when the fiber volume fraction is increased to 0.55. This fiber volume fraction is close to the maximum allowable for the RVE with the elliptical fiber, which is limited by the contact of fibers along the major axis in two adjacent RVE.This contact occurs when the fiber volume fraction is 0.59 in the case of fibers with an aspect ratio of 4/3. From

Figure 9(b), it can be seen that the composites response is affected by the fiber crosssection shape when the off-axis angle is just 45° , which is smaller than the preceding cases. In additionally, for both of the two kinds of strain rates, the difference of the three kinds of fibers is more obvious with the increase of the off-axis angle. In Figure 9(d), the difference between the composites with square fibers and the composites with circular fibers is very big, and the response of the composites with square fibers at the strain rate of 10^{-5}/sec is almost the same as the response of the composites with circular fibers at the strain rate of 0.1/sec.

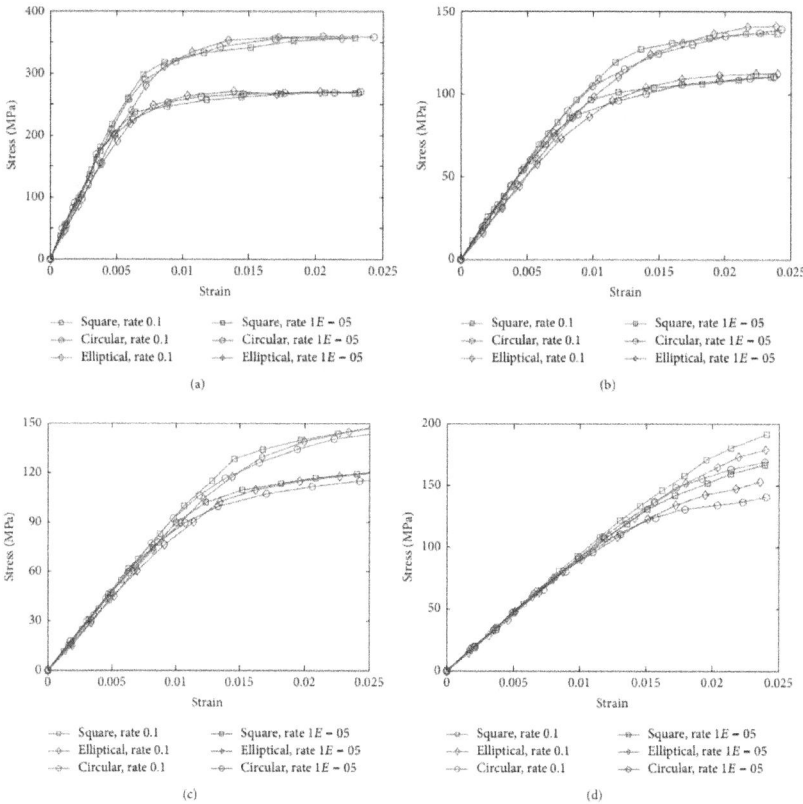

Figure 9: Off-axis responses of AS4/PEEK laminate (V_f = 0.55) at strain rate of 0.1/ sec and 10^{-5}/sec: (a) 15° , (b) 45° , (c) 60° , and (d) 75° .

Stress-Strain Response of Composites with the Same Fiber Off-Axis Orientation but Different Fiber Shapes and Fiber Volume Fractions at Different Strain Rates. Figure 10 shows the stress-strain response for 0.15, 0.30, 0.45, and 0.55 fiber volume fractions in the case that composites fiber off-axis angle is 90° . It can be seen that when the fiber volume fraction is less than 0.30, the stress-strain response is barely affected by the fiber cross-section shapes. When the fiber volume fraction is more than 0.30, the difference between different fiber cross-section shapes can be obtained. With the increase of the fiber volume fraction, the difference becomes larger and the stiffness of composites will increase, which is due to the bigger stiffness of fiber. When the fiber volume is increased to 0.45, the response of the composites with square fibers at the strain rate of 10^{-5}/sec is almost the same as the response rate of 0.1/ sec. When the fiber volume is increased to 0.55, the response of the composites with square fibers at the strain rate of 10^{-5}/sec is even higher than the response of the composites with circular fibers at the strain rate of 0.1/sec.

In Figure 10(d), the square fiber provides a 33% increase in the transverse flow stress of the composite relative to that of the elliptical and circular fibers when the strain rate is 10^{-5}/sec, while the square fiber provides a 15% increase when the strain rate is 0.1/sec.

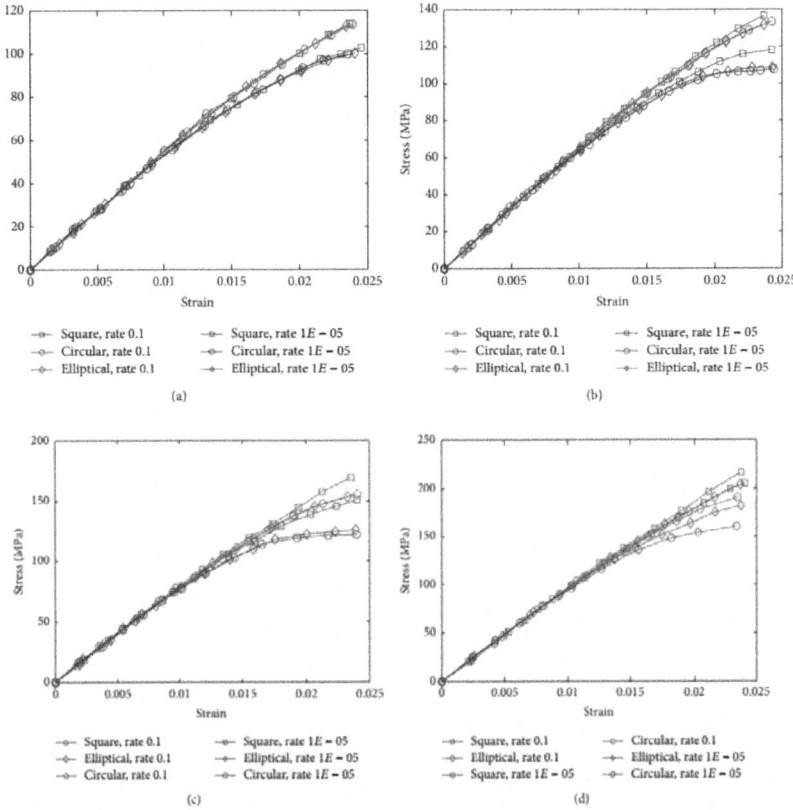

Figure 10: Stress-strain response of AS4/PEEK [90°] laminate at strain rate of 0.1/sec and 10^{-5}/sec: (a) $V_f = 0.15$, (b) $V_f = 0.30$, (c) $V_f = 0.45$, and (d) $V_f = 0.55$.

CONCLUSIONS

A viscoplastic constitutive model has been employed in the micromechanical method based on generalized model of cells to analyze the inelastic, rate dependent stress-strain response of fiber-reinforced polymer matrix composites with three different microstructures at different fiber off-axis angles condition. The acceptable agreement between the model predictions and experimental results shows that the proposed model can well predict the behaviors of AS4/PEEK composite. At the same time, from the predicted results, the following conclusions are obtained.

(1)   The AS4/PEEK composite is a kind of rate dependent material. When the strain rate changes from 10^{-5}/sec to 0.1/sec, the composites provide an effective increase in the flow stress while the elastic behavior almost remain unchanged.

(2)   The effects of fiber cross-sectional shape on the behavior of AS4/PEEK composite are related to the fiber volume fraction and fiber off-axis orientation. When the fiber volume fraction is smaller than 0.15, it can be seen that the composites response is hardly affected by the fiber cross-section shape; with the increasing of fiber volume fraction and fiber off-axis orientation, the effects of fiber cross-sectional shape become more obvious. Among the three kinds of fiber shapes, the stiffest response is obtained for the composites with the square fibers and the most compliant response for the composites with the circular fibers.

(3)   The increasing of fiber volume fraction can improve the stiffness of AS4/PEEK composite. However, for the elliptical fiber, the maximum allowable fiber volume fraction is 0.59 in the case of fibers with an aspect ratio of 4/3, so it should be noted that the elliptical fiber may not be chosen when the fiber volume fraction needed is big.

(4)   The influence of fiber off-axis orientation on the stress-strain curves of AS4/PEEK composite is very large. The response of composites decreases obviously when the off-axis orientation changes from 15° to 45° and then increases from 60° to 90°. So when the composites have been chosen to bear the load, the fiber off-axis orientation should be paid attention to.

CONFLICT OF INTERESTS

The authors declare that there is no conflict of interests regarding the publication of this paper.

ACKNOWLEDGMENTS

This work was supported by the National Natural Science Foundation of China (nos. 51175401 and 51335006), the Research Fund for the Doctoral Program of Higher Education of China (no. 20120201110028), and the Program for Changjiang Scholars and Innovative Research Team in University.

REFERENCES

1. D. Motamedi, A. Milani, M. Komeili, M. Bureau, F. Thibault, and D. Trudel-Boucher, "A stochastic XFEM model to study delamination in PPS/glass UD composites: effect of uncertain fracture properties," Applied Composite Materials, vol. 21, no. 2, pp. 341–358, 2014. · View at Google Scholar

2. P. W. R. Beaumont, "Targeting problems of composite failure," Key Engineering Materials, vol. 417-418, pp. 37–40, 2010.

3. Y. Zhou and Z. Huang, "A bridging model prediction of the ultimate strength of composite laminates subjected to triaxial loads," Journal of Composite Materials, vol. 46, no. 19-20, pp. 2343–2378, 2012.

4. Riccio, A. Raimondo, and F. Scaramuzzino, "A study on skin delaminations growth in stiffened composite panels by a novel numerical approach," Applied Composite Materials, vol. 20, no. 4, pp. 465–488, 2013.

5. R. K. Goldberg, G. D. Roberts, and A. Gilat, "Incorporation of mean stress effects into the micromechanical analysis of the high strain rate response of polymer matrix composites," Composites B: Engineering, vol. 34, no. 2, pp. 151–165, 2003.

6. G. M. Pearce, A. F. Johnson, R. S. Thomson, and D. W. Kelly, "Experimental investigation of dynamically loaded bolted joints in carbon fibre composite structures," Applied Composite Materials, vol. 17, no. 3, pp. 271–291, 2010.

7. C. A. Weeks, Nonlinear rate dependent response of thick-section composite laminates [Ph.D. thesis], Purdue University, 1995.

8. S. V. Thiruppukuzhi and C. T. Sun, "Models for the strain-rate-dependent behavior of polymer composites," Composites Science and Technology, vol. 61, no. 1, pp. 1–12, 2001.

9. Haque and M. Ali, "High strain rate responses and failure analysis in polymer matrix composites—an experimental and finite element study,"Journal of Composite Materials, vol. 39, no. 5, pp. 423–450, 2005.

10. M. M. Shokrieh and M. J. Omidi, "Tension behavior of unidirectional glass/epoxy composites under different strain rates," Composite Structures, vol. 88, no. 4, pp. 595–601, 2009. · ·

11. J. Ye, Y. Qiu, Z. Zhai, and X. Chen, "Strain rate influence on nonlinear response of polymer matrix composites," Polymer Composites, 2014. · View at Google Scholar

12. Z. Zhai, Z. He, X. Chen, J. Ye, and X. Zhu, "Fiber cross-section shape effect on rate-dependent behavior of polymer matrix composites with FBGs sensors," Sensors and Materials, vol. 25, no. 6, pp. 403–410, 2013. ·

13. C. A. Weeks and C. T. Sun, "Modeling non-linear rate-dependent behavior in fiber-reinforced composites," Composites Science and Technology, vol. 58, no. 3-4, pp. 603–611, 1998.

14. S. V. Thiruppukuzhi and C. T. Sun, "Testing and modeling high strain rate behavior of polymeric composites," Composites B: Engineering, vol. 29, no. 5, pp. 535–546, 1998.

15. H. D. Espinosa, H. Lu, P. D. Zavattieri, and S. Dwivedi, "A 3-D finite deformation anisotropic visco-plasticity model for fiber composites," Journal of Composite Materials, vol. 35, no. 5, pp. 369–410, 2001.

16. E. Kontou and A. Kallimanis, "Thermo-visco-plastic behaviour of fibre-reinforced polymer composites," Composites Science and Technology, vol. 66, no. 11-12, pp. 1588–1596, 2006. · ·

17. W. Hufenbach, A. Hornig, B. Zhou, A. Langkamp, and M. Gude, "Determination of strain rate dependent through-thickness tensile properties of textile reinforced thermoplastic composites using L-shaped beam specimens," Composites Science and Technology, vol. 71, no. 8, pp. 1110–1116, 2011.

18. L. Raimondo, L. Iannucci, P. Robinson, and P. T. Curtis, "Modelling of strain rate effects on matrix dominated elastic and failure properties of unidirectional fibre-reinforced polymer-matrix composites," Composites Science and Technology, vol. 72, no. 7, pp. 819–827, 2012.

19. L. Xing, K. L. Reifsnider, and X. Huang, "Progressive damage modeling for large deformation loading of composite structures," Composites Science and Technology, vol. 69, no. 6, pp. 780–784, 2009.

20. J. Ye, X. Chen, Z. Zhai, B. Li, Y. Duan, and Z. He, "Predicting the elastoplastic response of fiber-reinforced metal matrix composites," Mechanics of Composite Materials, vol. 46, no. 4, pp. 405–416, 2010.

21. D. D. Robertson and S. Mall, "Micromechanical relations for fiber-reinforced composites using the free transverse shear approach," Journal of Composites Technology and Research, vol. 15, no. 3, pp. 181–192, 1993.

22. R. K. Goldberg and D. C. Stouffer, "Strain rate dependent analysis of a polymer matrix composite utilizing a micromechanics approach," Journal of Composite Materials, vol. 36, no. 7, pp. 773–793, 2002.

23. Tabiei and S. B. Aminjikarai, "A strain-rate dependent micro-mechanical model with progressive post-failure behavior for predicting impact response of unidirectional composite laminates," Composite Structures, vol. 88, no. 1, pp. 65–82, 2009.

24. Z. Huang, "Simulation of the mechanical properties of fibrous composites by the bridging micromechanics model," Composites A: Applied Science and Manufacturing, vol. 32, no. 2, pp. 143–172, 2001.

25. Z.-M. Huang, "Inelastic and failure analysis of Laminate structures by ABAQUS incorporated with a general constitutive relationship," Journal of Reinforced Plastics and Composites, vol. 26, no. 11, pp. 1135–1181, 2007.

26. M. Paley and J. Aboudi, "Micromechanical analysis of composites by the generalized cells model," Mechanics of Materials, vol. 14, no. 2, pp. 127–139, 1992.

27. S. Ogihara, S. Kobayashi, and K. L. Reifsnider, "Characterization of nonlinear behavior of carbon/epoxy unidirectional and angle-ply laminates," Advanced Composite Materials, vol. 11, no. 3, pp. 239–254, 2003.

28. J. Tsai and K. Chen, "Characterizing nonlinear rate-dependent behaviors of graphite/epoxy composites using a micromechanical approach," Journal of Composite Materials, vol. 41, no. 10, pp. 1253–1273, 2007.

29. Gilat, R. K. Goldberg, and G. D. Roberts, "Strain rate sensitivity of epoxy resin in tensile and shear loading," Journal of Aerospace Engineering, vol. 20, no. 2, pp. 75–89, 2007.

30. M.-J. Pindera and B. A. Bednarcyk, "An efficient implementation of the generalized method of cells for unidirectional, multi-phased composites with complex microstructures," Composites B: Engineering, vol. 30, no. 1, pp. 87–105, 1999.

31. D. C. Stouffer and L. T. Dame, Inelastic Deformation of Metals: Models, Mechanical Properties, and Metallurgy, John Wiley & Sons, New York, NY, USA, 1996.

32. R. K. Goldberg, G. D. Roberts, and A. Gilat, "Implementation of an associative flow rule including hydrostatic stress effects into the high strain rate deformation analysis of polymer matrix composites," Journal of Aerospace Engineering, vol. 18, no. 1, pp. 18–27, 2005.

Chapter 7

SELF-CONSISTENT MICROMECHANICAL ENHANCEMENT OF CONTINUOUS FIBER COMPOSITES

Andrew Ritchey[1], Joshua Dustin[1], Jonathan Gosse[2] and R. Byron Pipes[1]
[1]Purdue University

[2]The Boeing Company USA

INTRODUCTION

Much of the previous work in developing analytical models for high performance composite materials has focused on representations of the heterogeneous medium as a homogenous, anisotropic continuum. The development of the equivalent properties of the homogenous medium from the geometry of the microstructure and the fiber and matrix properties has been come to be known as "micromechanics" (Daniel & Ishai, 2006). The term "homogenization" has been applied to the process of determining the effective properties of the homogenous medium and for much of the past half century homogenization was the only task of micromechanics. However, increases in computational capability has allowed for the use of micromechanics as a "de-homogenization" tool as well. The dehomogenization method that is the focus of the current study has been come to be known as "micromechanical enhancement" (Gosse & Christensen, 2001; Buchanan et al., 2009). Here the deformation of the homogeneous medium is enhanced by influence functions derived from unit cell micromechanical models representing extremes in the packing efficiencies of fiber arrangements. The motivation for development of the de-homogenization step is the need for an increase in the robustness and fidelity of failure theories used for these material systems wherein the deformation fields within the homogenized solutions are enhanced to reflect the actual strain field topologies within the fiber and matrix constituents. It is these enhanced strain fields that are used to determine the onset of damage initiation within the medium. There are several categories of models which have been proposed to perform the homogenization step of micromechanics including: mechanics of materials (Voigt, 1887; Reuss, 1929); self-consistent field (Hill,

1965); bounding methods based on variation principals (Paul, 1960; Hashin & Rosen 1964); semi-empirical (Halpin & Tsai, 1967); numerical finite element methods (Sun & Vaidya, 1996) and experimental methods such as uniaxial coupon tests. A significant amount of work has been devoted to this topic and more complete reviews are found elsewhere (Christensen 1979; Pindera et al., 2009). Although any analysis method used should be vetted against a rigorous testing program, accurate micromechanics models can provide a cost effective method for a priori material evaluation and ranking of composite systems. In the traditional composite analysis workflow, homogenized material properties are used in a laminate analysis of a structural member to determine lamina level stresses and strains. Stresses and strains at the lamina level are then used directly in a failure criterion to determine the ultimate performance of the member. Some success has been achieved with this approach but the analysis fails to take into account the actual state of stress and strain within the constituent phases. In addition, residual thermal stresses resulting from a mismatch in the coefficient of thermal expansion between the fiber and matrix phases are usually neglected. Others have noted that non-physical singularities may arise in homogenized solutions containing free-edges (Pagano & Rybicki, 1974; Pagano & Yuan, 2000). Several methods for recovery of the state of stress/strain from a homogenized solution have been proposed as well. Analytic methods have been proposed base on phase averaging methods (Hill, 1963; Hashin 1972). More recently, numerical methods have been employed. One method is to perform a global-local finite element analysis. In this approach the forces or displacements obtained from a homogenized solution are applied to a domain in which the fiber and matrix phases are modelled explicitly (Wang et al., 2002). With this method one must first determine an appropriate size for the local region, typically containing several fibers, using the so-called "local domain test." It has been suggested that a single fiber local region is feasible for determining fiber-matrix interface stresses if the continuum is modelled using the micro-polar theory of elasticity (Hutapea et al., 2003). Others have suggested the use of a multilevel analysis that models a homogenized region, a transition region and a region containing the explicit microstructure in a single finite element analysis (Raghavan et al., 2001). A more computationally efficient method for recovering the stress and strain in the fiber and matrix phases is to use an influence function formulation (Gosse & Christensen, 2001). In this method, also referred to as mircomechanical enhancement, a set of six canonical states of deformation and a separate thermal load are applied to a unit cell prior to performing an analysis of the homogeneous medium. The influence functions extracted from the unit cells are then used to relate the state of homogenous strain in each lamina to the state of strain within the representative volume element through the use of the enhancement matrix.

Microscopic residual thermal strains can also be recovered with a superposition vector (Buchanan et al., 2009). In a previous study (Gosse & Christensen, 2001), the homogenization step was an experimental one wherein the effective properties of the homogenous medium employed in the analysis were determined by experiments while the de-homogenization (micromechanical enhancement) step was carried out by a finite element analysis of a representative volume element. In addition to this procedure, an alternative method has been developed to utilize the derived effective elastic and thermal lamina properties from the same micromechanical models developed to assessed the strain fields within the unit cells. In this paper the latter approach is investigated exclusively in order to provide the consistency of utilizing the same method for both homogenization and de-homogenization. In the current chapter, the micromechanical enhancement method is investigated in more detail and a self-consistent method for determining the microscopic strain field is presented. By using a self-consistent analysis, the inherent approximations of the method are present in both steps while no new uncertain quantities, such as experimental test variables, are introduced. Self-consistency is assured by utilizing the same micromechanical models for both the homogenization and de-homogenization steps in the method. The goal is to provide an efficient link in a multi-scale analysis of a composite structure and to elucidate the analysis steps used in the current method.

HOMOGENIZATION

The homogenization process seeks to obtain equivalent homogenous continuum properties for a medium composed of multiple phases of varying constitutive properties. For the current discussion, we will limit ourselves to a heterogeneous medium consisting of collimated, continuous fibers within an isotropic matrix. Many methods and closed-form expressions have been developed to achieve this goal (Pindera et al., 2009). Among these, the most accurate in predicting the average response of an orthotropic medium is the finite element method (Daniel & Ishai, 2006). In the finite element approach, one would like to determine the relationship between the average stress and average strain as expressed in Equation 1.

$$\bar{\sigma}_i = \bar{C}_{ij}\left(\bar{\varepsilon}_j - \bar{\alpha}_j\Delta T\right)\ (i,j = 1-6)$$

(1)

The overbar indicates an average or homogenized quantity. From the homogenous stiffness matrix $(\bar{C}_{ij})$, the effective lamina engineering constants $(E_1, E_2, v_{12}, G_{12}, \text{etc.})$ can be calculated. Alternatively, the engineering constants can be determined directly by systematically performing finite element analysis corresponding to the definitions of the engineering constants (Sun & Vaidya,

1996). In this approach, the average stress is related to the average strain through the strain energy density. Typically, a representative volume element, as shown in Figure 1, is used to simplify the analysis. Equation 2 gives the stressstrain relation for the case when $\Delta T = 0$.

$$\bar{\sigma}_i = \bar{C}_{ij}\bar{\varepsilon}_j \; \left(i,j = 1-6\right)$$

(2)

Where "1" coincides with the fiber direction, "2" is transverse to the fiber direction and "3" is normal to the 1 and 2 directions. Also, note the use of a contracted notation such that i=1- 3 are the three normal components of stress and strain, 11, 22, and 33, respectively, while i=4-6 are the three engineering shear components, 23, 13 and 12, respectively. The average stress is shown in Equation 3 for the case when a canonical state of deformation is applied so that the only active strain component is $\bar{\varepsilon}_1$.The superscript indicates the isolated average mechanical strain component. As shown, the relationship of can be rearranged to determine the first column of the homogenous stiffness matrix.

$$\bar{\sigma}_i^{(\bar{\varepsilon}_1)} = \bar{C}_{i1}^k\bar{\varepsilon}_1 \text{ or } \bar{C}_{i1}^k = \bar{\sigma}_i^{(\bar{\varepsilon}_1)}\big/\bar{\varepsilon}_1 \; \left(i = 1-6\right)$$

(3)

The other five columns of the effective homogenous stiffness matrix are determined by applying the remaining five other states of canonical deformation such that each strain component is isolated. A total of six finite element analyses are required to fully determine the effective homogenous stiffness matrix for the general anisotropic solid. A seventh finite element analysis is required to determine the thermal response of the effective medium. For this case, the domain is subjected to a thermal loading of ΔT with the average mechanical strain, $\{\bar{\varepsilon} - \bar{\alpha}\Delta T\}$, constrained to be zero. Equation 4 gives the calculation of the average coefficients of thermal expansion. As seen from Equation 1, constraining the average mechanical strain to be zero requires that the average stress be zero. This condition can be used to check the validity of the boundary conditions.

$$\{\bar{\alpha}_i\} = \bar{\varepsilon}_i\big/\Delta T \; \left(i = 1-6\right)$$

(4)

Equation 5 shows the average strain in a unit cell determined from the surface displacement, u_i and u_j with i,j = 1-3, by using Gauss' theorem and written in index notation (Sun & Vaidya, 1996).

$$\bar{\varepsilon}_{ij} = \frac{1}{2V}\int_S \left(u_i n_j + u_j n_i\right) dS$$

(5)

Where S is the boundary surface of the representative volume element and ni is the unit surface normal in the ith direction. Thus, the average strain in the unit cell can be calculated for a set of displacement boundary conditions. Specifying the components of deformation on the surface of the representative volume element will, in general, induce average stress components. The average stress is calculated in Equation 6 using the reaction forces obtained on the boundaries of the unit cell and the definition of stress.

$$\bar{\sigma}_{ij} = F_j n_i / A \quad (i, j = 1 - 3) \tag{6}$$

Where index notation is used with F_j the jth component of the total force applied to a face with a total area of A oriented in the ith direction. In this way, all six components of the stress tensor that may result for a state of deformation applied to the representative volume element are determined.

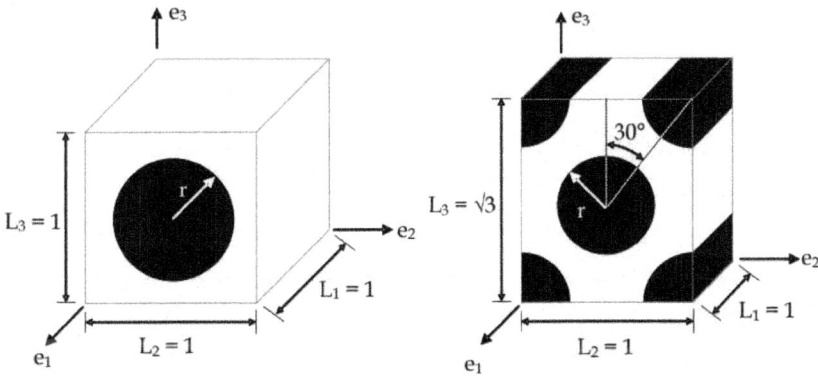

Figure. 1: Schematics of the square and hexagonal representative volume elements (Buchanan et al., 2009)

DE-HOMOGENIZATION

Utilizing classical laminate theory in the analysis of a composite laminate represented as a homogenous, anisotropic solid provides accurate predictions of structural deformations resulting from applied forces and moments. Although analysts have successfully used this approach, there are several shortcomings, which, if overcome, may provide increasingly accurate predictions of ultimate properties. The most apparent shortcomings of a homogenized analysis are: the modeling of fictitious interfaces; stresses and strains in the homogenized continuum exist in neither the fiber phase nor the matrix phase and the loss of the residual micromechanical thermal stress field due to a temperature change. The current chapter will focus on the latter two shortcomings by predicting the strain state within the fiber and matrix phases using a process referred to as micromechanical enhancement (Gosse & Christensen, 2001; Buchanan et al., 2009). The role of micromechanical enhancement is to provide a computationally efficient micromechanics analysis that includes congruent homogenization and de-homogenization steps. The current approach uses a single finite element model subjected to canonical states of deformation to provide the information needed for both homogenization (micromechanics) and de-homogenization (micromechanical enhancement) and is thus considered to be a self-consistent approach. This chapter is primarily focused on building a general framework required to obtain self-consistent results and transferring information between micro and macro scale composite models. Through the use of a simple example problem we will address the process used to recover strains at the micro-scale resulting from both mechanical loading and residual thermal stresses. First, consider a representative volume element subjected to an arbitrary state of average mechanical strain, $\{\bar{\varepsilon} - \bar{\alpha}\Delta T\}$, where $\bar{\varepsilon}$, where ε is the average total strain and $\bar{\alpha}$, the vector of effective coefficients of thermal expansion for the homogenized medium. As shown in Equation 7, the state of strain at a prescribed point within the representative volume element is calculated using an influence function formulation.

$$\varepsilon_i^k - \alpha_i^k \Delta T = M_{ij}^k \left(\bar{\varepsilon}_j - \bar{\alpha}_j \Delta T \right) + A_i^k \Delta T \quad (i, j = 1 - 6)$$

(7)

The matrix M_{ij}^k and the vector A_i^k are the influence function matrix and the thermal superposition vector of strain for the kth point in the representative volume element, respectively (Gosse & Christensen, 2001; Buchanan et al., 2009). The components of the influence function matrix can be determined uniquely, in a fashion similar to determining the stiffness matrix for the effective homogeneous medium, by prescribing a canonical state of deformation in the representative volume element and carrying out three-dimensional finite element analyses to determine the components of the strain tensor at the specified point, k. For example, let $\bar{\varepsilon}_1 \neq 0$, in the absence of the other five strain components and with no thermal loading. Shown in Equation 8, the first column of the influence matrix can be determined by relating the local strain to the average axial strain by using Equation 7.

$$\varepsilon_i^{k(\bar{\varepsilon}_1)} = M_{i1}^k \bar{\varepsilon}_1 \text{ or } M_{i1}^k = \varepsilon_i^{k(\bar{\varepsilon}_1)} / \bar{\varepsilon}_1 \quad (i = 1 - 6)$$

(8)

Note that a single finite-element analysis with boundary conditions that meet the condition, $\bar{\varepsilon}_1 \neq 0$ with $\bar{\varepsilon}_{2-6} = 0$, with $\bar{\varepsilon}_{2-6} = 0$, yields six of the 36 coefficients in the influence function matrix at any point within the representative volume element. A total of six finite-element analyses are required to completely determine terms of M_{ij}^k at any point within the domain for a given representative volume element geometry.

Calculation of the thermal superposition vector requires an additional finite element analysis in which the unit cell is subjected to a temperature change with the constraint that the average mechanical strain vanishes, i.e. $\{\bar{\varepsilon} - \bar{\bar{\alpha}}\Delta T\} = 0$. Equation 9 gives the thermal superposition vector obtained by inserting this constraint into Equation 7.

$$A_i^k = \frac{\varepsilon_i^k - \alpha_i^k \Delta T}{\Delta T} \quad (i = 1-6)$$

(9)

REPRESENTATIVE VOLUME ELEMENT BOUNDARY CONDITIONS

The imposition of canonical states of strain upon the representative volume element utilizing finite-element analyses requires the development of a corresponding set of displacement boundary conditions. The representative volume element principal directions, (e_1, e_2, e_3) are shown in Figure 1. Equation 10 defines the appropriate displacement boundary conditions for the prescribed extensional strain in the "1" direction with ui representing the displacement vector and xi the position vector.

$$u(0, x_2, x_3) = \tau_{xy}(0, x_2, x_3) = \tau_{xz}(0, x_2, x_3) = 0$$

(10a)

$$u(L_1, x_2, x_3) = \bar{\varepsilon}_1 L_1; \quad \tau_{xy}(L_1, x_2, x_3) = \tau_{xz}(L_1, x_2, x_3) = 0$$

(10b)

$$v(x_1, 0, x_3) = \tau_{yx}(x_1, 0, x_3) = \tau_{yz}(x_1, 0, x_3) = 0$$

(10c)

$$v(x_1, L_2, x_3) = \tau_{yx}(x_1, L_2, x_3) = \tau_{yz}(x_1, L_2, x_3) = 0$$

(10d)

$$w(x_1, x_2, 0) = \tau_{zx}(x_1, x_2, 0) = \tau_{zy}(x_1, x_2, 0) = 0$$

(10e)

$$w(x_1, x_2, L_3) = \tau_{zx}(x_1, x_2, L_3) = \tau_{zy}(x_1, x_2, L_3) = 0$$

(10f)

As an example, Equation 11 gives the canonical shearing displacements for shearing in the 2-3 plane.

$$u(0, x_2, x_3) = \tau_{xy}(0, x_2, x_3) = \tau_{xz}(0, x_2, x_3) = 0$$

(11a)

$$u(L_1, x_2, x_3) = \tau_{xy}(L_1, x_2, x_3) = \tau_{xz}(L_1, x_2, x_3) = 0$$

(11b)

$$w(x_1, 0, x_3) = \sigma_{yy}(x_1, 0, x_3) = \tau_{yx}(x_1, 0, x_3) = 0$$

(11c)

$$w(x_1, L_2, x_3) = \bar{\gamma}_{23} L_2 / 2; \quad \sigma_{yy}(x_1, L_2, x_3) = \tau_{yx}(x_1, L_2, x_3) = 0$$

(11d)

$$v(x_1,x_2,0) = \sigma_{zz}(x_1,x_2,0) = \tau_{zx}(x_1,x_2,0) = 0 \qquad (11e)$$

$$v(x_1,x_2,L_3) = \overline{\gamma}_{23}L_3/2; \quad \sigma_{zz}(x_1,x_2,L_3) = \tau_{zx}(x_1,x_2,L_3) = 0 \qquad (11f)$$

This simple set of displacement boundary conditions are only valid for doubly periodic representative volume elements. In general, further constraints are required on the displacement field to maintain periodicity between adjacent unit cells. However, periodicity is satisfied automatically in the symmetric unit cells studied here. The desired average strain is recovered by inserting the boundary conditions shown in Equations 10 and 11 into Equation 6. The nodal forces taken from each face of the representative volume element can then be used in Equation 6 to determine the average stress and thereby provide the homogenized material properties as discussed in Section 2. For the case of a uniform temperature change of the representative volume element, boundary conditions are imposed to allow for free expansion of the representative volume element with the constraint that all faces must remain planar. This condition results from Equation 8 that requires the representative volume element to exhibit the free thermal deformation $\overline{\alpha}\Delta T$ in order that the homogenized mechanical strains vanish. In this case, free thermal deformation of representative volume element is equal to that defined by the coefficients of thermal expansion of the homogenized unidirectional lamina, $\overline{\alpha}\Delta T$. The planar constraint is required to maintain periodicity between adjacent volume elements. Procedures to implement these constraints are implemented in both Abaqus (© Dassault Sytémes) and StressCheck® (ESRD). For StressCheck® when the analysis is performed for any given load, the program will create constraint equations for all the degrees of freedom associated with the selected faces. The internal degrees of freedom (faces and edges) are eliminated at the element level (local constraint equations), while the equation for the nodal variables are written in compact form at the global level.

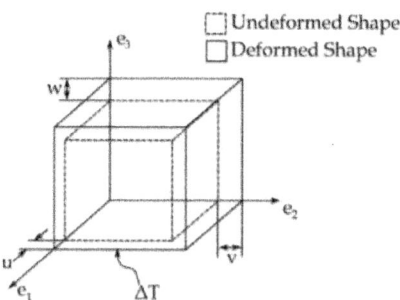

Figure. 2: Constrained deformation due to thermal loading. All faces remain planar to maintain periodicity

Since each face is constrained to remain planar, no average shearing strain will be obtained. The normal strain components are given in Equation 12. Using the normal strain components, the homogenized coefficients of thermal expansion can be determined with Equation 4, where U, V and W are the displacement components shown in Figure 2. It should be noted that these displacements are unknown prior to performing the analysis.

$$\bar{\varepsilon}_1 = \frac{U}{L_1}, \ \bar{\varepsilon}_2 = \frac{V}{L_2}, \ \bar{\varepsilon}_3 = \frac{W}{L_3}$$

(12)

EXAMPLE

First, consider a general laminate to be analysed using a self-consistent micromechanics method. Presented graphically in Figure 3 is the self-consistent micromechanics method used herein. A complete set of material properties for both the fiber and matrix phase are required. Six canonical states of deformation, extending from Equations 10 and 11, are applied as boundary conditions to two representative volume elements, a square array and a hexagonal array shown in Figure 1. The fiber volume fraction of the representative volume elements is 60 percent. For this step the finite element program Abaqus is utilized. Six canonical states of deformation provide both the homogenized stiffness matrix $(\bar{C}_{ij})$ and the enhancement matrix (M_{ij}) for the two micro-geometries. Both domains are subjected to a uniform change in temperature in a seventh finite element analysis. This thermal loading case provides the homogenized coefficients of thermal expansion (αi) and the thermal superposition vector (A_i). In total, seven finite element analyses are required for each representative volume element of interest. The homogenized material properties become the input for the laminate level analysis. For illustration, the boundary conditions are limited to in-plane force resultants and a uniform change in temperature ΔT applied to symmetric, balanced laminates.

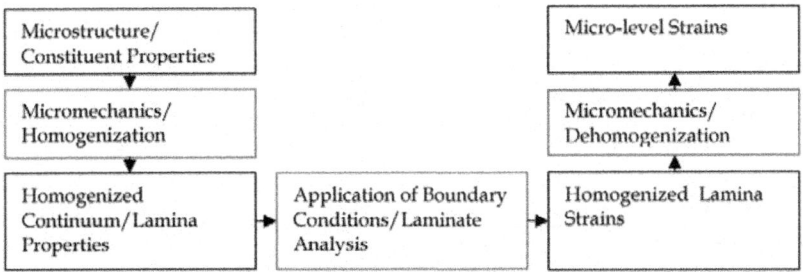

Figure. 3: Flow chart of self-consistent micromechanical enhancement. The analysis steps are boxed in red while the inputs and outputs to each step are boxed in black

The laminate level calculation is performed twice, once for both sets of homogenized material properties corresponding to the representative volume elements modelled.

The strains in each lamina of the laminate are calculated with a classical laminated plate theory analysis and become inputs to Equation 7. From this step, we obtain two sets of selfconsistent states of strain at the micro-level, i.e. in the fiber and matrix phases. This method is considered to be a highly efficient way to obtain micro-level information because laminate geometry and loading conditions can be changed independently of the micromechanics step. Therefore, the initial set of seven finite element analyses only need to be carried out once for each representative volume element. This decoupling of micro and macro level analysis is the characteristic that is responsible for the flexibility and computational efficiency of the method described herein. The alternative approaches described in the introduction require explicit modelling of the fiber and matrix phases for each loading condition and laminate geometry.

Prediction of Homogenized Properties

In the current example, two representative volume elements are considered, the square and hexagonal arrays. However the self-consistent micromechanics method can be applied to other representative volume element geometries that meet the doubly periodic condition. A schematic of each geometry is given in Figure 1. The first two columns in Table 1 are the input constituent material properties for an IM7/8552 carbon fiber, epoxy matrix composite. The final two columns in Table 1 give the predicted homogenized composite properties for the two representative volume elements.

Table 1: Fiber, matrix and equivalent homogenized medium material properties

Property	Matrix	Fiber	Square Cell	Hex Cell
E_1 (GPa)	4.76	276.0	167.5	167.5
E_2 (GPa)	4.76	19.5	11.5	10.7
E_3 (GPa)	4.76	19.5	11.5	10.7
G_{12} (GPa)	1.74	70.0	6.78	6.30
G_{13} (GPa)	1.74	70.0	6.78	6.30
G_{23} (GPa)	1.74	5.74	3.10	3.34
v_{12}	0.37	0.28	0.31	0.31
v_{13}	0.37	0.28	0.31	0.31
v_{23}	0.37	0.70	0.57	0.60
α_1 (10^{-6}/°C)	64.8	-0.4	0.41	0.41
α_2 (10^{-6}/°C)	64.8	5.6	34.7	35.1
α_3 (10^{-6}/°C)	64.8	5.6	34.7	35.1

The homogenzied stiffness matrix $(\bar{C}_{ij})$ is first calculated from Equation 3. Equation 13 shows the calculations used to determine the homogenized engineering elastic constants from the homogenzied stiffness matrix. Shown in Tables 2 and 3 are the homogenized stiffness matrix and the homogenized compliance matrix $(\bar{S}_{ij})$, respectively. The predicted engineering constants are used as inputs in the laminate level analysis.

$$[\bar{S}] = [\bar{C}]^{-1} \qquad (13a)$$

$$\bar{E}_1 = 1/\bar{S}_{11}, \ \bar{E}_2 = 1/\bar{S}_{22}, \ \bar{E}_3 = 1/\bar{S}_{33} \qquad (13b)$$

$$\bar{\nu}_{23} = -\bar{S}_{32}/\bar{S}_{22}, \ \bar{\nu}_{13} = -\bar{S}_{31}/\bar{S}_{11}, \ \bar{\nu}_{12} = -\bar{S}_{21}/\bar{S}_{11} \qquad (13c)$$

$$\bar{G}_{23} = 1/\bar{S}_{44}, \ \bar{G}_{13} = 1/\bar{S}_{55}, \ \bar{G}_{12} = 1/\bar{S}_{66} \qquad (13d)$$

Table 2: Homogenized stiffness matrix representative volume elements

Square Array, $\bar{C}_{ij}$ (GPa)						Hexagonal Array, $\bar{C}_{ij}$ (GPa)					
172.8	8.7	8.7	0.0	0.0	0.0	172.8	8.6	8.6	0.0	0.0	0.0
8.7	17.6	10.3	0.0	0.0	0.0	8.6	17.1	10.5	0.0	0.0	0.0
8.7	10.3	17.6	0.0	0.0	0.0	8.6	10.5	17.1	0.0	0.0	0.0
0.0	0.0	0.0	3.1	0.0	0.0	0.0	0.0	0.0	3.3	0.0	0.0
0.0	0.0	0.0	0.0	6.8	0.0	0.0	0.0	0.0	0.0	6.3	0.0
0.0	0.0	0.0	0.0	0.0	6.8	0.0	0.0	0.0	0.0	0.0	6.3

Table 3: Homogenized compliance matrix representative volume elements

Square Array, $\bar{S}_{ij}$ (1/GPa 10^{-3})						Hexagonal Array, $\bar{S}_{ij}$ (1/GPa 10^{-3})					
5.97	-1.86	-1.86	0.00	0.00	0.00	5.97	-1.86	-1.86	0.00	0.00	0.00
-1.86	87.24	-50.16	0.00	0.00	0.00	-1.86	93.48	-56.07	0.00	0.00	0.00
-1.86	-50.16	87.24	0.00	0.00	0.00	-1.86	-56.07	93.48	0.00	0.00	0.00
0.00	0.00	0.00	322.27	0.00	0.00	0.00	0.00	0.00	299.00	0.00	0.00
0.00	0.00	0.00	0.00	147.54	0.00	0.00	0.00	0.00	0.00	158.69	0.00
0.00	0.00	0.00	0.00	0.00	147.54	0.00	0.00	0.00	0.00	0.00	158.69

Determination of Influence Matrix and Thermal Superposition Vector

The influence matrix and thermal superposition vector can be extracted from the same set finite element analyses used to determine the effective lamina properties. It should be noted that both the influence matrix and the thermal superposition vector are field variables. That is, each specific geometric point within a representative volume element yields a unique influence matrix.

Presented as field variables, the terms of the influence matrices and thermal superposition vectors are illustrated graphically in Figures 4 and 5, respectively. The micro-strain field can be extracted at every node or integration point within the representative volume element. The enhanced strain field at every point within a volume element can be used in a point failure criteria. Alternatively, a smaller set of points can be selected for examination in order to increase computational efficiency.

Table 4: Influence matrix for both representative volume elements at the selected location

Square Array, M_{ij} $(x,y,z) = (0.5,1.0,0.5)$						Hexagonal Array, M_{ij} $(x,y,z) = \left(0.5,1.0,\sqrt{(3)}/2\right)$					
1.0	0.0	0.0	0.0	0.0	0.0	1.0	0.0	0.0	0.0	0.0	0.0
0.8	3.2	1.0	0.0	0.0	0.0	0.6	2.8	0.3	0.0	0.0	0.0
-0.3	-0.6	0.5	0.0	0.0	0.0	-0.2	-0.5	0.6	0.0	0.0	0.0
0.0	0.0	0.0	2.0	0.0	0.0	0.0	0.0	0.0	1.4	0.0	0.0
0.0	0.0	0.0	0.0	0.1	0.0	0.0	0.0	0.0	0.0	0.3	0.0
0.0	0.0	0.0	0.0	0.0	6.8	0.0	0.0	0.0	0.0	0.0	4.7

The same method is applicable for a stress based criteria whereby the stress state at a point is determined from the strain state through the appropriate constitutive relationships. In the current example, a single point is used to provide a numerical illustration of the micromechanical enhancement process. Tables 4 and 5 contain the influence matrix and thermal superposition vectors respectively. The data is extracted for both the square and hexagonal representative volume elements at the point $(e_1, e_2, e_3) = (L_1/2, L_2, L_3/2)$. Although both points represent locations that are midway between two fiber centers, the influence matrix and thermal superposition vectors are different for the two representative volume elements. This shows the effect of packing geometry on the local strain fields and the need for a comprehensive understanding of the underlying geometry contained in the composite material.

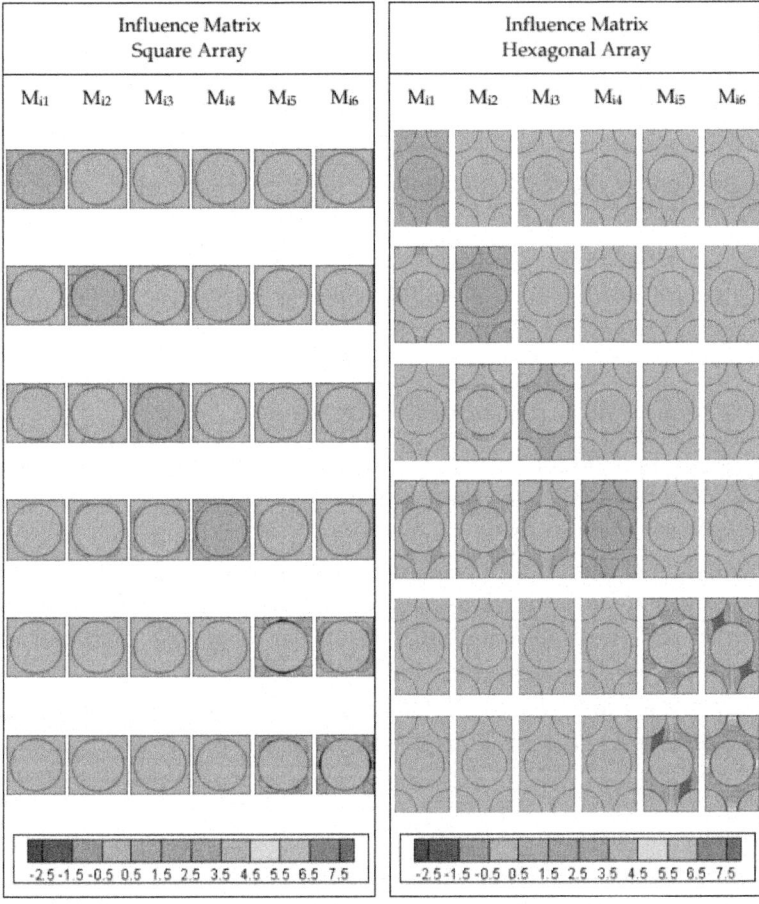

Figure. 4: Influence matrix fields for both representative volume elements, square and hexagonal

Table 5: Thermal superposition vector both representative volume elements at a selected location

Square Array, A_i (10⁻⁶/°C), $(x,y,z) = (0.5, 1.0, 0.5)$						Hexagonal Array, A_i (10⁻⁶/°C), $(x,y,z) = (0.5, 1.0, \sqrt{(3)}/2)$					
-64	129	-83	0	0	0	-64	90	-60	0	0	0

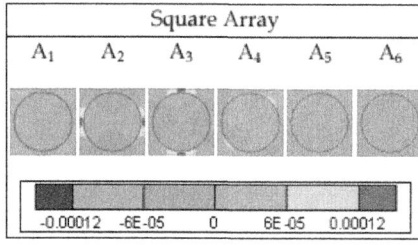
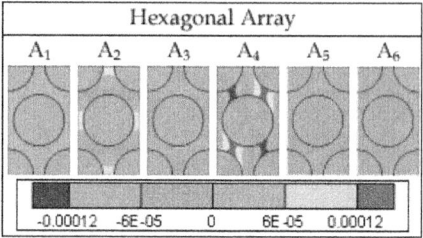

Figure. 5: Thermal superposition vector fields for both representative volume elements, square and hexagonal

Review of Classical Laminated Plate Theory

To illustrate the full process of using the self-consistent micromechanics method described herein, two laminate stacking sequences are investigated, the [0/90/90/0] cross-ply laminate and the [45/-45/-45/45] angle-ply laminate. For this example, the laminate level analysis is preformed using classical laminate plate theory but, finite element methods can also be used for more complex geometries and loading conditions. Consider only the in-plane resultant forces, $[N_x, N_y, N_{xy}]$ as defined in Figure 6, and thermal loading. Under these conditions, Equation 14 gives relationship between the lamina stresses and strains referenced to the principal material axis (1,2).

$$\bar{\sigma}_i = \bar{Q}_{ij}\varepsilon_j \ (i,j = 1,2,6) \tag{14}$$

Here, $\bar{Q}_{ij}$ is the reduced stiffness matrix in the material principal coordinate system. Equation 15 shows the entries in the reduced stiffness matrix written in terms of the homogenized stiffness matrix.

$$\bar{Q}_{ij} = \bar{C}_{ij} - \frac{\bar{C}_{i3}\bar{C}_{j3}}{\bar{C}_{33}} \ (i,j = 1,2,6) \tag{15}$$

The stresses and strains can be written in the laminate coordinate system (x,y) obtained by rotating the material coordinate system through an angle, θ, about the material 3 axis, see Figure 6. Equation 16 gives the stress strain relation of Equation 14 written in the transformed coordinate system.

$$\bar{\sigma}_i = \bar{Q}_{ij}(\varepsilon_j - \alpha_j\Delta T) \ (i,j = 1,2,6) \tag{16}$$

A primed quantity is referenced to the laminate coordinate system (x,y). As given in Equation 17, the reduced stiffness matrix referenced to the laminate coordinate system (x,y) is obtained by applying a transformation to the reduced stiffness matrix of Equation 14.

$$\left[\bar{Q}\right]=\left[T\right]^{T}\left[\bar{Q}\right]\left[T\right]\text{ with }\left[T\right]=\begin{bmatrix} c^2 & s^2 & sc \\ s^2 & c^2 & -sc \\ -2sc & 2sc & c^2-s^2 \end{bmatrix}$$

(17a,b)

$$s = \sin\theta \text{ and } c = \cos\theta$$

(17c,d)

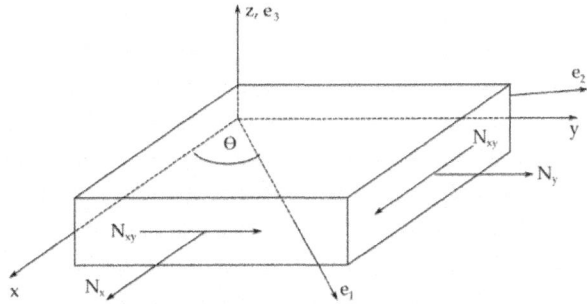

Figure. 6: Definition of laminate resultant forces

For both the cross-ply and the angle-ply laminates being considered, the in-plane loading is decoupled from any out of plane curvatures. As shown in Equation 18, the laminate strains are related to the force resultants, [N_x, N_y, N_{xy}], through the extensional stiffness matrix A_{ij}.

$$\{N\} = A_{ij}\bar{\varepsilon}_j - \{N^T\}$$

(18a)

$$A_{ij} = \sum_{l=1}^{L}\left(\bar{Q}_{ij}^{'}\right)^{(l)} t^{(l)} \text{ and } \{N^T\} = \Delta T \sum_{l=1}^{L}\left[\bar{Q}^{'}\right]^{(l)}\{\bar{\alpha}^{'}\}^{(l)} t^{(l)}$$

(18b,c)

Where the l index sums over the total number of lamina in the laminate and t is the thickness of each lamina. The laminate strains are determined in Equation 19 by rearranging Equation 18a.

$$\bar{\varepsilon}_i^{'} = A_{ij}^{-1}\left[\{N\}+\{N^T\}\right]$$

(19)

For the de-homogenization step, the components of homogenized strain field must be referenced to the principal material axis. This is accomplished by applying the transformation matrix of Equation 17b to the laminate strain in Equation 19. The transformation is applied for each lamina within the laminate. The plane stress condition is then inserted into Equation 2 to determine the out-of-plane strain component, $\bar{\varepsilon}_3$. Equation 20 gives the relationships obtained for an orthotropic material system. The homogenized strain state is now fully specified and the strain state within each representative volume element can be determined.

$$\bar{\varepsilon}_3 = -\frac{\bar{C}_{13}\bar{\varepsilon}_1 + \bar{C}_{23}\bar{\varepsilon}_2}{\bar{C}_{33}}$$

(20)

Application of Uniaxial Force Resultant, N_x

First, a purely mechanical loading is considered. The lamina level homogenized material properties calculated in Section 5.3 are used to determine the macroscopic state of strain in a both laminates subjected an axial force resultant, N_x, in the absence of other two force resultants, Ny and Nxy. The laminate resultants are defined in Figure 4. Consider the cross-ply and angle-ply laminates under a loading case in which the resulting edge forces are [N_x, N_y, N_{xy}] = [100 kN/m, 0, 0] without a temperature change. The macroscopic strain state in each lamina is determined using the process described in Section 5.3 with a lamina thickness, t, of 0.2 mm. The homogenized mechanical strains in the cross-ply and angle-ply laminates are given in Tables 6 and 7 respectively. From these homogenized states of strain, the state of strain within both representative volume elements is found by applying the influence matrices according to Equation 8. For this example, the influence matrices shown in Table 4 are used to determine the strain in the matrix phase at the location $(e_1, e_2, e_3) = (L_1/2, L_2, L_3/2)$. Tables 8 and 9 list the strains within the matrix at the selected point for the cross-ply and angle-ply laminates, respectively. The results show that the state of strain at the point of inquiry in each representative volume element can be very different from the homogenized state of strain in each lamina.

Application of Uniform Temperature Change, T

Next, consider the cross-ply and angle-ply laminates subjected to a uniform temperature change of $\Delta T = -100$ °C. The macroscopic strain state in each lamina is determined using the classical laminate theory analysis outlined in Section 5.3 with a lamina thickness, t, of 0.2 millimetres. The homogenized mechanical strains in the cross-ply and angle-ply laminates are given in Tables 10 and 11 respectively. The strain within both representative volume element is found from the homogenized states of strain by applying the influence matrices and thermal superposition vectors according to Equation 8. Mechanical strains are present at the lamina level and at the micro-level for the thermal loading case. This is due to a mismatch in the homogenized coefficients of thermal expansion of the lamina and a mismatch in the coefficients of thermal expansion of the constituents. Again, the influence matrices shown in Table

4 and the thermal superposition vectors of Table 5 are used to determine the strain in the matrix phase at the location (x, y, z) = (Lx/2, Ly, Lz/2) for both representative volume elements. Tables 12 and 13 list the strain state at the selected location within the matrix for the cross-ply and angle-ply laminates, respectively. It is obvious that identical results are obtained for the [0/90/90/0] and [45/-45/-45/45] laminates because the two laminates thermally identical after a rotation of 45°. As such, both laminates have identical states of strain in the material principal coordinates.

Table 6: Homogenized mechanical strains due to a force resultant, N_x = 100 k N/m in the [0/90/90/0] laminate

	Square Array, 10^{-6}		Hexagonal Array, 10^{-6}	
	0° Ply	90° Ply	0° Ply	90° Ply
$\bar{\varepsilon}_1 - \bar{\alpha}_1 \Delta T$	1,390	-55	1,396	-52
$\bar{\varepsilon}_2 - \bar{\alpha}_2 \Delta T$	-55	1,390	-52	1,396
$\bar{\varepsilon}_3 - \bar{\alpha}_3 \Delta T$	-655	-786	-670	-825

Table 7: Homogenized mechanical strains due to a force resultant, N_x = 100 kN/m in the [45/-45/-45/45] laminate

	Square Array, 10^{-6}		Hexagonal Array, 10^{-6}	
	45° Ply	-45° Ply	45° Ply	-45° Ply
$\bar{\varepsilon}_1 - \bar{\alpha}_1 \Delta T$	667	667	672	672
$\bar{\varepsilon}_2 - \bar{\alpha}_2 \Delta T$	667	667	672	672
$\bar{\varepsilon}_3 - \bar{\alpha}_3 \Delta T$	-720	-720	-748	-748
$\bar{\varepsilon}_6 - \bar{\alpha}_6 \Delta T$	-9221	9221	-9921	9921

Table 8: Micro-level mechanical strain due to a force resultant, N_x = 100 kN/m at the selected location in the [0/90/90/0] laminate

	Square Array, 10^{-6} $(x,y,z) = (0.5, 1.0, 0.5)$		Hexagonal Array, 10^{-6} $(x,y,z) = \left(0.5, 1.0, \sqrt{(3)}/2\right)$	
	0° Ply	90° Ply	0° Ply	90° Ply
$\varepsilon_1 - \alpha_1 \Delta T$	1,390	-55	1,396	-52
$\varepsilon_2 - \alpha_2 \Delta T$	281	3618	491	3,630
$\varepsilon_3 - \alpha_3 \Delta T$	-712	-1211	-655	-1,183

Table 9: Micro-level mechanical strain due to a shell edge traction, $N_x = 100$ k N/m at the selected location in the [45/-45/-45/45] laminate

	Square Array, 10^{-6} $(x,y,z)=(0.5,1.0,0.5)$		Hexagonal Array, 10^{-6} $(x,y,z)=(0.5,1.0,\sqrt{(3)}/2)$	
	45° Ply	-45° Ply	45° Ply	-45° Ply
$\varepsilon_1 - \alpha_1 \Delta T$	667	667	672	672
$\varepsilon_2 - \alpha_2 \Delta T$	1,948	1,948	2060	2060
$\varepsilon_3 - \alpha_3 \Delta T$	-960	-960	-919	-919
$\varepsilon_6 - \alpha_6 \Delta T$	-62,703	62,703	-46,629	46,629

Table 10: Homogenized mechanical strains due to thermal loading, $\Delta T = -100$ °C in the [0/90/90/0] laminate

	Square Array, 10^{-6}		Hexagonal Array, 10^{-6}	
	0° Ply	90° Ply	0° Ply	90° Ply
$\bar{\varepsilon}_1 - \bar{\alpha}_1 \Delta T$	-277	-277	-263	-263
$\bar{\varepsilon}_2 - \bar{\alpha}_2 \Delta T$	3,152	3,152	3,206	3,206
$\bar{\varepsilon}_3 - \bar{\alpha}_3 \Delta T$	-1,708	-1,708	-1,823	-1,823

Table 11: Homogenized mechanical strains due to thermal loading, $\Delta T = -100$ °C in the [45/-45/-45/45] laminate

	Square Array, 10^{-6}		Hexagonal Array, 10^{-6}	
	45° Ply	-45° Ply	45° Ply	-45° Ply
$\bar{\varepsilon}_1 - \bar{\alpha}_1 \Delta T$	-277	-277	-263	-263
$\bar{\varepsilon}_2 - \bar{\alpha}_2 \Delta T$	3,152	3,152	3,206	3,206
$\bar{\varepsilon}_3 - \bar{\alpha}_3 \Delta T$	-1,708	-1,708	-1,823	-1,823

Table 12: Micro-level mechanical strain due to thermal loading, $\Delta T = -100$ °C at the selected location in the [0/90/90/0] laminate

	Square Array, 10^{-6} $(x,y,z)=(0.5,1.0,0.5)$		Hexagonal Array, 10^{-6} $(x,y,z)=(0.5,1.0,\sqrt{(3)}/2)$	
	0° Ply	90° Ply	0° Ply	90° Ply
$\varepsilon_1 - \alpha_1 \Delta T$	6123	6123	6137	6137
$\varepsilon_2 - \alpha_2 \Delta T$	-4743	-4743	-728	-728
$\varepsilon_3 - \alpha_3 \Delta T$	5638	5638	3356	3356

Table 13: Micro-level mechanical strain due to thermal loading, $\Delta T = -100$ °C at the selected location in the [45/-45/-45/45] laminate

	Square Array, 10^{-6} $(x,y,z)=(0.5,1.0,0.5)$		Hexagonal Array, 10^{-6} $(x,y,z)=\left(0.5,1.0,\sqrt{(3)}/2\right)$	
	45° Ply	-45° Ply	45° Ply	-45° Ply
$\varepsilon_1 - \alpha_1\Delta T$	6123	6123	6137	6137
$\varepsilon_2 - \alpha_2\Delta T$	-4743	-4743	-728	-728
$\varepsilon_3 - \alpha_3\Delta T$	5638	5638	3356	3356

CONCLUSIONS

A computationally efficient method for estimating the microscopic strain field within the discrete phases of a heterogeneous medium consisting of collimated, continuous fibers within an isotropic matrix has been developed. The goal of the development has been to provide an essential link in a multi-scale analysis of a composite structure. The structural loading and deformations at the macro-scale can be related to the state of strain within the fiber and matrix phases at the micro scale by using the self-consistent micromechanics method. The model utilizes a conventional influence function formulation and considers thermo-mechanical deformations. Results have been presented that illustrate the utility of the approach in determining microscopic state of strain in the [0/90/90/0] and [45/-45/- 45/45] laminates. Enhanced strain components within each of the lamina were calculated for both uniaxial loading and a uniform change in temperature. The present example showed results extracted for a single point within the representative volume element. As shown in Figure 3, the influence matrix and thermal superposition vector are field values. Therefore, the state of strain within the representative volume element can also be represented as a field value. As such, the analysis is not limited to analysis of the state of strain at a single point. However, the reader may choose as many interrogation points as are required in order to address de-homogenization of all of the phases or to meet a specific need with a minimum computational cost. Microstructures found in fiber reinforced composites typically consist of an irregular array of fibers which differ from the representative volume elements analysed herein. An efficient method for dealing with variability is through the use of a statistically equivalent periodic unit cell. With this approach, a computational step is used to generate an equivalent representative volume element that replaces the actual complex geometry. This method has been applied at several length scales including a unidirectional fiber tow (Zeman and Sejnoha, 2007). The main assumption implicit in the analysis is that the representative volume element is subjected to a uniform state of strain. Certainly, this is not true at all locations within a laminate. Examples include areas with large strain gradients or locations with discontinuities in the assumed periodicity such as ply interfaces. In light of these limitations, the described method provides a reasonable first order approximation of the state of strain within the constitutive phases of an ordered heterogeneous medium.

REFERENCES

1.   Buchanan, D. L.; Gosse, J. H.; Wollschlager, J. A.; Ritchey, A. & Pipes, R. B. (2009). Micromechanical enhancement of the macroscopic strain state for advanced composite materials. Composites Science and Technology, Vol. 69, (month and year of the edition) page numbers (1974–1978), 0266-3538

2.   Christensen, R. M. (1979). Mechanics of Composite Materials, John Wiley & Sons, 0-471-05167- 5, New York, New York.

3.   Daniel, I. M. & Ishai, O. (2006). Engineering Mechanics of Composite Materials, Oxford University Press, 0-19-515097, New York, New York.

4.   Gosse, J.H. & Christensen, S. (2001). Strain Invariant Failure Criteria for Polymers in Composites. Proceedings of 42nd AIAA/ASME/ASCE/AHS/ASC Structures, Structural Dynamics, and Materials Conference and Exhibit, Seattle WA, April 2001

5.   Halpin, J. C. & Tsai, S. W. (1967). Effects of Environmental Factors on Composite Materials. Air Force Technical Report AFML-TR-67-423, Wright Aeronautical Labs, Dayton, OH Hashin, Z. & Rosen, B. W. (1964). The Elastic Moduli of Fiber-Reinforced Materials. Journal of Applied Mechanics, Vol. 21 (1964) page numbers (233-242), 0021-8936

6.   Hashin, Z. (1972). Theory of Fiber Reinforced Materials. NASA CR-1974, Langley Research Center Hill, R. (1963). Elastic Properties of Reinforced Solids: Some Theoretical Principles. Journal of Mechanics and Physics of Solids, Vol. 11, No. 5, (September 1963) page numbers (357-372), 0022-5096

7.   Hill, R. (1965). Theory of Mechanical Properties of Fibre-Strengthened Materials: III, SelfConsistent Model. Journal of the Mechanics and Physics of Solids, Vol. 13, No. 4,(August 1965) page numbers (189-198), 0022-5096

8.   Hutapea, P.; Yuan, F.G.; Pagano, N.J. (2003) Micro-stress prediction in composite laminateswith high stress gradients. International Journal of Solids and Structures, Vol. 40, No.9, (May 2003) page numbers (2215–2248), 0020-7683

9.   Pagano, N. J. & Rybicki, E. F. (1974). On the significance of effective modulus solutions for fibrous composites. Journal of Composite Materials, Vol. 8, No. 3, (July 1974) page numbers (214-228), 0021-9983

10. Pagano, N. J. & Yuan, F. G. (2000). The significance of effective modulus theory (homogenization) in composite laminate mechanics. Composites Science and Technology, Vol. 60, No. 12-13, (September 2000) page numbers (2471-2488), 0266-3538

11. Paul, B. (1960). Prediction of Elastic Constants of Multiphase Materials. Transactions of AIME, Vol. 218, page numbers (36-41), 0096-4778

12. Pindera, M. J.; Khatam, H.; Drago, A. S. & Bansal, Y. (2009). Micromechanics of spatially uniform heterogeneous media: A critical review and emerging approaches.

13. Composites Part B: Engineering, Vol. 40, No. 5, (July 2009) page numbers (349-378),1359-8368

14. Raghavan, P.; Moorthy, S.; Ghosh, S. & Pagano, N. J. (2001). Revisiting the composite laminate problem with and an adaptive multi-level computational model.

15. Composite Science and Technology, Vol. 61, 2001, (June 2001) page numbers (1017- 1040), 0266-3538

16. Reuss, A. (1929). "Berechnung der Fließgrenze von Mischkristallen auf Grund der Plastizitätsbedingung für Einkristalle". Zeitschrift für Angewandte Mathematik und Mechanik,Vol. 9 No. 1 (February 1929) page numbers (49–58), 1521-4001

17. Sun, C. T. & Vaidya, R. S. (1996). Prediction of composite properties from a representative volume element. Composites Science and Technology, Vol. 56, No. 2, (1996) page numbers (171-179), 0266-3538

18. Voigt, W. (1887). Theoretische Studien über die Elasticitätsverhältnisse der Krystalle.Abhandlungen der Gesellschaft der Wissenschaften zu Göttingen, Vol. 34, (August 1887) page numbers (3–51)

19. Wang, Y.; Sun, C.; Sun, X. & Pagano, N. J. (2002). Principles for Recovering Micro-Stress in Multi-Level Analysis, In: Composite Materials: Testing, Design and Acceptance Criteria, ASTM STP 1416, A. Zureick and A. T. Nettles, (Ed.), page numbers (200-211),

20. American Society for Testing and Materials International, 0-8031-2893-2, West Conshohocken, PA

21. Zeman, J; Sejnoha, M. (2007). From random microstructures to representative volume elements. Modeling and Simulation in Materials Science and Engineering, Vol. 15, No. 4,(2007) page numbers (S325-S335), 0965-0393.

Chapter 8

MICROMECHANICS MODELING ON ELECTRICAL CONDUCTIVITY OF CNT-POLYMER COMPOSITES

INTRODUCTION

A mixed micromechanics model was developed to predict the overall electrical conductivity of carbon nanotube (CNT)–polymer nanocomposites. Two electrical conductivity mechanisms, electron hopping and conductive networks, were incorporated into the model by introducing an interphase layer and considering the effective aspect ratio of CNTs. It was found that the modeling results agree well with the experimental data for both single-wall carbon nanotube and multi-wall carbon nanotube based nanocomposites. Simulation results suggest that both electron hopping and conductive networks contribute to the electrical conductivity of the nanocomposites, while conductive networks become dominant as CNT volume fraction increases. It was also indicated that the sizes of CNTs have significant effects on the percolation threshold and the overall electrical conductivity of the nanocomposites. This developed model is expected to provide a more accurate prediction on the electrical conductivity of CNT–polymer nanocomposites and useful guidelines for the design and optimization of conductive polymer nanocomposites. Recently, the addition of carbon nanotubes (CNTs) in compliant polymers to form conductive nanocomposites has stimulated a surge of scientific interests from the research communities due to their potential applications in stretchable electronics (Subramanian et al., 2006; Yu et al., 2009; Li et al., 2010). Such interests stem from CNT's excellent electrical conductivity, which is several orders of magnitude larger than that of all neat polymers (Ebbesen et al., 1996). Existing experiments have shown the remarkable improvement of electrical conductivity of polymer medium with the addition of CNTs (Kim et al., 2005; Gojny et al., 2006; Hu et al., 2008). Due to the large disparity of electrical conductivity between the polymer matrix and CNTs, the CNT-polymer composite demonstrates a percolation like behavior, and its electrical conductivity increases abruptly when the CNT concentration reaches a certain threshold. Explanations for this percolation behavior have centered on two conductivity mechanisms: electron hopping (or quantum tunneling) at the nanoscale and conductive networks at the microscale (Qunaies et al., 2003; Du et al., 2004; Chang et al., 2009; Zhang et al., 2009). From quantum mechanics, electrons have the probability of hopping intra-tube or from one CNT to another, but the probability is highly dependent on the separation distance between CNTs (Seidel and Lagoudas, 2009). When the CNT concentration in the composite is extremely low with larger separation distance between CNTs, electron hopping governs the electrical conductivity of the composite. However, when the separation distance between CNTs decreases with the increase of CNT concentration, some adjacent CNTs may be electrically connected resulting in microscale conductive networks. With the CNT concentration getting larger and larger, conductive networks are believed to be dominant over electron hopping. Quantitative prediction on the overall electrical conductivity of CNT-polymer composites which can capture the percolation behavior is essential for the design and optimization of conductive nanocomposites. However, accurate modeling of the electrical conductivity of nanocomposites is challenging due to the multi-scale nature of the problem as indicated by the conductivity mechanisms, i.e., phases of CNT-polymer nanocomposites range from the microscale down to the nanoscale.

Efforts have been devoted to predicting the electrical conductivity of CNT-polymer nanocomposites. Monte Carlo (MC) simulation has commonly been considered as an effective way to predict the electrical conductivity of the nanocomposites (Ma and Gao, 2008; Zhang and Yi, 2008; Lu et al., 2010) by incorporating the nanoscale features of the materials. However, MC simulation is numerically expensive for large-scale systems and does not provide an explicit formulation for materials design and optimization, thus analytical models have naturally been pursued as alternative tools. Regardless of the conductivity mechanisms, a three-parameter power law equation, which is based on classical percolation theory, has been adopted to predict the electrical conductivity of nanocomposites after percolation (Kirkpatrick, 1973; Grimmett, 1999). Comparisons showed that the power law equation successfully predicted the trend of some experimental data (Ramasubramaniam et al., 2003; McLachlan et al., 2005), however, such power law equation is phenomenological and the parameters in this power law equation need to be fitted from experimental data. In addition, the power law equation cannot capture the variation of electrical conductivity prior to percolation and is not able to distinguish the two conductivity mechanisms. Alternatively, some micromechanics theory based models have also been extended to predict the overall electrical conductivity of CNT-polymer nanocomposites. Deng and Zheng (2008) developed a simplified micromechanics model to evaluate the effective electrical conductivity for CNT composites by accounting for the percolation, conductive networks, conductivity anisotropy and non-straightness of CNTs. It was found that the non-straightness of CNTs had significant influence on the effective electrical conductivity of the composites, and model predictions successfully predicted the trend of some measured experimental data in literature. Based on Deng and Zheng's model, Takeda et al. (Takeda et al., 2011) considered the effect of electron hopping among CNTs by taking CNTs as effective fibers to predict the overall electrical conductivity of the composites.

The influence of electron hopping and the formation of conductive networks on the electrical conductivity of CNT-polymer composites has been investigated by Seidel and Lagoudas (2009) using a Mori-Tanaka micromechanics model (Mori and Tanaka, 1973; Hatta and Taya, 1985). In their work, electron hopping was assessed through an interphase layer surrounding the CNT, while the effect of conductive networks was captured by changing the CNT aspect ratio. The developed micromechanics model was successful in qualitatively identifying the potential causes for low percolation concentrations. However, large discrepancy was observed between Seidel and Lagoudas predicted results and experimental data after percolation. It should be mentioned that in Seidel's work (2009) the thickness and the electrical conductivity of the interphase layer were kept constant, and the two electrical conductivity mechanisms were considered separately in the simulations, i.e., either electron hopping or conductive networks solely dominated electrical conductivity. From a statistical mechanics point of view, it is believed that some of CNTs in the polymer form conductive networks, while others contribute to the effective electrical conductivity of the composites through electron hopping (Deng and Zheng, 2008). The probability of the formation of conductive networks will increase with the increase of CNT concentration. In addition, the thickness and the electrical conductivity of the interphase layer accounting for electron hopping will also vary with CNT concentration. In order to more accurately predict the overall electrical conductivity of CNT-polymer composites, all these factors should be incorporated into a micromechanics model.

It is the objective of the current work to develop a mixed micromechanics model for CNT-polymer composites with the incorporation of electron hopping and conductive networks simultaneously. According to the percolation process, electron hopping is believed to be dominant below the percolation threshold. In this model, hollow CNTs were modeled as effective solid fibers, in which an interphase layer outside the CNT was used to capture nanoscale effects of electron hopping among CNTs as shown in literature. After percolation, the thickness and the electrical conductivity of the interphase layer were defined by considering electric tunneling effect, which vary with CNT concentration. The developed micromechanics model was validated by comparing the prediction results with experimental data.

Nanoscale composite cylinder model for CNT

As mentioned before, electron hopping exists among CNTs distributed in the polymer matrix, which results in the formation of a continuum interphase layer surrounding the CNT. In order to capture this interphase layer and the hollow nature of CNTs, an effective composite cylinder model well accepted by researchers (Hashin, 1990; Yan et al., 2007; Seidel and Lagoudas, 2009), was used to determine the effective electrical conductivity of the CNT together with the surrounding interphase. Such an effective composite cylinder assemblage was further homogenized as an equivalent solid filler based on the application of a set of homogeneous boundary conditions as shown in Figure 2.1. In this way, the composite itself is composed of two phases: the polymer matrix and the effective solid filler. For this two-phase composite, a micromechanics model can further be applied to determine its overall electrical conductivity.

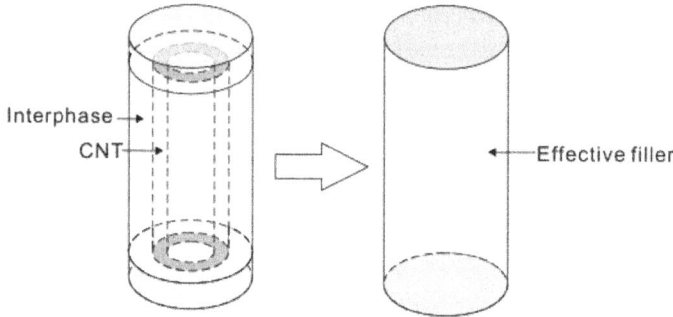

Figure 2.1: Schematic illustration of the composite cylinder and the effective filler.

Effective electrical conductivity of composite cylinder

In order to apply micromechanics theory, the effective electrical conductivity of the solid filler phase needs to be determined first. Figure 2.2 is a sketch of the composite cylinder assemblage (Yan et al., 2007), which consists of a CNT and the surrounding interphase with thickness t. A local coordinate system $(\tilde{x}_1, \tilde{x}_2, \tilde{x}_3)$ is used to describe the cylinder assemblage, in which the electrical isotropic plane (cross-section) can also be described by a polar coordinate system (r, β). Here the CNT is treated as a solid cylinder instead of a hollow tube due to the difficulty of obtaining the actual electrical conductivity of the CNT considering the nanoscale structures (Ebbesen et al., 1996; Takeda et al., 2011). According to the structure shown in Figure 2.2, the composite cylinder can be divided into three parts: isotropic parts 1 and 3 are interphases; transversely isotropic part 2 is a combination of the CNT with the surrounding interphase. Knowing the properties of these three parts, the law-of-mixture rule (Taya, 2005; Yan et al., 2007), which is based on equating the total electric energy of the composite cylinder consisting of three parts to the electric energy of a homogeneous effective solid filler transferred by the same electric field, can be applied to determine the effective longitudinal and transverse electrical conductivity of the effective filler. For part 2, which consists of a CNT (length L, radius rc) and an annual interphase (thickness t), the effective longitudinal electrical conductivity, L σ 2 % in the local coordinate system, is first determined from the law-ofmixture as (Taya, 2005; Yan et al., 2007):

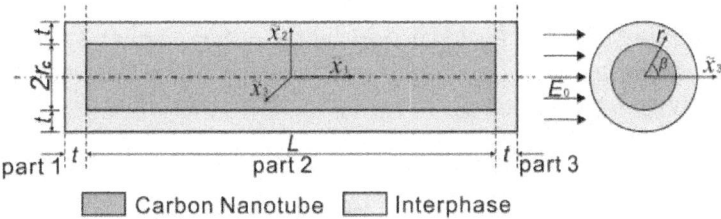

Figure 2.2: Sketch of a composite cylinder.

$$\tilde{\sigma}_2^L = \frac{\sigma_c^L r_c^2 + \sigma_{Int}\left(2r_c t + t^2\right)}{\left(r_c + t\right)^2} \, ,$$

(2.1)

where σ_c^L and σ_{Int} are the electrical conductivities of the CNT along the longitudinal direction and the interphase, respectively. The superscript "L" represents the longitudinal direction in the local coordinate system.

Due to the variation of the cylindrical surface area in the transverse direction, the law-ofmixture rule is not applicable for calculating the electrical conductivity in this direction. In order to determine the transverse electrical conductivity $\tilde{\sigma}_2^T$ in the isotropic plane of part 2, Maxwell's equation in the local polar coordinates (r, β) is applied when the system is subjected to an uniform electric field E0 along 3 x% at a large distance R sufficiently far away from the CNT (Yan et al., 2007), i.e.,

$$\nabla^2\phi = \frac{1}{r}\frac{\partial}{\partial r}\left(r\frac{\partial\phi}{\partial r}\right) + \frac{1}{r^2}\frac{\partial^2\phi}{\partial\beta^2} = 0 \, ,$$

(2.2)

where φ(r, β) is the electric potential. Note the (r, β) plane is the isotropic plane and $\tilde{x}_3$ is an arbitrary radial direction along which the electrical conductivity is the same. The boundary conditions for this problem are prescribed as

$$\phi_c\big|_{r=0} = \text{constant}, \quad E_m\big|_{r=R} = -\frac{\partial \phi_m}{\partial r}\bigg|_{r=R} = E_0$$

$$\phi_c\big|_{r=r_c} = \phi_{\text{Int}}\big|_{r=r_c}, \quad -\sigma_c^T \frac{\partial \phi_c}{\partial r}\bigg|_{r=r_c} = -\sigma_{\text{Int}} \frac{\partial \phi_{\text{Int}}}{\partial r}\bigg|_{r=r_c}$$

$$\phi_{\text{Int}}\big|_{r=r_c+t} = \phi_m\big|_{r=r_c+t}, \quad -\sigma_{\text{Int}} \frac{\partial \phi_{\text{Int}}}{\partial r}\bigg|_{r=r_c+t} = -\sigma_m \frac{\partial \phi_m}{\partial r}\bigg|_{r=r_c+t}$$

$$(2.3)$$

The subscripts "c", "m" and "Int" represent the CNT, the polymer matrix and the interphase, respectively, while superscript "T" represents the transverse direction in the local coordinate system. Solving Eq. (2.2) with the corresponding boundary conditions from Eq. (2.3), the electric potential distribution can be derived as

$$\phi_c(r, \beta) = 2A\sigma_{\text{Int}} E_0 r \cos \beta \quad 0 \le r \le r_c$$

$$\phi_{\text{Int}}(r, \beta) = A\left[\sigma_{\text{Int}} + \sigma_c^T + \left(\frac{r_c}{r}\right)^2 \left(\sigma_{\text{Int}} - \sigma_c^T\right)\right] E_0 r \cos \beta \quad r_c < r < r_c + t$$

$$\phi_m(r, \beta) = \left\{ A\left[\left(\frac{r_c+t}{r}\right)^2 \left(\sigma_{\text{Int}} + \sigma_c^T\right) + \left(\frac{r_c}{r}\right)^2 \left(\sigma_{\text{Int}} - \sigma_c^T\right)\right] + \left(\frac{r_c+t}{r}\right)^2 - 1 \right\} E_0 r \cos \beta \quad r_c + t \le r = R$$

$$(2.4)$$

Where

$$A = \frac{2\sigma_m}{\left(\sigma_{\text{Int}} - \sigma_c^T\right)\left(\sigma_{\text{Int}} - \sigma_m\right)\left(\dfrac{r_c}{r_c+t}\right)^2 - \left(\sigma_{\text{Int}} + \sigma_c^T\right)\left(\sigma_{\text{Int}} + \sigma_m\right)}$$

$$(2.5)$$

Correspondingly, the electric field along the 3 x% -axis for the composite assemblage can be determined as

$$E_{c,\tilde{x}_3} = -\frac{\partial \phi_c}{\partial \tilde{x}_3} = -\frac{\partial \phi_c}{\partial r}\frac{\partial r}{\partial \tilde{x}_3} = -\frac{1}{\cos \beta}\frac{\partial \phi_c}{\partial r}$$

$$E_{\text{Int},\tilde{x}_3} = -\frac{\partial \phi_{\text{Int}}}{\partial \tilde{x}_3} = -\frac{\partial \phi_{\text{Int}}}{\partial r}\frac{\partial r}{\partial \tilde{x}_3} = -\frac{1}{\cos \beta}\frac{\partial \phi_{\text{Int}}}{\partial r}$$

$$(2.6)$$

with the electric current density in the CNT and the interphase being derived as

$$j_{c,\tilde{x}_3} = \sigma_c^T E_{c,\tilde{x}_3} \quad \text{and} \quad j_{\text{Int},\tilde{x}_3} = \sigma_{\text{Int}} E_{\text{Int},\tilde{x}_3} .$$

$$(2.7)$$

From Eqs. (2.4)–(2.7), it can be seen that the electric field and the electric current density of the CNT and the interphase are dependent on the radius. To obtain the equivalent transverse electrical conductivity of part 2, a spatial average expression is introduced as (Yan et al., 2007)

$$\left\langle j_{\tilde{x}_3} \right\rangle = \tilde{\sigma}_2^T \left\langle E_{\tilde{x}_3} \right\rangle,$$

(2.8)

Where $\left\langle j_{\tilde{x}_3} \right\rangle = \frac{1}{V} \int_v j_{\tilde{x}_3} dv$ and $\left\langle E_{\tilde{x}_3} \right\rangle = \frac{1}{V} \int_v E_{\tilde{x}_3} dv$ are the spatial averages of the electric current density and the electric field along the $\tilde{x}_3$-axis, respectively. Combining Eqs. (2.4)–(2.8) the equivalent transverse electrical conductivity of part 2, $\tilde{\sigma}_2^T$ in the local coordinate system, can be derived as

$$\tilde{\sigma}_2^T = \frac{\left\langle j_{\tilde{x}_3} \right\rangle}{\left\langle E_{\tilde{x}_3} \right\rangle} = \frac{\int_0^{r_c} \sigma_c^T E_{c,\tilde{x}_3} 2\pi r dr + \int_{r_c}^{r_c+t} \sigma_{Int} E_{Int,\tilde{x}_3} 2\pi r dr}{\int_0^{r_c} E_{c,\tilde{x}_3} 2\pi r dr + \int_{r_c}^{r_c+t} E_{Int,\tilde{x}_3} 2\pi r dr} = \frac{2r_c^2 \sigma_c^T \sigma_{Int} + \sigma_{Int} \left(\sigma_{Int} + \sigma_c^T\right)\left(t^2 + 2r_c t\right)}{2r_c^2 \sigma_{Int} + \left(\sigma_{Int} + \sigma_c^T\right)\left(t^2 + 2r_c t\right)}.$$

(2.9)

Then by using the law-of-mixture rule again for the composite cylinder composed of parts 1, 2, and 3, the effective electrical conductivity of the solid filler in both the transverse and longitudinal directions in the local coordinate system can be derived as (Yan et al., 2007):

$$\tilde{\sigma}^T = \frac{\sigma_{Int}}{L + 2t}\left[L \frac{2r_c^2 \sigma_c^T + \left(\sigma_c^T + \sigma_{Int}\right)\left(t^2 + 2r_c t\right)}{2r_c^2 \sigma_{Int} + \left(\sigma_c^T + \sigma_{Int}\right)\left(t^2 + 2r_c t\right)} + 2t\right]$$

$$\tilde{\sigma}^L = \frac{(L + 2t)\sigma_{Int}\left[\sigma_c^L r_c^2 + \sigma_{Int}\left(2r_c t + t^2\right)\right]}{2\sigma_c^L r_c^2 t + 2\sigma_{Int}\left(2r_c t + t^2\right)t + \sigma_{Int} L\left(r_c + t\right)^2}.$$

(2.10)

Thickness and conductivity of interphase It is clearly indicated in Eq. (2.10) that the properties of the interphase, i.e., the thickness (t) and the electrical conductivity (σ_{Int}), may also play a significant role in the effective electrical conductivity of the solid filler. In addition, the percolation behavior is governed by two mechanisms as discussed before: nanoscale electron hopping (EH) (or quantum tunneling) and microscale conductive networks (CN). Therefore, definitions of the thickness and the electrical conductivity of the interphase accounting for electron hopping and conductive networks are essential. Figure 2.3 shows an example of an inplane contact configuration between two adjacent CNTs (Takeda et al., 2011), in which the separation distance naturally decreases with the increase of CNT volume fraction. When the CNT volume fraction reaches the percolation threshold f_c, the formation of conductive networks starts. In literature, several works have demonstrated that the average separation distance da between adjacent CNTs when conductive networks are formed after percolation rate follows a power law description (Allaoui et al., 2008; Deng and Zheng, 2008; Takeda et al., 2011). Here we take the power law relation suggested in (Deng and Zheng, 2008), i.e.,

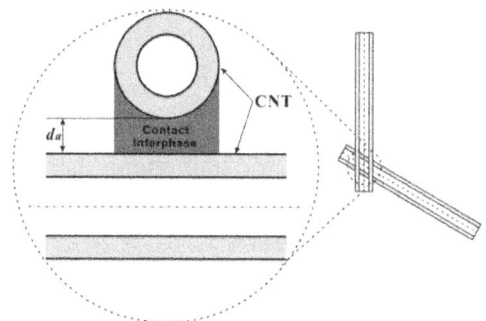

Figure 2.3: Contact configuration between two SWCNTs.

$$d_a = \frac{\alpha}{f^{1/3}},$$

(2.11)

where f is the volume fraction of CNTs and α is a constant to be determined. The upper limit separation distance for CNTs in conductive networks was suggested as $d_c = 1.8$ nm (Li et al., 2007; Takeda et al., 2011), which is the maximum possible thickness of the medium separating two adjacent CNTs that allows the tunneling penetration of electrons. Therefore, α can be approximately determined as

$$\alpha = d_c f_c^{1/3}.$$

(2.12)

Electron hopping is highly dependent on the separation distance between CNTs. When the separation distance of CNTs is larger than 1.8 nm with CNT volume fraction less than the percolation threshold f_c, it is regarded that CNTs are more independent rather than electrically connected to each other. Under this situation, electron hopping dominates the electrical conductivity of the composite. When the CNT volume fraction reaches the percolation threshold f_c, conductive networks start to form and the separate distance between percolated CNTs is taken as 1.8 nm. With the increase of CNT volume fraction after percolation, it is understood that the separation distance of CNTs in conductive networks is getting smaller than 1.8 nm. However, these percolated CNTs are still in electrical contact rather than physical contact due to van der Waals forces. For each CNT forming a conductive network, the thickness of the interphase can be defined as

$$t = \frac{1}{2}d_a = \frac{1}{2}\left(\frac{f_c}{f}\right)^{1/3} d_c.$$

(2.13)

It should be noted that the interphase thickness of part 1 (or 3) and part 2 of the composite cylinder was taken the same as assumed in (Takeda et al., 2011). By using Simmons' derivation for electron tunneling (Simmons, 1963; Takeda et al., 2011), the tunneling-type contact resistance between two CNTs can be expressed as

$$R_{Int}(d_a) = \frac{d_a \hbar^2}{ae^2 (2m\lambda)^{1/2}} \exp\left(\frac{4\pi d_a}{\hbar}(2m\lambda)^{1/2}\right),$$

(2.14)

where λ is the potential barrier height, which is approximate 5.0 eV for CNTs dispersed in most polymers (Shiraishi and Ata, 2001; Takeda et al., 2011); m (9.10938291×10^{-31} kg) and e ($-1.602176565\times10^{-19}$ C) are the mass and the electric charge of an electron, respectively; a is the contact area of CNTs and $\hbar$ (6.626068×10-34 $m^2 \cdot kg/s$) is the Planck constant. The electrical conductivity of the interphase surrounding a CNT in conductive networks can then be obtained correspondingly,

$$\sigma_{Int} = \frac{d_a}{aR_{Int}(d_a)}.$$

(2.15)

Due to the lack of information in literature for estimating the average separation distance between CNTs without electrical contact, an assumption of d_c is given on this quantity in the current work, which is the upper limit separation distance between CNTs allowing electron to tunnel through. Therefore, the thickness and the electrical conductivity of the interphase surrounding CNTs without forming conductive networks are determined as:

$$t = \frac{1}{2}d_c$$

(2.16)

And

$$\sigma_{Int} = \frac{d_c}{aR_{Int}(d_c)}.$$

(2.17)

It is obvious that such an assumption on the average separation distance between CNTs may result in an overestimated electrical conductivity of the nanocomposites, particularly for the composite with CNT volume fraction below and around the percolation threshold. For example, when the CNT volume fraction is f = 0.003% and the CNT aspect ratio is L/d = 1000, the electrical conductivity of the interphase was calculated as 8.6776×10^{-14} S/m from Eq. (2.17), which is comparable to the interphase electrical conductivity 9.95×10^{-14} S/m calculated from Eq. (2.15). However, with the increase of CNT volume fraction, the contribution of conductive networks becomes more dominant. For example, when CNT volume fraction f reaches 0.01%, the electrical conductivity for the interphase surrounding CNTs is calculated as 7.922×10^{-8} S/m from Eq. (2.15), which is much larger than the one calculated from Eq. (2.17). Therefore, such an overestimation can be neglected after percolation.

Mixed micromechanics model

For the purpose of modeling simplicity, CNTs were assumed to be straight, and uniformly and randomly dispersed in the polymer matrix. The CNT volume fraction is denoted as f. As discussed in the previous section, CNTs together with their surrounding interphases can be simulated as equivalent solid fillers due to electron hopping. Therefore, the composite can be considered as a two-phase system with solid fillers dispersed in the polymer matrix. The volume fraction f_{eff} of the effective solid fillers can

be obtained in terms of the volume fraction of CNTs in the polymer according to Figure 2.2 (Seidel and Lagoudas, 2009; Yan et al., 2007) as

$$f_{\text{eff}} = \frac{\left(r_c + t\right)^2 \left(L + 2t\right)}{r_c^2 L} f .$$

(2.18)

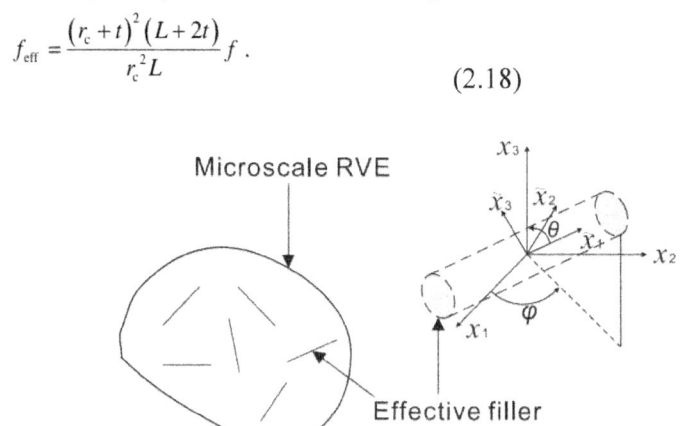

Figure 2.4: Sketch of a microscale RVE containing effective filler.

To determine the effective electrical conductivity of the composites with two phases, a representative volume element (RVE) containing enough effective fillers with random orientations, which is able to statistically represent overall properties of the material, as shown in Figure 2.4 was chosen in the micromechanics model to study the overall electrical conductivity of the composite. In micromechanics theory, each orientation of a given effective filler in the polymer can be treated as a separate phase, and the effective electrical conductivity of the nanocomposite can be determined by averaging over all possible orientations in the RVE according to the following expression (Odegard et al., 2003; Taya, 2005):

$$\sigma_{eff} = \sigma_m + \frac{\int_0^{2\pi} \int_0^{\pi} \rho(\varphi,\theta) f_{eff}(\sigma - \sigma_m) A \sin\theta \, d\theta \, d\varphi}{\int_0^{2\pi} \int_0^{\pi} \rho(\varphi,\theta) \sin\theta \, d\theta \, d\varphi}$$

(2.19)

where φ and θ are Euler angles defining effective filler orientation as shown in Figure 2.4; $\rho(\varphi, \theta)$) is the orientation distribution function, which equals unity for a random

distribution; σ and σ_m are electrical conductivity tensors of the effective filler and the polymer, respectively; and A is the electric field concentration tensor. Since all quantities in Eq. (2.19) are in the global coordinate system (x_1, x_2, x_3), quantities obtained in the local coordinate system need to be transferred to the global coordinate system through a transformation, i.e.,

$$\sigma = Q^T \tilde{\sigma} Q,$$

(2.20)

where $\tilde{\sigma}$ is the electrical conductivity tensor of the effective filler in the local coordinate system, which is given as:

$$\tilde{\sigma} = \begin{bmatrix} \tilde{\sigma}^L & 0 & 0 \\ 0 & \tilde{\sigma}^T & 0 \\ 0 & 0 & \tilde{\sigma}^T \end{bmatrix},$$

(2.21)

with $\tilde{\sigma}^L$ and $\tilde{\sigma}^T$ being the longitudinal and transverse electrical conductivity of the effective filler as obtained in Eq. (2.10), and the transformation matrix is (Seidel and Lagoudas, 2009)

$$Q = \begin{bmatrix} \sin\theta\cos\varphi & -\cos\theta\cos\varphi & \sin\varphi \\ \sin\theta\sin\varphi & -\cos\theta\sin\varphi & -\cos\varphi \\ \cos\theta & \sin\theta & 0 \end{bmatrix}.$$

(2.22)

As a case study, the Mori-Tanaka method (Hatta and Taya, 1985; Mori and Tanaka, 1973) was selected as the micromechanics model. Based on the assumption of uniform and random distribution of CNTs, the electric field concentration tensor A in the global coordinate system can be expressed as (Seidel and Lagoudas, 2009; Taya, 2005):

$$A = Q^T \tilde{T} Q \left\{ (1 - f_{\text{eff}}) \delta + \frac{f_{\text{eff}}}{4\pi} \int_0^{2\pi} \int_0^{\pi} \{Q^T \tilde{T} Q\} \sin\theta \, d\theta \, d\varphi \right\}^{-1} , $$

(2.23)

$$\tilde{T} = \left\{ \delta + S (\sigma_m)^{-1} (\tilde{\sigma} - \sigma_m) \right\}^{-1}$$

(2.24)

with δ and S being the Kronecker delta tensor and the Eshelby tensor of the effective filler, respectively. The Eshelby tensor of the effective filler is given as

$$S = \begin{bmatrix} S_{11} & 0 & 0 \\ 0 & S_{22} & 0 \\ 0 & 0 & S_{33} \end{bmatrix} ,$$

(2.25)

Where

$$S_{22} = S_{33} = \frac{A_{re}}{2 (A_{re}^2 - 1)^{3/2}} \left[A_{re} (A_{re}^2 - 1)^{1/2} - \cosh^{-1} A_{re} \right]$$

(2.26)

with Are being the aspect ratio of the effective filler, i.e., Are=(L+2t)/(2rc+2t), and S11=1− S22, respectively (Taya, 2005; Seidel and Lagoudas, 2009).

As mentioned before, several experiments and simulations have demonstrated that CNTpolymer nanocomposites have a percolation-like behavior displaying a sharp increase in electrical conductivity after the CNT volume fraction reaches a certain threshold. For a two-phase nanocomposite with a uniformly random distribution of CNTs, the percolation threshold, fc, is approximately determined by the following analytical formulation (Gao and Li, 2003; Deng and Zheng, 2008)

$$f_c(H) = \frac{9H(1-H)}{2 + 15H - 9H^2} ,$$

(2.27)

Where $H(A_r) = \frac{1}{A_r^2 - 1} \left[\frac{A_r}{\sqrt{A_r^2 - 1}} \ln \left(A_r + \sqrt{A_r^2 - 1} \right) - 1 \right]$ and A_r is the aspect ratio of the CNT, i.e., $A_r = L/2r_c$. Here it should be noted that the percolation threshold defined in Eq. (2.27)

is the critical CNT volume fraction denoting the onset of the percolation process (a process of forming conductive networks). As argued by Deng and Zheng (2008), the percolation process of CNT composites would last for a volume fraction range due to the random distribution nature of CNTs. Before the CNT volume fraction reaches the percolation threshold, i.e., $f < f_c$, no CNTs are percolated due to the very small CNT concentration and large separation distance between CNTs. Under this condition, only electron hopping is believed to contribute to the electrical conductivity of the nanocomposite. However, once percolation starts ($f > f_c$), certain amount of CNTs are percolated to form conductive networks, i.e., a percentage ξ of CNTs are percolated while ($1-\xi$) are not. Therefore, in the percolation process, both electron hopping and conductive networks contribute to the overall electrical conductivity of the composite. It is obvious that with the CNT volume fraction increasing from f_c to 1 (a limiting case for material consisting of CNTs only), the percentage of percolated CNTs, ξ, will change from 0 to 1. According to Deng and Zheng (2008), the percentage of percolated CNTs, ξ, can be approximately estimated as

$$\xi = \frac{f^{1/3} - f_c^{1/3}}{1 - f_c^{1/3}}, \quad (f_c \le f < 1).$$

(2.28)

From the above analysis, it is concluded that both electron hopping and conductive networks govern the electrical conductivity of the composite after percolation, while only electron hopping contributes to the electrical conductivity of the composite prior to percolation. Therefore, the overall electrical conductivity of CNT-polymer nanocomposites can be determined in a mixed form from the general micromechanics formulation of Eq. (2.19), i.e.,

$$\sigma_{eff} = \begin{cases} \sigma_m + \dfrac{1}{4\pi} \int_0^{2\pi} \int_0^{\pi} \left\{ f_{eff} \left(\sigma_{EH} - \sigma_m \right) A_{EH} \right\} \sin\theta \, d\theta \, d\varphi, \ f < f_c \\[2mm] \sigma_m + \left(1 - \xi\right) \dfrac{1}{4\pi} \int_0^{2\pi} \int_0^{\pi} \left\{ f_{eff} \left(\sigma_{EH} - \sigma_m \right) A_{EH} \right\} \sin\theta \, d\theta \, d\varphi \\[2mm] \quad + \xi \dfrac{1}{4\pi} \int_0^{2\pi} \int_0^{\pi} \left\{ f_{eff} \left(\sigma_{CN} - \sigma_m \right) A_{CN} \right\} \sin\theta \, d\theta \, d\varphi, \ f \ge f_c \end{cases},$$

(2.29)

where σ_{EH} and σ_{CN} denote the electrical conductivity tensors of the effective filler contributed by electron hopping and conductive networks, respectively. These two tensors in the global coordinate system are determined from Eq. (2.20) with EH σ% and CN σ% being derived by E_{qs}. (2.10) and (2.21). For electron hopping, the interphase thickness and electrical conductivity are determined from E_{qs}. (2.16) and (2.17), while for conductive networks, these quantities are determined from E_{qs}. (2.13) and (2.15). AEH and A_{CN} are the corresponding electric field concentration tensors determined from Eq. (2.23) with the Eshelby tensor given in Eq. (2.25). For conductive networks, several CNTs will be electrically connected to each other (not in physical contact because of van der Waals forces between CNTs), and the effective aspect ratio of formed networks can thus be taken as infinite due to the large aspect ratio of the CNTs (Seidel and Lagoudas, 2009). Therefore, the quantities associated with the electron hopping corresponds to the real CNTs aspect ratio, while the quantities related to conductive networks corresponds to an infinite aspect ratio.

This mixed micromechanics model with considering electron hopping and conductive networks was employed to predict the effective electrical conductivity of CNT-polymer nanocomposites.

Results and discussion

To validate the micromechanics model developed in the current work, the modeling results were compared with the experimental data for a single walled carbon nanotube (SWCNT)-epoxy nanocomposite (Gojny et al., 2006) and a multiple walled carbon nanotube (MWCNT)-epoxy nanocomposite (Kim et al., 2005), respectively. In these experimental works the electrical conductivities of the nanocomposites were measured by means of a dielectric spectroscopy analyzer. It should be noted that for the current theoretical modeling, CNTs were assumed to be uniformly dispersed in the polymer. Conventionally, it is accepted that such a perfect dispersion without any agglomeration of CNTs is desirable to enhance the electrical properties of composites as validated in experiments (Song and Youn, 2008; Golosova et al., 2012).

However, some other experiments showed that a less than perfect dispersion with CNT agglomerations could favor the formation of conductive networks in composites (Martin et al., 2004; Li et al., 2007). In order to get a better uniform dispersion of CNTs in the polymer, in Gojny's (2006) experiments SWCNTs were dispersed in the epoxy matrix by a three-roll mill (with roller speeds 20, 30 180 rpm, respectively). The suspensions were fully mixed before hardening and then cured for 48 hours to reduce agglomerations. For the MWCNT-epoxy nanocomposite in (Kim et al., 2005), surfaces of MWCNTs were chemically treated by acetone/surfactant solution and sonication to reduce van der Waals interactions among MWCNTs. MWCNTs were dispersed and fully mixed with epoxy matrix by a two-roll mill. After approximate five hour's cure, it was observed that MWCNTs were well dispersed in the epoxy. Figure 2.5 shows the comparison between the analytical results from the current micromechanics model and the experimental data for the SWCNT-epoxy nanocomposite (Gojny et al., 2006), in which the SWCNTs were dispersed in a modified DGEBA-based epoxy resin. In this work, the length and diameter of the CNTs and the electrical conductivity of the epoxy were selected as L = 2 μm, d = 2 nm and σ_m = 9×10^{-9} S/m, respectively. The electrical conductivity of the CNT was taken as 5×10 S/m from (Ebbesen et al., 1996). By comparison, it was observed that the analytical result derived from the mixed micromechanics model of Eq. (2.29) successfully predicts the trend of the experimental data. Without considering conductive networks, the micromechanics modeling result was also provided in this figure for comparison. It was noticed that when the volume fraction is small, little discrepancy exists between the modeling results considering conductive networks and the results without considering conductive networks. However, with the increase of CNT volume fraction, a significant discrepancy appears between these two modeling results. It can be concluded that both electron hopping and conductive networks contribute to the overall electrical conductivity of the composite, while conductive networks become dominant conductivity mechanism for the composite with higher CNT volume fraction. Therefore, the overestimation of electron hopping caused by definitions in Eqs. (2.16) and (2.17) as mentioned in section 2 can be neglected after percolation due to the dominant contribution of conductive networks to the overall electrical conductivity of the nanocomposites. The same conclusion can be obtained for the MWCNT case as shown in Figure 2.6.

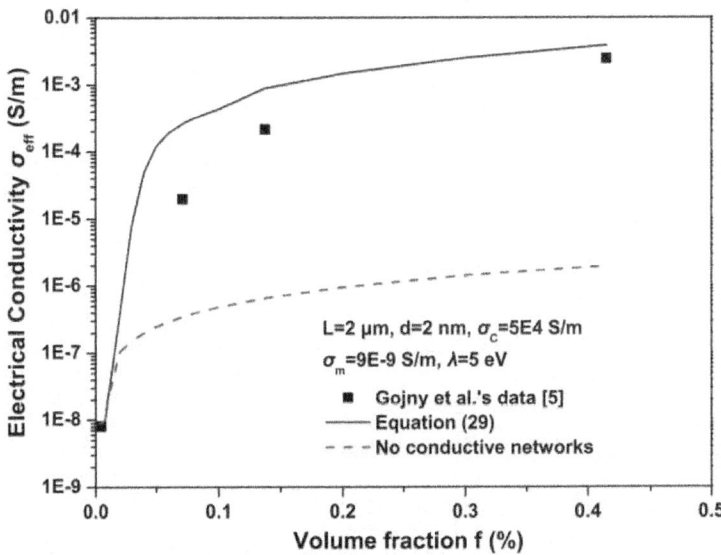

Figure 2.5: Comparison between modeling results and experimental data of a SW-CNT– epoxy nanocomposite.

Figure 2.6 shows another comparison between the micromechanics results and the experimental data of a MWCNT-epoxy nanocomposite. In simulation, the length and diameter of the CNTs and the electrical conductivity of the epoxy were selected as L = 20 μm, d = 20 nm and $\sigma_m = 1 \times 10^{-13}$ S/m, respectively (Kim et al., 2005). According to literature Deng's work (2008), the electrical conductivity of MWCNTs ranges from 10 S/m to 10000 S/m, which will be used in the simulation for comparison. It was found that the analytical results successfully predict the trend of the experimental data when the electrical conductivity of the CNT was set between 100 S/m and 1000 S/m. It was also observed that for MWCNTs with the same aspect ratio the increasing in electrical conductivity of the CNT significantly enhances the overall electrical conductivity of the nanocomposite. It was also found that the overall electrical conductivity of the nanocomposite increases with the increase of CNT volume fraction. Without considering conductive networks in the micromechanics model, significant discrepancy from the experimental data was observed in Figure 2.6. Therefore, the formation of conductive networks is a significant factor in governing the percolation behavior of nanocomposites, and a micromechanics model without considering conductive networks may lead to significant error in predicting the overall electrical conductivity of CNT-polymer

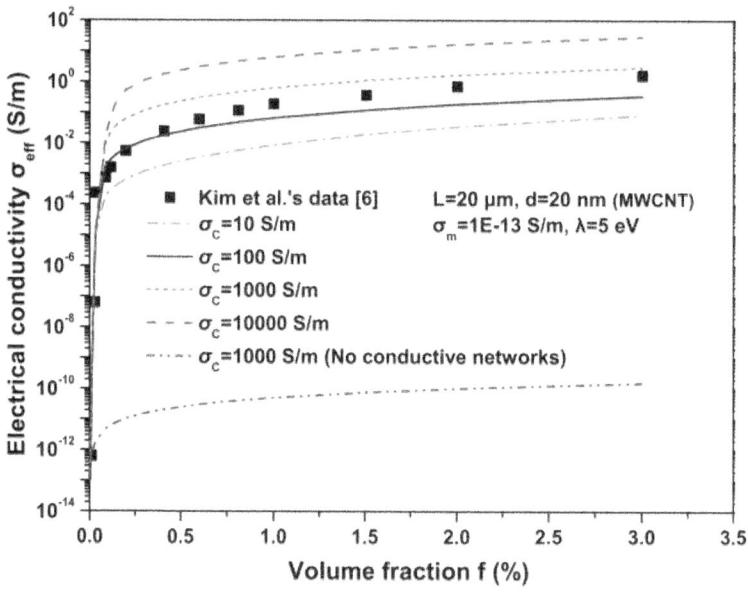

Figure 2.6: Comparison between modeling results and experimental data for a MW-CNT– epoxy nanocomposite.

Figure 2.7 and Figure 2.8 demonstrate the effects of CNT sizes, i.e. length and diameter, on the overall electrical conductivity of MWCNT-epoxy nanocomposites. It was found that CNT sizes have a significant effect on the percolation rate of the composite, while having a moderate effect on the overall electrical conductivity after percolation. For the same CNT volume fraction, the overall electrical conductivity increases with the increase in CNT length or the decrease in CNT diameter. These results suggest that the electrical conductivity of CNT-polymer nanocomposite can be improved by increasing the CNT aspect ratio. Such a phenomenon on the improved electrical conductivity can be explained by the existence of more conductive networks for CNTs with larger aspect ratio at the same CNT concentration. Moreover, the percolation rate was found to decrease with the increase of CNT length, while increasing with the increase of CNT diameter. Similar trends of the aspect ratio effect on the percolation rate have also been observed in existing literature (Yan et al., 2007).

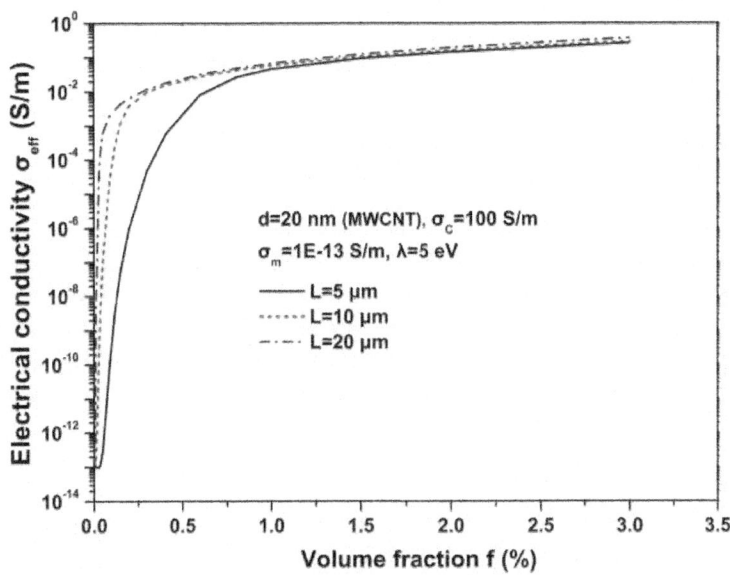

Figure 2.7: Effect of the CNT length on the electrical conductivity of the MWCNT–epoxy nanocomposite.

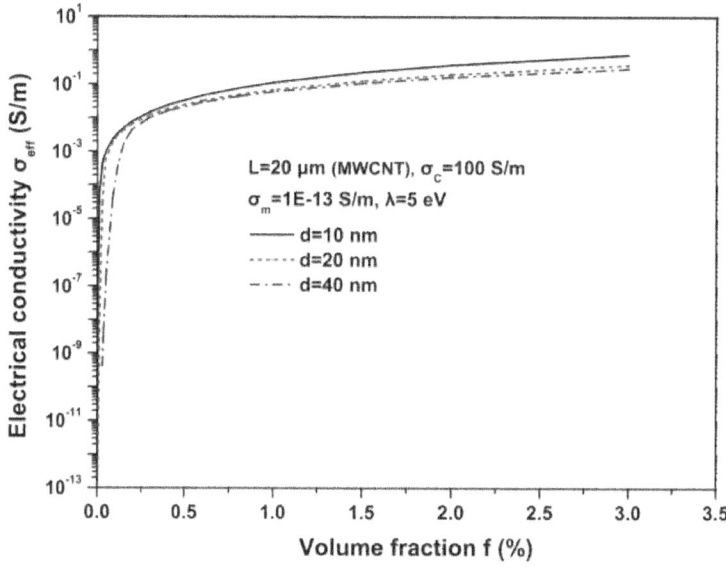

Figure 2.8: Effect of the CNT diameter on the electrical conductivity of the MWCNT–epoxy nanocomposite. It should be mentioned that the mixed micromechanics model in the current work was developed based on the assumption of uniformly random distribution of straight CNTs in polymers. However, due to the large aspect ratio of CNTs and van der Waals forces among CNTs, CNTs distributed in polymer media usually exist in agglomerates and in a curved state. A more accurate model with the consideration of the effects of curvature and agglomerates should be further developed to analyze the electrical properties of nanocomposites, which is our future research concentration.

CONCLUSIONS

In this work, two conductivity mechanisms, electron hopping and conductive networks, have been considered to develop a mixed micromechanics model to predict the overall electrical conductivity of CNT-polymer nanocomposites. An interphase surrounding the CNT was introduced to capture the electron hopping effect, and its properties such as thickness and conductivity, were defined through electron hopping theory. By comparison with experimental data, the mixed micromechanics model was validated to provide accurate quantitative predictions on the overall electrical conductivity of the nanocomposite. It was found that both electron hopping and conductive networks contribute to the electrical conductivity of the nanocomposite, while conductive networks become dominant after percolation. Meanwhile, effects of CNT sizes on the electrical conductivity were also investigated. It was observed that both CNT length and diameter significantly affect the percolation concentration of nanocomposites, while having moderate effects on the overall electrical conductivity of the nanocomposites after percolation. This work with quantitative prediction on the electrical conductivity of CNTpolymer composites is envisaged to be helpful for the design and optimization of the CNT-polymer nanocomposites with desirable electrical properties.

UNI-AXIAL STRETCHING EFFECTS ON ELECTRICAL CONDUCTIVITY OF CNT-POLYMER COMPOSITES

In recent years, conductive polymer composites have been attracting extensive interests from scientific and industrial communities due to their unique features, such as high stretchability, low cost, easy processability and good compatibility (Yu et al., 2009; Nambiar and Yeow, 2011; Shang et al., 2011). Compared with traditional conductive or semi-conductive materials, such as metals or silicon, conductive polymer composites possess conductive properties from the filler materials while keeping the intrinsic flexibility of the polymers. Thus, the unique combination of conductivity and flexibility enables conductive polymer composites the most promising materials in a variety of applications including flexible and stretchable electronics, conductive coatings, electromagnetic shielding, solar cells, etc, which cannot be achieved by traditional rigid conductive materials (Yang et al., 2005; Berson et al., 2007; Yu et al., 2009; Shang et al., 2011).

Due to their extraordinary physical properties, particularly electrical properties, carbon nanotubes (CNTs) are widely adopted as promising candidates of conducting fillers in soft polymers for the potential fabrication of conductive polymer composites. Experiments have shown that the addition of a very small amount of CNTs into polymers can remarkably improve the electrical conductivity of the composites with a percolationlike behavior (Ounaies et al., 2003; Kim et al., 2005; Gojny et al., 2006). Interpretation on the conductivity of these CNT–polymer composites is attributed to two mechanisms: nanoscale electron hopping (or quantum tunneling) and microscale conductive networking (Chang et al., 2009; Zhang et al., 2009; Lu et al., 2010). It is well explained

in literature (Deng and Zheng, 2008; Seidel and Lagoudas, 2009; Takeda et al., 2011) that the contribution of the electron hopping and conductive networking depends on the CNT concentration, such that, with increasing CNT concentration, conductive networks are formed and believed to be dominant over the electron hopping, while electrons always have the probability of hopping intra-tube or among different CNTs, which governs the overall electrical conductivity of the composites particularly with extremely low CNT concentration.

Understanding on the overall electrical conductivity of CNT–polymer composites is essential for their design and full potential applications. In literature, efforts have been devoted to investigating the electrical properties of these composites through experiments, numerical simulations and theoretical modeling (Chang et al., 2009; Feng and Jiang, 2013; Gojny et al., 2006; Kim et al., 2005; Li et al., 2007; Seidel and Lagoudas, 2009; Takeda et al., 2011; Yan et al., 2007). However, most of these studies were focused on preparing composites with well dispersed conductive fillers or predicting the electrical behavior of the as-received composites without considering stretching effects. For the potential application of conductive polymer composites as stretchable electronics, it is necessary to investigate the stretching effects upon the overall electrical conductivity of the composites in order to accurately predict and control the device performance. Existing studies have demonstrated that stretching may significantly influence the electrical behavior of conductive polymer composites. For example, Bao et al. (2011) experimentally examined the morphology and the electrical conductivity of carbon nanofibre composites before and after mechanical stretching. Their results showed that mechanical stretching could lead to a sharp decrease in the electrical conductivity of the composites due to the breakdown of conductive networks. Miao and his co-workers (2011a, 2011b and 2012) examined the piezoresistive response of CNT–polymer composites by experiments and percolation theory and their results indicated that stretching decreases the electrical conductivity of the composites. In addition, Park et al. (2008) and Hu et al.'s (2010) works demonstrated an abrupt increase in resistance of CNT–polymer composites when subjected to stretching. However, other experimental results (Cheng et al., 2009; Shang et al., 2011; Wang et al., 2011) showed an opposite trend, i.e., stretching increased the electrical conductivity of CNT–polymer composites.

This phenomenon was explained as resulting from the substantial alignment enhancement of conductive fillers along the stretching direction of the composites. From the aspect of numerical simulations, Taya et al. (1998) and Lin et al. (2010) applied the fibre percolation model and Monte Carlo method, respectively, to investigate the stretching/compression effects upon the electrical properties of fibre-filled composites and their results showed that the deformation could shift the percolation threshold of the composites. A recent study by Tallman and Wang (2013) also indicated that stretching could alter the percolation threshold due to the CNTs' re-orientation. Despite the fact that different results have been observed in literature, it is suggested that there are three major expected changes occurred during the stretching, including composite volume expansion, re-orientation of conductive fillers and change in conductive networks, which may contribute to the variation of the overall electrical properties of the polymer composites.

Traditional compressible materials expand when subjected to stretching. Since CNTs are much stiffer than polymers in the CNT–polymer composites, the stretching deformation is mainly sustained by the polymer matrix. As a result, the concentration of the conductive fillers decreases due to the very low concentration and the nearly unchanged volume of conductive CNTs. Consequently, the electrical conductivity of the composites will be altered by the decrease in CNT concentration. Some researchers argued that such effect of volume expansion of the CNT–polymer composites could be neglected for small deformation due to their limited compressibility Taya (1998). After stretching, experiments have shown that CNTs tend to be more re-aligned along the stretching direction rather than randomly distributed as observed before stretching (Cheng et al., 2009; Wang et al., 2011). Due to the highly anisotropic properties of CNTs, the electrical properties of the CNT–polymer composites strongly depend upon the orientation of the CNTs, which has been demonstrated by several experiments (Du et al., 2005; Cheng et al., 2009; Wang et al., 2011). Conductive networking is one of the conductivity mechanisms in conductive composites as discussed before. Researchers have argued that due to the re-alignment of CNTs certain existing conductive networks may break down causing an increase in the percolation threshold, which is believed to be responsible for the observed decrease in the electrical conductivity of composites (Taya et al., 1998; Du et al., 2005;

Bao et al., 2011). In addition, some simulations and analytical studies (Taya et al, 1998; Lin et al., 2010; Tallman and Wang, 2013;) also showed that stretching/compression could change the percolation threshold, a critical concentration of conductive fillers identifying the formation of conductive networks.

Despite the fact that efforts have been made to investigate the stretching effects on the electrical conductivity of CNT–polymer composites experimentally and numerically, there is limited work on theoretical modeling with the consideration of the three expected changes mentioned before. Therefore, the objective of the current work is to investigate the effects of the stretching induced volume expansion, CNT re-orientation and conductive network change on the overall electrical conductivity of the CNT–polymer composites following our previous work (Feng and Jiang, 2013).

Micromechanics model for CNT-polymer composites

For a two-phase composite, in which conductive fillers with volume fraction f are assumed as straight and uniformly dispersed in the polymer matrix, micromechanics model is commonly adopted by researchers to quantitatively predict the overall effective physical properties of the composite, i.e., mechanical and electrical properties (Entchev and Lagoudas, 2002; Seidel and Lagoudas, 2009). Following the routine procedure of the micromechanics theory, a representative volume element (RVE) containing enough fillers with random orientations as shown in Figure 3.1 (Seidel and Lagoudas, 2009; Feng and Jiang, 2013) is chosen to study the overall electrical conductivity of the composite in the current work. In the RVE $(\tilde{x}_1, \tilde{x}_2, \tilde{x}_3)$ is selected as the local coordinate system describing the position of an individual filler and ϕ and θ are Euler angles identifying its orientation with respect to the global coordinate system (X_1, X_2, X_3). Based on the micromechanics theory, the effective electrical conductivity of this two-phase composite can be determined in terms of the electrical conductivity of the fillers and the polymer by averaging the contribution of the fillers from all possible orientations in the RVE (Taya, 1995; Entchev and Lagoudas, 2002; Odegard et al., 2003; Taya, 2005; Seidel and Lagoudas, 2009), i.e.,

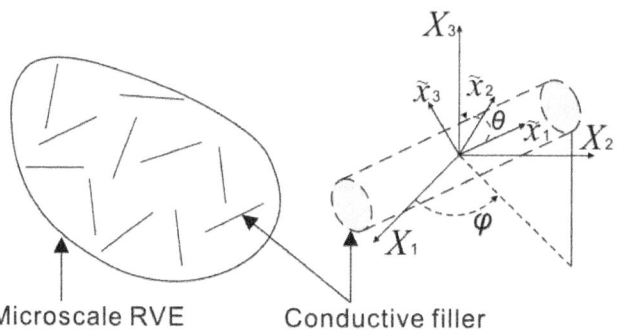

Figure 3.1: Sketch of a RVE containing conductive fillers.

$$\sigma_{eff} = \sigma_m + \frac{\int_0^{2\pi}\int_0^{\pi} \rho(\varphi,\theta) f(\sigma - \sigma_m) A \sin\theta d\theta d\varphi}{\int_0^{2\pi}\int_0^{\pi} \rho(\varphi,\theta)\sin\theta d\theta d\varphi} ,$$

(3.1)

where A is the electric field concentration tensor, which will be defined later, σ and σ_m are the electrical conductivity tensors of the fillers and the polymer, respectively and $\rho(\phi,\theta)$ is the orientation distribution function (ODF) representing the probability density of the distribution of the fillers with a given orientation, particularly $\rho(\phi,\theta)$ equals unity for a random distribution of fillers.

In CNT–polymer composites, electron hopping among CNTs results in the formation of a continuum interphase layer surrounding a CNT. In order to capture the effect of this interphase layer and the hollow nature of the CNT, a well-accepted effective composite cylinder model (Yan et al., 2007; Seidel and

Lagoudas, 2009; Feng and Jiang, 2013) as shown in Figure 3.2, which consists of a CNT (length L, radius r_c) and a surrounding interphase layer (thickness t), is adopted in the current work to determine the effective electrical conductivity of the CNT with the consideration of the electron hopping effect. Knowing the electrical conductivity of the interphase (σ_{Int}) and the CNT (σ_c) and the interphase thickness (t), the effective longitudinal and transverse electrical conductivity of the effective solid filler in its own local coordinate system was determined by applying the law-of-mixture rule as (Taya, 2005; Yan et al., 2007; Feng and Jiang, 2013):

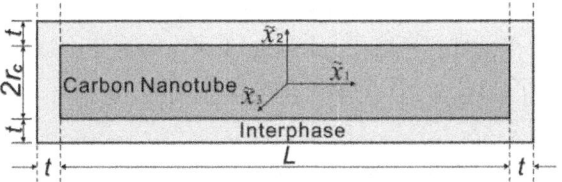

Figure 3.2: Sketch of an effective composite cylinder.

$$\tilde{\sigma}^L = \frac{(L+2t)\sigma_{Int}\left[\sigma_c^L r_c^2 + \sigma_{Int}\left(2r_c t + t^2\right)\right]}{2\sigma_c^L r_c^2 t + 2\sigma_{Int}\left(2r_c t + t^2\right)t + \sigma_{Int}L\left(r_c + t\right)^2}$$

$$\tilde{\sigma}^T = \frac{\sigma_{Int}}{L+2t}\left[L\frac{2r_c^2\sigma_c^T + \left(\sigma_c^T + \sigma_{Int}\right)\left(t^2 + 2r_c t\right)}{2r_c^2\sigma_{Int} + \left(\sigma_c^T + \sigma_{Int}\right)\left(t^2 + 2r_c t\right)} + 2t\right]$$

$$(3.2)$$

where superscripts "L" and "T" denote the longitudinal and the transverse directions in the local coordinate system. Thus, the electrical conductivity tensor of the effective cylinder assembly in its local coordinate system can be written as:

$$\tilde{\sigma} = \begin{bmatrix} \tilde{\sigma}^L & & \\ & \tilde{\sigma}^T & \\ & & \tilde{\sigma}^T \end{bmatrix}.$$

$$(3.3)$$

It should be mentioned that the electrical conductivity tensor σ in Eq. (3.1) is the one in the global coordinate system, which can be obtained from its counterpart $\tilde{\sigma}$ in the local coordinate system through a transformation (Seidel and Lagoudas, 2009), i.e.,

$$\sigma = Q\tilde{\sigma}Q^T$$

$$(3.4)$$

with the transformation matrix Q being given in terms of the Euler angles (ϕ, θ) as (Entcheve and Lagoudas, 2002):

$$Q = \begin{bmatrix} \sin\theta\cos\varphi & -\cos\theta\cos\varphi & \sin\varphi \\ \sin\theta\sin\varphi & -\cos\theta\sin\varphi & -\cos\varphi \\ \cos\theta & \sin\theta & 0 \end{bmatrix}.$$

(3.5)

For the CNT–polymer composite with an initial volume fraction f_{CNT} of hollow CNTs before stretching, the homogenization of the solid filler from the CNT with the consideration of the interphase results in the change of the filler volume fraction, i.e., the volume fraction of the effective solid fillers is re-defined as (Yan et al., 2007; Seidel and Lagoudas, 2009):

$$f_{eff} = \frac{(r_c + t)^2 (L + 2t)}{r_c^2 L} f_{CNT}.$$

(3.6)

From the above analysis, it is obvious that the properties of the interphase, i.e., the electrical conductivity (σ_{Int}) and the thickness (t), must be defined for determining the electrical conductivity tensor and the volume fraction of the effective solid filler. Since the electrical behavior of the CNT–polymer composite is governed by both the electron hopping and the conductive networks, it is naturally believed that the interphase properties are correlated to these two conductivity mechanisms. When the CNT concentration is small, CNTs are more independent rather than electrically connected to each other with electron hopping dominating the conductivity. However, when CNT concentration reaches a critical value f_c, i.e., the percolation threshold, the formation of conductive networks starts. According to existing experiments and simulations (Li et al., 2007; Allaoui et al., 2008; Takeda et al., 2011), the upper limit of the critical separation distance between adjacent CNTs for the formation of conductive networks is suggested as dc = 1.8 nm, which is the maximum possible thickness of the separating medium between adjacent CNTs allowing the tunneling penetration of electrons. Obviously, such a separation distance varies with the CNT concentration, which was suggested by some studies (Allaoui et al., 2007; Deng and Zheng, 2008; Takeda et al., 2011) to follow a power law relation after percolation. Accordingly, in our previous work (Feng and Jiang, 2013), the separation distance between the adjacent CNTs in the conductive networks was defined as:

$$d_a = \left(\frac{f_c}{f_{CNT}}\right)^{1/3} d_c.$$

(3.7)

In contrast, when the CNT concentration is under the threshold, the separation distance between adjacent CNTs should be larger than dc. Under this situation, CNTs are more independent without the formation of conductive networks. Due to the difficulty of estimating the average separation distance between independent CNTs before the percolation, the separation distance was assumed as a constant, i.e., $d_a = d_c$ (Feng and Jiang, 2013). The thickness of the interphase for each CNT was taken as half of the separation distance between the adjacent CNTs, i.e., $t = d_{a/2}$ (Feng and Jiang, 2013).

As discussed above, the electrical conductance between adjacent CNTs is governed by the tunneling penetration of electrons. Accounting for such electrical tunneling effect, Simmons (1963) has derived a formula for the tunneling-type resistance for the separation interphase between CNTs, ie

$$R_{Int}(d_a) = \frac{d_a \hbar^2}{ae^2 (2m\lambda)^{1/2}} \exp\left(\frac{4\pi d_a}{\hbar} (2m\lambda)^{1/2} \right).$$

Accordingly, the electrical conductivity of the interphase was derived by the tunnelingtype contact resistance R_{Int} as (Takeda et al., 2011)

$$\sigma_{Int} = \frac{d_a}{aR_{Int}(d_a)}$$

(3.8)

$\lambda = 5.0$ eV is the potential barrier height for CNTs dispersed in most polymers (Celzard et al., 1996; Takeda et al., 2011); $m = 9.10938291 \times 10^{-31}$ kg and $e = -1.602176565 \times 10^{-19}$ C are mass and electric charge of an electron, respectively; a is the contact area of CNTs and $\hbar = 6.626068 \times 10^{-34}$ m^2 kg/s is the Planck constant.

As argued by Deng and Zheng (2008), before percolation, i.e., $f_{CNT} < fc$, no CNTs are percolated and only electron hopping is believed to contribute to the overall electrical conductivity of the composites. However, once the percolation starts, i.e., $f_{CNT} > f_c$, both electron hopping and conductive networks contribute to the conductivity. In literature, there were several approximations on this critical concentration (percolation threshold) of conductive CNTs. In order to consider the stretching induced CNT re-orientation effect on the percolation threshold or the formation of conducting networks, the excluded volume method (Balberg et al., 1984; Shiraishi and Ata, 2001; Tallman and Wang, 2013), which accounts for fillers' orientation through an averaging expression in terms of the angles between fillers, is adopted in the current work in which the percolation threshold was defined as:

$$f_c = 1 - \exp\left(\frac{-\langle V_{ex} \rangle V_{CNT}}{\langle V_e \rangle} \right),$$

(3.9)

Where $\langle V_e \rangle = \frac{4\pi}{3}D^3 + 2\pi D^2 L + 2 \cdot D \cdot L^2 \langle \sin \gamma \rangle_\mu$ is the average excluded volume of CNT with D being the diameter of the CNT, $\langle V_{ex} \rangle$ is a dimensionless parameter denoting the total average excluded volume of CNT, which particularly equals 1.4 and 2.8 for randomly oriented CNTs and perfectly aligned CNTs, respectively; V_{CNT} is the volume of the spherocylinder CNT. The term $\langle \sin \gamma \rangle_\mu$ in the expression of $\langle V_e \rangle$ is an averaging term accounting for the filler orientation, with γ being the angle between the ith and the jth CNTs. The detailed interpretation of the parameters in Eq. (3.9) will be discussed later. Obviously, with CNT concentration increasing from fc to 1, the percentage of the percolated CNTs, ξ, will change from 0 to 1. Deng and Zheng (2008) estimated this percentage of percolated CNTs as:

$$\xi = \frac{f_{CNT}^{1/3} - f_c^{1/3}}{1 - f_c^{1/3}}, \quad (f_c \leq f_{CNT} < 1).$$

(3.10)

According to the percolation process with the involvement of both the electron hopping and conductive networking, a mixed form of micromechanics model was developed in our previous work (Feng and Jiang, 2013) to determine the overall electrical conductivity of CNT–polymer composites, which was modified from the general formulation (3.1) as:

$$\sigma_{eff} = \begin{cases} \sigma_m + \dfrac{\int_0^{2\pi}\int_0^{\pi}\{f_{eff}(\sigma_{EH} - \sigma_m)A_{EH}\}\rho(\varphi,\theta)\sin\theta d\theta d\varphi}{\int_0^{2\pi}\int_0^{\pi}\rho(\varphi,\theta)\sin\theta d\theta d\varphi}, & f_{CNT} < f_c \\[4ex] \sigma_m + (1-\xi)\dfrac{\int_0^{2\pi}\int_0^{\pi}\{f_{eff}(\sigma_{EH} - \sigma_m)A_{EH}\}\rho(\varphi,\theta)\sin\theta d\theta d\varphi}{\int_0^{2\pi}\int_0^{\pi}\rho(\varphi,\theta)\sin\theta d\theta d\varphi} \\[4ex] + \xi\dfrac{\int_0^{2\pi}\int_0^{\pi}\{f_{eff}(\sigma_{CN} - \sigma_m)A_{CN}\}\rho(\varphi,\theta)\sin\theta d\theta d\varphi}{\int_0^{2\pi}\int_0^{\pi}\rho(\varphi,\theta)\sin\theta d\theta d\varphi}, & f_{CNT} \geq f_c \end{cases}$$,

(3.11)

where the subscripts "EH" and "CN" denote the terms that are associated with electron hopping and conductive networking, respectively.

Uni-axial stretching induced changes

As discussed in the introduction of this paper, a mechanical stretching may induce volume expansion of the composites, re-orientation of CNTs and change in conductive networks. Accordingly, certain parameters in Eq. (3.11) should be re-defined, for example, the volume fraction f, Euler angles φ and

θ, the orientation distribution function $\rho(\phi,\theta)$, the percolation threshold, the electrical conductivity tensors σ_{EH} and σ_{CN} in the global coordinate system, and the electric field concentration tensors A_{EH} and A_{CN}. In the following, the three stretching induced changes are investigated and incorporated into the micromechanics model.

Volume expansion and re-orientation of CNTs

Following the fiber re-orientation model in literature (Kuhn and Grün, 1942; Taya et al., 1998), a cell containing an effective solid filler before and after uni-axial stretching as shown in Figure 3.3 is used here to describe the change of volume expansion and reorientation of CNTs. The original lengths of the cell in the X_1, X_2 and X_3 directions are h_0, w_0 and l_0, respectively. With any stretching strain ε in the X_3 direction, the lengths of the cell after stretching can be determined by the following general formula:

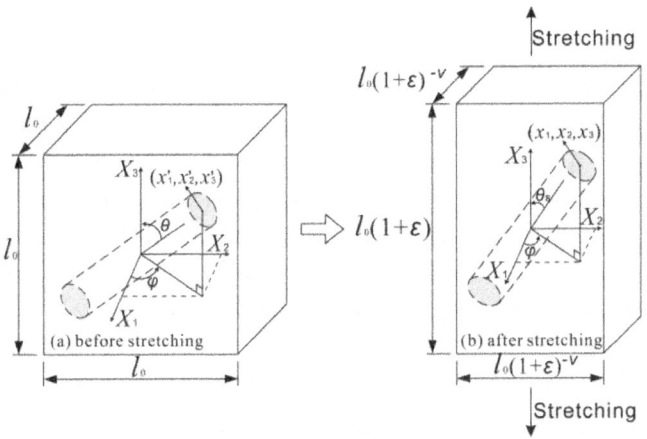

Figure 3.3: Orientation description of a conductive filler in a cell (a) before stretching; (b) after stretching.

$$l = l_0(1+\varepsilon), \quad w = w_0(1+\varepsilon)^{-\nu} \text{ and } h = h_0(1+\varepsilon)^{-\nu},$$

(3.12)

where ν is the Poisson's ratio of the composite. For small deformation, the width and height of the cell defined in Eq. (3.12) can be reduced to $w = w_0(1-\nu\varepsilon)$ and $h = h_0(1-\nu\varepsilon)$, , respectively. Due to the much larger stiffness of CNTs in comparison to the polymer, the deformation of the composite is mainly sustained by the polymer. Therefore, the volume fraction of the effective solid fillers after the uni-axial stretching can be approximately estimated by neglecting the volume change of the CNTs, i.e.,

$$f_{\text{update}} = \frac{V_0 f_{\text{eff}}}{V} = \frac{f_{\text{eff}}}{\left(1+\varepsilon\right)^{1-2v}},$$

where V_0 and V are the volumes of the cell in Figure 3.3 before and after stretching, respectively. Particularly, there is no change for the CNT volume fraction after stretching if the composite is incompressible, i.e., $v = 0.5$.

In addition to the volume expansion, after stretching the filler in the cell as shown in Figure 3.3 also tends to re-align along the stretching direction, which results in the decrease of the polar angle from θ to θ_s. However, the change of the azimuth angle, ϕ, can be neglected for this uni-axial stretching condition (Kuhn and Grün, 1942).

After the stretching of strain ε along the X_3 direction, the new polar angle θ_s can be found in terms of the initial polar angle θ as (Kuhn and Grün, 1942):

$$\tan\theta_s = \left(1+\varepsilon\right)^{-(1+v)} \cdot \tan\theta .$$

(3.14)

Such a variation of the polar angle will change the orientation distribution function (ODF) from $\rho(\phi, \theta)$ to $\rho(\phi, \theta_s)$. It should be mentioned that the ODF which describes the orientation distribution of fillers, always satisfies the following conditions (Perez et al., 2008; Vangurp, 1995):

$$\rho(\varphi,\theta) \geq 0 \text{ and } \frac{1}{4\pi}\int_0^{2\pi}\int_0^{\pi}\rho(\varphi,\theta)\sin\theta d\theta d\varphi = 1 .$$

(3.15)

Before stretching, CNTs are assumed as uniformly distributed in the polymer matrix. Thus, the ODF is a constant, exhibiting a uniform distribution of CNTs along any orientation. However, after stretching, the decrease of the polar angle θ of any individual filler means that these fillers are no longer randomly distributed and tend to be more aligned along the stretching direction. In order to determine this new ODF, it is assumed that there are G fillers distributed in the RVE of the micromechanics model. Following Eq. (3.15), the total number of fillers falling in the ranges of (θ, θ+dθ) and (ϕ, ϕ+dϕ) in the RVE can be determined as (Kuhn and Grün, 1942):

$$dN_{\substack{\theta,\theta+d\theta \\ \varphi,\varphi+d\varphi}} = \frac{1}{4\pi}G\rho(\varphi,\theta)\sin\theta d\theta d\varphi .$$

(3.16)

These fillers will be re-oriented within the ranges of $(\theta_s,\ \theta_s+d\theta_s)$ and $(\varphi,\ \varphi+d\varphi)$ after a stretching, i.e.,

$$dN_{\substack{\theta_s,\theta_s+d\theta_s \\ \varphi,\varphi+d\varphi}} = \frac{1}{4\pi} G\rho(\varphi,\theta_s)\sin\theta_s d\theta_s d\varphi = dN_{\substack{\theta,\theta+d\theta \\ \varphi,\varphi+d\varphi}}.$$

(3.17)

Substituting Eq. (3.14) into Eq. (3.17), the ODF $\rho(\varphi, \theta_s)$ after stretching is determined as:

$$\rho(\varphi,\theta_s) = \frac{(1+\varepsilon)^{\frac{1+v}{2}}}{\left[(1+\varepsilon)^{-(1+v)}\cos^2\theta_s + (1+\varepsilon)^{1+v}\sin^2\theta_s\right]^{3/2}}$$

(3.18)

Without stretching, i.e., $\varepsilon = 0$, the ODF ρ (φ, θ_s) is reduced to unity for a random distribution as expected. Figure 3.4 demonstrates an example of the variation of the ODF with the strain and the polar angle θ_s. From Figure 3.4 (a), it is observed that with the increase of strain, more conductive fillers tend to be re-aligned along the stretching direction ($\theta_s = 0°$ and $180°$). Figure 3.4(b) suggests that stretching has a more significant effect on the re-alignment of the conductive fillers for the composites with larger Poisson's ratio than that with smaller Poisson's ratio, i.e., with the increase of Poisson's ratio, more CNTs tend to be re-aligned along the stretching direction. Here it should be noted that due to the re-alignment of the CNTs, the Poisson's ratio of the composite is theoretically dependent on the ODF instead of a fixed value. However, according to the evaluation in (Pan, 1996) the variation of the Poisson's ratio due to the re-orientation of fillers can be neglected. From the derived formula in that work, it is calculated that for a CNT composite with volume fraction of 3%, $v = 0.46$ for a random CNT distribution while $v \approx 0.45$ for a perfectly aligned distribution. Therefore, in the current work the dependency of the Poisson's ratio on the re-orientation of the fillers due to the stretching is neglected and the value of Poisson's ratio is taken as constant regardless of strain.

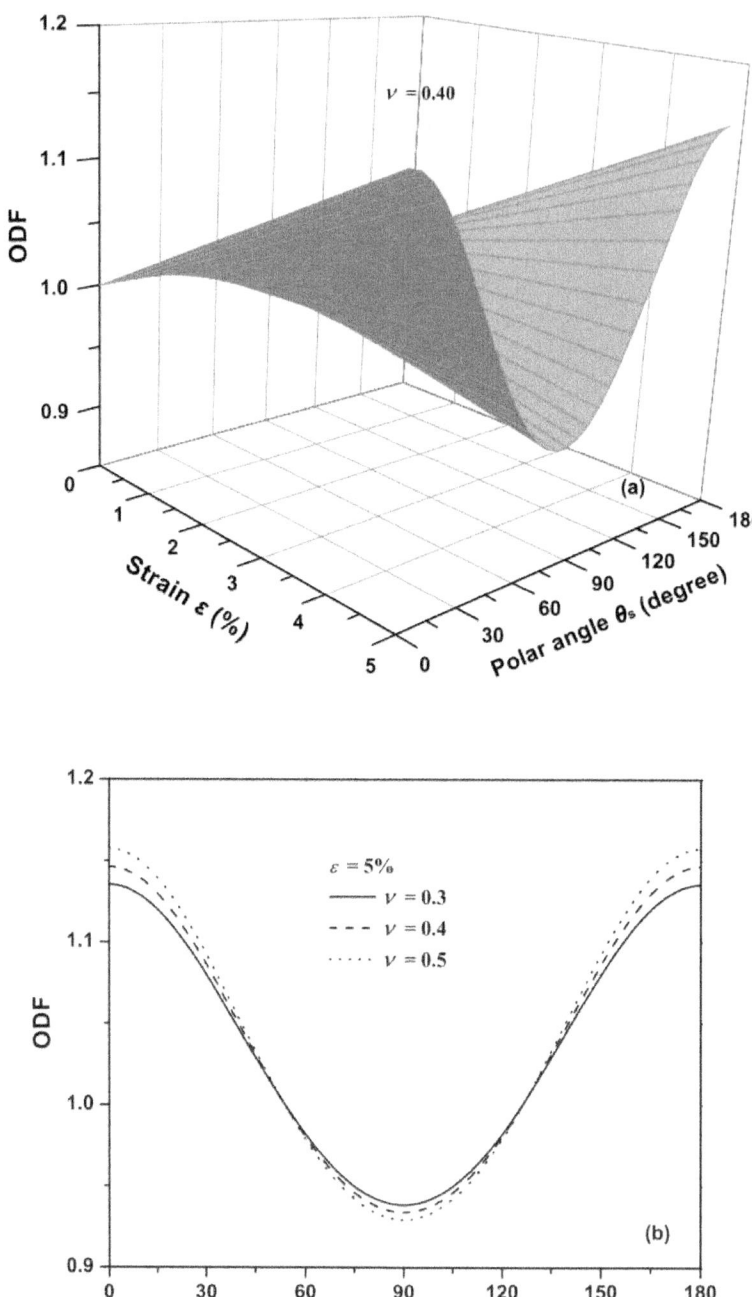

Figure 3.4: Variation of ODF with strain and polar angle θ_s (a) $v = 0.4$; (b) $\varepsilon = 5\%$.

Change in conductive networks

In addition to composite volume expansion and CNT re-orientation, stretching also induces change in the conductive networks, i.e., stretching can induce an increase in the percolation threshold (Tallman and Wang, 2013). As demonstrated in Figure 3.1, the orientation of the CNT is identified by the orientation angles φ and θ. The azimuth angle φ ranges from 0 to 2π while the polar angle ranges from 0 to π for the composites with random CNT distribution. As discussed before, CNTs re-align in a manner which favors the direction of the uni-axial stretching. As a result, the randomness of the CNT distribution decreases with the increase of the stretching strain. Such an assumption on the randomness of the distribution of CNTs can be verified from Figure 3.4, i.e., with the increase of the stretching strain, CNTs tend to more re-align along the X_3 direction. It was understood from (Tallman and Wang, 2013) that more CNTs would concentrate within the range of $[0, \theta_\mu]$ and $[\pi-\theta_\mu, \pi]$ in which θ_μ was related to the strain as $\theta_\mu = \arcsin[(1+\varepsilon)^{-1-\nu}]$ to represent the randomness of the distribution of the CNTs. According to the fitting from Monte Carlo simulation in (Tallman and Wang, 2013), the term $\langle \sin \gamma \rangle_\mu$ in Eq. (3.9), which accounts for the CNTs' orientation, can be related to θ_μ as:

$$\langle \sin \gamma \rangle_\mu = 0.018\theta_\mu^5 + 0.021\theta_\mu^4 - 0.234\theta_\mu^3 - 0.015\theta_\mu^2 + 0.909\theta_\mu.$$

(3.19)

In addition, Tallman and Wang (2013) also proposed a linear relation between Vex in Eq. (3.9) and $\langle V_{ex} \rangle$ in Eq. (3.9) and $\langle \sin \gamma \rangle_\mu$ as:

$$\langle V_{ex} \rangle = 2.8 - 5.6 \frac{\langle \sin \gamma \rangle_\mu}{\pi}.$$

(3.20)

Then with the consideration of the stretching effect the percolation threshold defined in Eq. (3.9) is modified and simplified as:

$$f_c = 1 - \exp\left[-\frac{\left(2.1\pi - 4.2\langle \sin \gamma \rangle_\mu\right)p}{4\pi + 6\pi p + 6p^2 \langle \sin \gamma \rangle_\mu} \right],$$

(3.21)

where $p = L/D$ is the aspect ratio of the CNTs. It should be noted here that due to the variation of the percolation threshold, the separation distance defined in equation (3.7) changes with the stretching strain, which would result in significant variation of the contact conductivity of the interphase due to its exponential dependency on the separation distance. In addition, the percentage of the percolated CNTs defined in Eq. (3.10) varies with the stretching strain due to the variation of the percolation threshold. Figure 3.5

plots the variations of the normalized percentage $\zeta_N = \zeta_s/\zeta_0$ of the percolated CNTs for a composite with the stretching strain for different CNT volume fractions and Poisson's ratios, where ζ_0 and ζ_s denote the percentages of the percolated CNTs before and after stretching, respectively. From Figure 3.5(a), it is found that the percentage of the percolated CNTs decreases with the increase of the stretching strain. It is also indicated that the percentage of the percolated CNTs is more sensitive to the stretching for the composites with lower volume fraction. Figure 3.5(b) informs us that the stretching induced re-alignment of CNTs with larger Poisson's ratio decreases the percentage of the percolated CNTs more significantly than that with smaller Poisson's ratio. It should be noted that in addition to the reduced percentage of percolated CNTs, the separation distance defined in Eq. (3.7) also varies with the strain, resulting in the change of interphase thickness and conductivity.

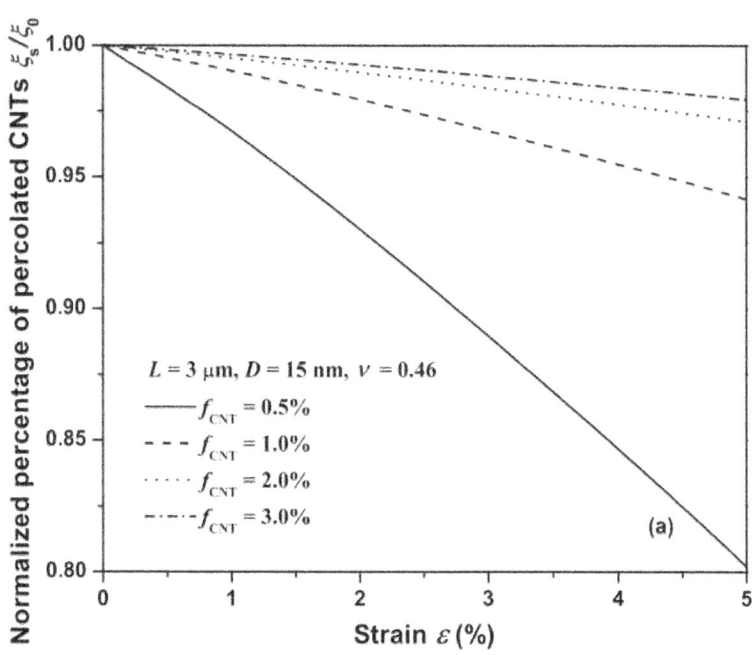

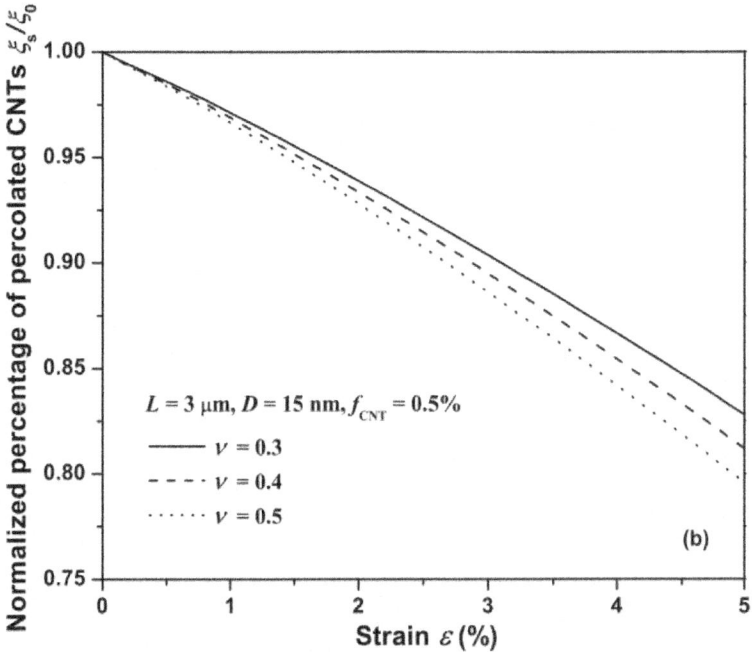

Figure 3.5: Variation of normalized percentage of percolated CNTs with strain (a) $v =$ 0.46; (b) $f_{CNT} = 0.5\%$.

In the current work, the Mori-Tanaka method (Taya, 2005; Mori and Tanaka, 1973), which accounts for interactions between dispersed inhomogeneities, is chosen as the micromechanics model. Accounting for the effect of the stretching, the electric field concentration tensor A in Eq. (3.1) is defined as (Taya, 1995; Entchev and Lagoudas, 2002; Odegard et al., 2003):

$$A = QA^{dil}Q^{T}\left\{(1-f_{update})\delta + f_{update}\frac{\int_{0}^{2\pi}\int_{0}^{\pi}f_{update}\{QA^{dil}Q^{T}\}\rho(\varphi,\theta_{s})sin\theta_{s}d\theta_{s}d\varphi}{\int_{0}^{2\pi}\int_{0}^{\pi}\rho(\varphi,\theta_{s})sin\theta_{s}d\theta_{s}d\varphi}\right\}^{-1},$$
(3.22)

where the transformation matrix Q is determined by replacing θ with θ_{s} and A^{dil} is defined as:

$$A^{dil} = \left\{\delta + S(\sigma_{m})^{-1}(\tilde{\sigma} - \sigma_{m})\right\}^{-1}$$
(3.23)

with δ being the Kronecker delta tensor. S is the Eshelby tensor of the effective filler, which is given by:

$$S = \begin{bmatrix} S_{11} & 0 & 0 \\ 0 & S_{22} & 0 \\ 0 & 0 & S_{33} \end{bmatrix},$$

(3.24)

where

$$S_{22} = S_{33} = \frac{A_{re}}{2\left(A_{re}^{2}-1\right)^{3/2}}\left[A_{re}\left(A_{re}^{2}-1\right)^{1/2} - \cosh^{-1} A_{re} \right]$$

(3.25)

with A_{re} being the aspect ratio of the effective filler, i.e $A_{re} = (L+2t)/(D+2t)$, and $S_{11} = 1- 2S_{22}$. When conductive networks are formed, several CNTs will be electrically connected to each other while not in physical contact due to van der Waals forces between CNTs. Thus, the effective aspect ratio of the formed networks can be taken as infinite due to the large aspect ratio of CNTs. However, quantities associated with the electron hopping correspond to the real effective filler aspect ratio as defined. Correspondingly, A_{EH} and A_{CN} in Eq. (3.11) can be determined from Eq. (3.22) by using different aspect ratios for CNTs. In the following Section, the mixed micromechanics model with the consideration of the uni-axial stretching effects will be employed to predict the effective electrical conductivity of a CNT–polymer nanocomposite through case study.

RESULTS AND DISCUSSION

To study the uni-axial stretching effects on the electrical conductivity of the CNT– polymer composites, a MWCNT/PEO (Polyethylene Oxide) composite under stretching in (Park and Kim, 2006; Park et al., 2008) is chosen as the example material. The length and diameter of the CNTs range from 1μm to 5μm and 10 nm to 20 nm, respectively. For the purpose of theoretically predicting the electrical conductivity of stretched composites in our simulation, the length and the diameter of the CNTs are taken as constants, i.e., L = 3 μm and D = 15 nm, respectively. According to the mechanical and physical properties of PEO, the Poisson's ratio and the electrical conductivity of the PEO are approximately set as $\nu = 0.46$ and $\sigma_m = 1\times10^{-13}$ S/m, respectively. While the electrical conductivity of the MWCNTs was not provided in this reference, other references suggest $\sigma_{CNT} = 100$ S/m as the effective conductivity of the MWCNTs (Deng and Zheng, 2008; Feng and Jiang, 2013). In the experiment of (Park and Kim, 2006; Park et al., 2008), MWCNTs were surface functionalized and well dispersed in the PEO matrix by stirring and sonication before stretching. Then the MWCNT/PEO composite strip was bonded to a gage section and stretched using a uni-axial test machine. A laser extensometer and a precision multimeter were used simultaneously to record strain and resistance of the composite. Two representative CNT volume fractions, 0.56% (near the percolation threshold before stretching) and 1.44% (away from the percolation threshold before stretching), were tested to investigate the stretching effect on the electrical resistance of the MWCNT/PEO composites. In order to compare the analytical predictions with the experimental results, the relative resistance change of the composite is expressed in terms of the stretching and the electrical conductivity of the composite as:

$$\frac{\Delta R}{R_0} = \frac{\sigma_{eff}^0}{\sigma_{eff}^L} (1 + \varepsilon)^{1+2\nu} - 1 \, ,$$

$$(3.26)$$

where σ_{eff}^0 and σ_{eff}^L are electrical conductivities of the composite before and after stretching in the stretching direction, respectively. Figure 3.6 shows the comparison between the analytical results and experimental results for the MWCNT/PEO composite. It is observed that the analytical prediction successfully predicts better the trend of the experimental result for the composite with higher volume fraction (f_{CNT} = 1.44%) and the variation trend of the resistance change is captured by the current model. However, a large discrepancy between the analytical prediction and the experimental results still exists which may be attributed to the assumption on the electrical conductivity of CNTs, the fixed length and diameter of the CNTs, etc.

For the composite with lower volume fraction (f_{CNT} = 0.56%), the sharp increase of the resistance change corresponding to a percolation behavior is not captured by the current model, which could be explained by the adoption of the excluded volume method for predicting the percolation threshold in Eq. (3.21). It is found that with the considered range of the strain, the change of the percolation threshold is negligible, which is always below 0.56% and thus no percolation is observed after stretching. Such a small change in the percolation threshold due to the stretching is also evidenced by the Monte Carlo simulation by Lin et al. (2010). However, it should be mentioned that the stretching effect on the percentage of the percolated CNTs is more significant for the composites with smaller volume fraction (see Figure 3.5(a)), which means that the stretching may significantly change the overall electrical conductivity of the composites. Due to the above mentioned limitation of the current model, this work can be claimed to provide a theoretical prediction on the trend of the stretching effect upon the electrical conductivity of the CNT-polymer composites. A more accurate model with the assistance of statistics on the distribution of the CNT size needs to be further pursued.

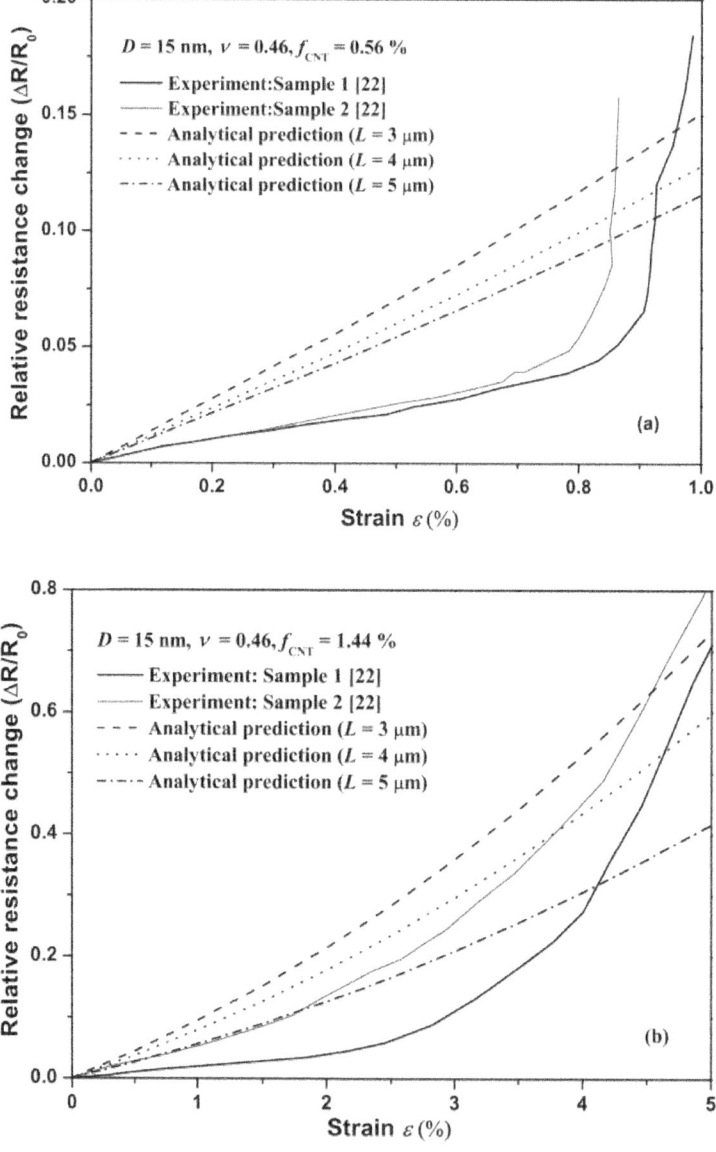

Figure 3.6: Comparisons between experimental results and analytical predictions (a) fCNT = 0.56% ; (b) fCNT = 1.44%.

From equations (3.7) and (3.8), it can be seen that the tunneling-type electrical conductivity of the interphase highly denpends on the critical separation distance dc, which was suggested by Monte Carlo simulation and experiments (Li et al., 2007; Allaoui et al., Takeda et al., 2011) that this critical separation distance might vary within a small range. In order to see the sensitivity of the overall electrical conductivity on the value of the critical separation distance, Figure 3.7 plots the variation of the electrical conductivity of the composite with the CNT volume fraction for different critical separation distances when the composite is under no stretching. From this figure, it can be seen that for the composite with lower CNT volume fraction, the electrical conductivity is very sensitive to the critical separtion distance, i.e., the increase of the critical separation distance significantly decreases the electrical conductivity. Such high sensitivity of the electrical conductivity to the critical separation distance may partly explain the large discrepancy between the modeling results and the experimental results for lower CNT volume fraction ($f_{CNT} = 0.56\%$). However, the sensitivity decreases as the CNT volume fraction increases.

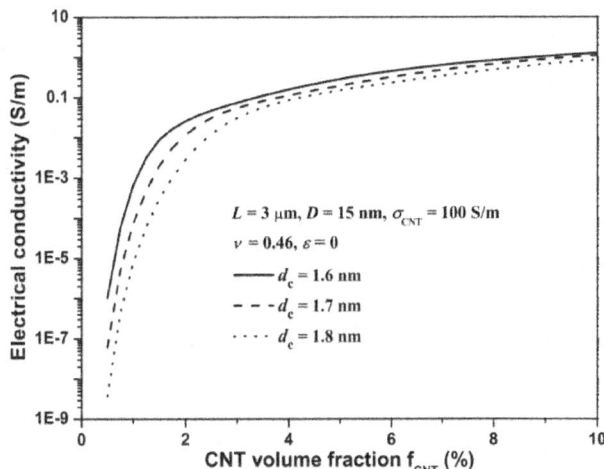

Figure 3.7: Variation of electrical conductivity with CNT volume fraction for different critical separtion distances.

Variation of the electrical conductivity with the CNT volume fraction for different stretching strains is plotted in Figure 3.8. It is observed that the electrical conductivity of the composite in both the longitudinal (along the stretching direction) and the transverse (perpendicular to the stretching direction) directions increases with the increase of CNT volume fractions as expected. In addition, for the composite with a fixed volume fraction of CNTs, stretching decreases the electrical conductivity in both the longitudinal and the transverse directions. Such a decrease can be attributed to the volume expansion and the conductive network change, i.e., increased percolation threshold due to the stretching.

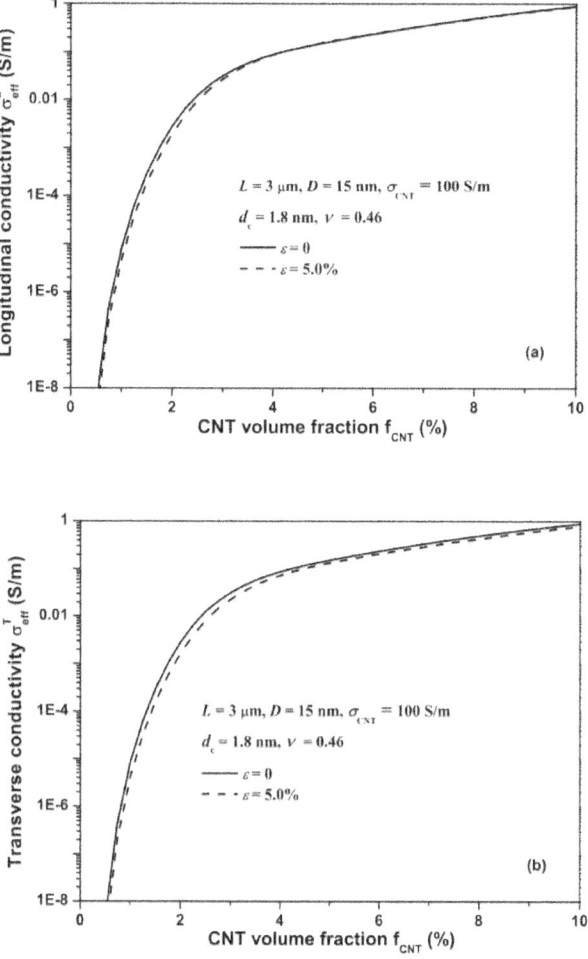

Figure 3.8: Variation of electrical conductivity with CNT volume fraction for different strains (a) Longitudinal direction; (b) Transverse direction.

Figure 3.9 shows the variation of the normalized electrical conductivity $\sigma_N = \sigma_{\text{eff}}/\sigma_{\text{eff}}^0$ with the stretching strain for different CNT volume fractions, where the quantities with the superscript "0" denotes the ones before stretching. It is again demonstrated in this figure that the stretching decreases the electrical conductivity and the stretching effect on the electrical conductivity along both directions is more significant for the composite with lower CNT concentration. This phenomenon can be interpreted by the effect of CNT volume fraction on the normalized percentage of the percolated CNTs (see Figure 3.5(a)), which was also commented by Jiang et al. (2007), i.e., for composites with lower filler concentration above the percolation threshold, only a few conductive networks are formed, so any small change in the networks, i.e., the separation distance between CNTs and re-alignment of CNTs, affects the magnitude of the electrical conductivity significantly. In contrast, a large number of conductive networks exist in the composites with higher filler concentration, for which any change in the networks has relatively limited effects on the electrical properties of the composites. The results in this figure suggest that the sensitivity of the electrical conductivity on stretching depends on the CNT volume fraction of the composites.

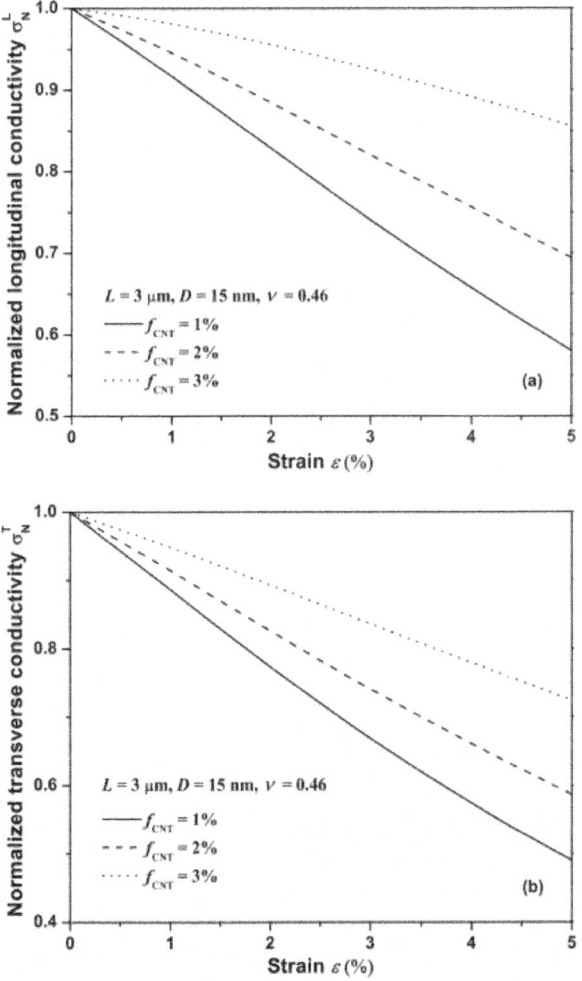

Figure 3.9: Variation of normalized electrical conductivity with strain for different CNT volume fractions (a) Longitudinal direction; (b) Transverse direction.

As indicated in the previous formulation and Figure 3.4(b), stretching effects also depend on the Poisson's ratio. In order to see the Poisson's ratio effect from a theoretical prediction perspective, the variation of the normalized electrical conductivity along the longitudinal and the transverse directions with the strain is plotted in Figure 3.10 for the composite with a fixed initial CNT volume fraction (f_{CNT} = 2%) but different Poisson's ratios. It is observed that the normalized electrical conductivity decreases with the increase of the strain and the decreasing rate is enhanced by the increase of the Poisson'sratio.

Such an enhancement of the decreasing rate can be explained by the combined effects of the Poisson's ratio on the electrical conductivity. For example, from Eq. (3.13) it is easily understood that with the increase of the Poisson's ratio volume expansion induced decrease in the electrical conductivity becomes more and more limited. From Figure 3.4(b), it is indicated that the increase of the Poisson's ratio enables CNTs to be more aligned along the stretching direction, which increases the electrical conductivity in the longitudinal direction while decreases the electrical conductivity in the transverse direction when the change in conductive networks is not taken into account. However, Eqs. (3.19)–(3.21) with the expression of θ_μ remind us that the re-alignment of the CNTs in the composites with larger Poisson›s ratio would always lead to a more significant increase in the percolation threshold and a decrease in the percentage of the percolated CNT (see Figure 3.5(b)), which could result in a more significant decrease in the electrical conductivity along the two directions.

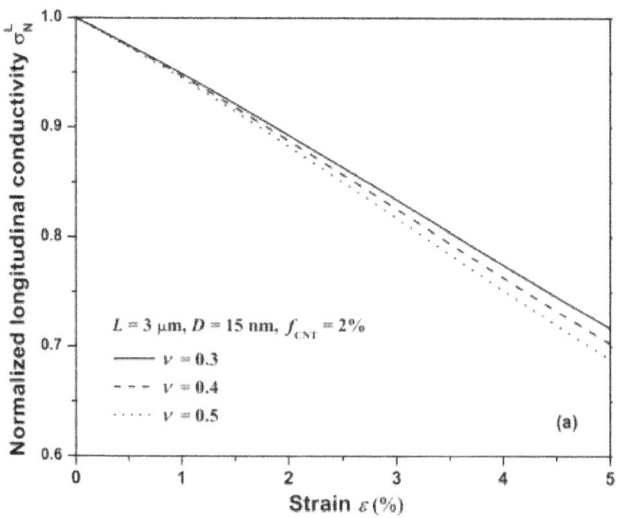

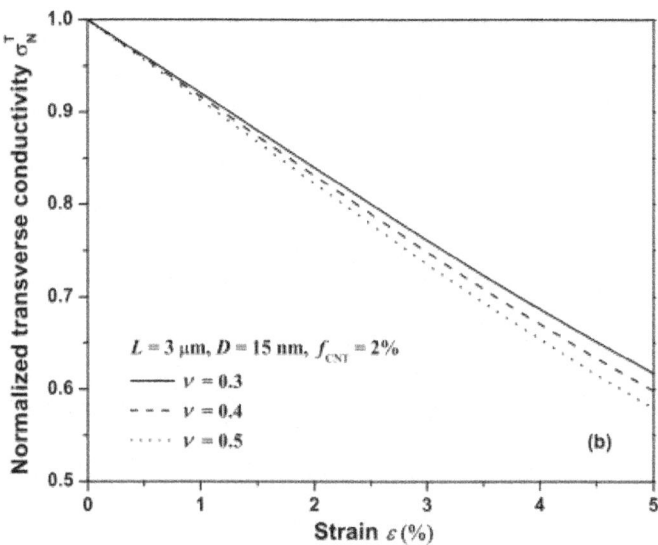

Figure 3.10: Variation of normalized electrical conductivity with strain for different Poisson's ratios (a) Longitudinal direction; (b) Transverse direction.

The observation of the combined effects of Poisson›s ratio on the overall electrical conductivity of the composites suggests that the conductive network change dominates the variation of electrical conductivity during the stretching over the re-alignment of the CNTs along the stretching direction.

Another appromixmation made in the current work is to fix the electrical conductivity of the CNTs in the simulation. However, it was suggested that the electrical conductivity of MWCNTs might vary within a big range (Deng and Zheng, 2008), i.e., from 10 S/m to 10000 S/m. Here we also test the sensitivity of the overall electrical conductivity of the composite to the electrical conductivity of the the CNTs in figure 11. From this figure, it can be seen that the electrical conductivity of the composite decreases more significantly with the stretching strain for CNTs with larger electrical conductivity. However, for the CNTs with sufficient large electrical conducitivy, σ_{CNT} = 100 S/m in the previous simulation for example, the selection value of the electrical conductivity of the CNTs will have slight influence on predicting the stretching effect upon the overall electrical conductivity of the composite.

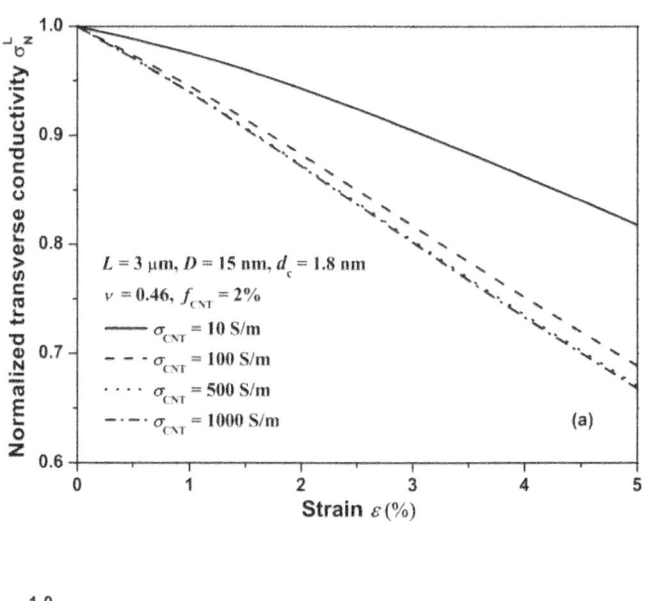

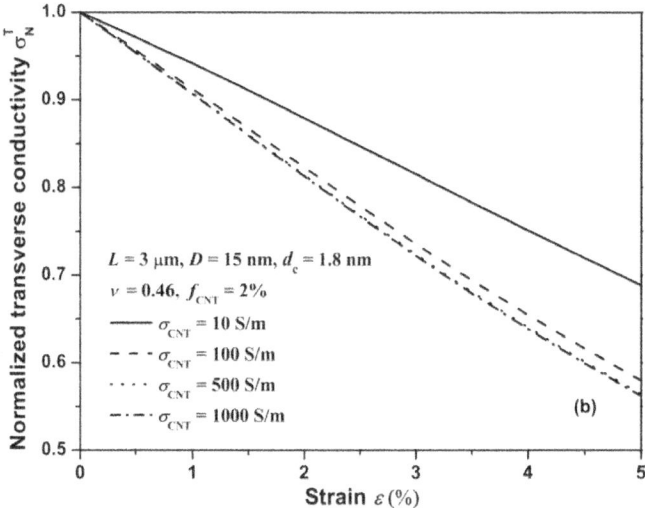

Figure 3.11: Variation of normalized electrical conductivity with strain for different electrical conductivity of CNTs (a) Longitudinal direction; (b) Transverse direction.

It should be mentioned that the modeling results in the current work could theoretically predict the stretching effects on the variation of the electrical conductivity of composites as reported in experiments to some extent (Park et al., 2008; Hu et al., 2010; Bao et al., 2011; Miao et al., 2011a, 2011b, 2012). However, the micromechanics model in the current work is based on small strain formulation. For certain CNT–polymer composites, such as CNT–rubber composites, which can undergo large deformation, linear kinematics assumed for micromechanics model with small deformation may be violated. In addition, the change of the azimuth angle φ and elongation of CNT length neglected in this paper.

may become significant under large deformation. Therefore, a more comprehensive micromechanics model based on finite deformation formulation with the consideration of the change of all Euler angles and the elongation of CNTs needs to be developed to predict the stretching effects on the electrical properties of conductive polymer composites.

CONCLUSIONS

In the current work, three expected changes induced by a uni-axial stretching, volume expansion, re-orientation of CNTs and change in conductive networks, have been considered and incorporated into a mixed micromechanics model to predict the stretching effects upon the overall electrical conductivity of CNT–polymer composites. Modeling results show that stretching decreases the electrical conductivity in both the longitudinal and transverse directions. It is found that the electrical conductivity of the composites highly depends on the critical separation distance. Also it is shown that the electrical conductivity of the composites is more sensitive to the stretching for composites with lower CNT concentration and shorter CNTs. Simulation results also indicate that the stretching induced change in conductive networks plays a dominant role for the variation of the electrical conductivity along the stretching direction. To some extent, the developed model in the current work is envisaged to be helpful in qualitatively understanding the trend of the stretching effects on the overall electrical conductivity of CNT–polymer composites.

BI-AXIAL STRETCHING EFFECTS ON ELECTRICAL CON-DUCTIVITY OF CNT-POLYMER COMPOSITES

Due to their broad spectrum of potential applications, conductive polymer composites have been attracting extensive interests over the past decades. In addition to the attribute of electrical conductivity as owned by traditional conductive or semi-conductive materials, such as metals or silicon, conductive polymer composites also possess a combined performance of flexibility, low cost, easy processability and good chemical and biological compatibility (Yu et al., 2009; Nambiar and Yeow, 2011; Shang et al., 2011), which makes them as promising material candidates for stretchable electronics, conductive coatings, electromagnetic shielding, solar cells, etc (Yang et al., 2005; Berson et al., 2007; Yu et al., 2009; Shang et al., 2011). Most traditional polymers are electrical insulators, while the desired electrical conductivity is commonly achieved by adding conductive fillers into compliant insulating polymers. Among different conductive fillers, CNTs have been widely adopted as doping materials due to their high aspect ratio and excellent intrinsic electrical conductivity. Some experiments (Qunaies et al., 2003; Kim et al., 2005; Gojny et al., 2006) have verified that the addition of a very small amount of CNTs into polymers can significantly improve the electrical conductivity of the nanocomposites, which demonstrate a percolation-like behavior with very low percolation threshold. Such a low percolation threshold is highly beneficial for conductive nanocomposite applications as the desired electrical conductivity could be achieved without any significant loss of the other inherent merits of the polymer matrix. It is well discussed in the literature that the electrical conductivity of the CNT-polymer composites attributes to two conductivity mechanisms, i.e., electron hopping at the nanoscale and conductive networks at the microscale (Chang et al., 2009; Zhang et al., 2009; Lu et al., 2010). As explained by Deng and Zheng (2008), the contribution of the electron hopping and the conductive networks depends on the CNT concentration. When the CNT concentration is below the percolation threshold, the electron hopping intra-tube or among different CNTs governs the electrical conductivity of the composites. However, when the CNT concentration exceeds the percolation threshold, some CNTs start to form conductive networks, which contribute to the electrical conductivity of the composites more significantly.

To fulfill the potential applications of CNT-polymer composites, understanding of the fundamental physics governing their overall electrical properties is of essential importance. Up to date, some efforts have been devoted to interpreting the electrical conductivity mechanisms and investigating how the overall electrical properties of the composites vary with their constituent features from both the theoretical and experimental perspectives (Kim et al., 2005; Gojny et al., 2006; Li et al., 2007; Yan et al., 2007; Chang et al., 2009; Seidel and Lagoudas, 2009; Takeda et al., 2011). It should be mentioned that most studies in the literature focused on investigating the electrical properties of the as-received composites without considering the stretching effects. However, for the particular applications of CNT-polymer composites in stretchable electronics, strain sensors (Park and Kim, 2006; Hu et al., 2010) for example, it is natural to investigate how the stretching influences the electrical behavior of the composites, which could particularly offer advice on the better design of the strain sensors. Under a uni-axial stretching, experimental data (Park et al., 2008; Hu et al., 2010; Bao et al., 2011) have demonstrated that such a stretching could lead to a sharp decrease/increase in the electrical conductivity/resistivity of the composites due to the breakdown of conductive networks. Meanwhile, an opposite trend was also observed (Cheng et al., 2009; Shang et al., 2011; Wang et al., 2011), i.e., the electrical conductivity of CNT-polymer composites increases with the increase of the stretching. Such a contrary observation was attributed to the fact that the CNT re-alignment along the stretching direction dominates over the breakdown of the conductive networks. By conducting numerial simulation, Taya et al. (1998) and Lin et al. (2010) investigated the stretching/compression effects upon the electrical conductivity of fibre-polymer composites and their results indicated that the deformation could increase the percolation threshold. A similar conclusion for the stretching effects on the percolation threshold was also captured by Tallman and Wang (2013) through applying excluded volume method.

Compared to the extensive explorations in studying the electrical properties of CNTpolymer nanocomposites and the uni-axial stretching effects, limited work has been found in investigating the bi-axial stretching effects on the electrical conductivity of the CNT-polymer composites in the literature. Nevertheless, bi-axial stretching is another typical stretching mode for CNT-polymer composites (Shen et al., 2012; Mayoral et al., 2013), which enables the CNTs to re-orientate along the two stretching directions in the polymer matrix. Such a stretching mode is expected to reduce the anisotropy of the electrical properties of the composites in the stretching plane compared to the uni-axial stretching case in which the CNTs are prone to get parallel along the uni-axial stretching direction. To complement the theoretical modeling of the uni-axial stretching effects upon the electrical conductivity of the CNT-polymer composites, the current work aims to investigate the bi-axial stretching effects on the electrical conductivity of the CNTpolymer composites with the incorporation of possible stretching-induced changes in a mixed micromechanics model.

Nanoscale and micromechanics modeling on electrical conductivity

Before applying the micromechanics modeling on the overall electrical behavior of the CNT-polymer composite, the nanoscale electrical conductivity mechanism, i.e., the electron hopping, is first considered. It was well-accepted in the literature (Yan et al., 2007; Seidel and Lagoudas, 2009; Feng and Jiang, 2013) that the electron hopping among CNTs results in the formation of an interphase layer surrounding the CNTs. Accordingly, an effective composite cylinder was developed in these studies to capture such a nanoscale electron hopping effect, which consists of a CNT with length L and diameter D and an interphase layer with thickness t. The effective volume fraction feff of the cylinder is re-defined in terms of the CNT concentration fCNT as (Yan et al., 2007; Seidel and

$$f_{\text{eff}} = \frac{(D+2t)^2(L+2t)}{D^2 L} f_{\text{CNT}}.$$

(4.1)

Provided the electrical conductivity of the CNT (σ_{CNT}) and the interphase (σ_{Int}), the effective longitudinal and transverse electrical conductivity of the cylinder can be obtained by applying the law-of-mixture rule as (Taya, 2005; Yan et al., 2007; Feng and Jiang, 2013):

$$\tilde{\sigma}^{\text{L}} = \frac{(L+2t)\sigma_{\text{Int}}\left[\sigma_{\text{CNT}}D^2 + 4\sigma_{\text{Int}}(Dt+t^2)\right]}{2\sigma_{\text{CNT}}D^2 t + 8\sigma_{\text{Int}}(Dt+t^2)t + \sigma_{\text{Int}}L(D+2t)^2},$$

$$\tilde{\sigma}^{\text{T}} = \frac{\sigma_{\text{Int}}}{L+2t}\left[L\frac{D^2\sigma_{\text{CNT}} + 2(\sigma_{\text{CNT}}+\sigma_{\text{Int}})(t^2+Dt)}{D^2\sigma_{\text{Int}} + 2(\sigma_{\text{CNT}}+\sigma_{\text{Int}})(t^2+Dt)} + 2t\right]$$

(4.2)

where the superscripts "L" and "T" represent the longitudinal and the transverse directions, respectively. Thus the electrical conductivity tensor of the effective cylinder in its local coordinate system can be expressed as $\tilde{\sigma} = \text{diag}\left(\tilde{\sigma}^{\text{L}}, \tilde{\sigma}^{\text{T}}, \tilde{\sigma}^{\text{T}}\right)$ Obviously, the interphase features, i.e., the electrical conductivity (σ_{Int}) and the thickness (t), are essential for determining the electrical property of the effective cylinder

According to the literature (Deng and Zheng, 2008; Seidel and Lagoudas, 2009; Takeda et al., 2011; Feng and Jiang, 2013), the electrical conductivity of the composites relies on the two mentioned conductivity mechanisms depending on the CNT concentration as mentioned in the Introduction Section. For example, when the CNT concentration is below the percolation threshold f_c, CNT fillers are more electrically independent and the electron hopping dominates the electrical property of the composite; when the CNT concentration is greater than the percolation threshold, some CNTs start to form conductive networks which govern the electrical conductivity over the electron hopping. For CNTs forming conductive networks, quite a few works suggested that the average separation distance between CNTs follows a power law relation with the CNT concentration (Allaoui et al., 2008; Deng and Zheng, 2008; Takeda et al., 2011). Here we adopt the following expression to determine the average separation distance between CNTs (Feng and Jiang, 2013):

$$d_a = \left(\frac{f_c}{f_{CNT}} \right)^{1/3} d_c,$$

where $d_c = 1.8$ nm is the critical separation distance between two adjacent CNTs that allows the tunneling penetration of electrons. It is easily understood that for the electrically independent CNTs without forming any conductive networks, the separation distance between the CNTs should be greater than d_c. Due to the lack of work on estimating the average separation distance between such CNTs, we assume the average separation distance between CNTs without forming conductive networks as the critical separation distance, i.e., $d_a = d_c$. It should be noted that such an assumption overestimates the electrical conductivity of the composite before the percolation. However, the overestimation in the electrical conductivity after the percolation can be neglected due to the fact that the conductive networks dominate over the electron hopping as argued in our previous work (Feng and Jiang, 2013). The thickness of the interphase was taken as half of the corresponding average separation distance between the CNTs, i.e., $t = d_a/_2$ (Feng and Jiang, 2013).

Following Simmons' formulation for electron tunneling (Simmons, 1963; Takeda et al., 2011), the electrical conductivity of the interphase was derived as:

$$\sigma_{Int} = \frac{d_a}{a R_{Int}(d_a)},$$

(4.4)

Where $R_{Int}(d_a) = \frac{d_a h^2}{ae^2(2m\lambda)^{1/2}} \exp\left(\frac{4\pi d_a}{h}(2m\lambda)^{1/2} \right)$ is the tunneling-type contact resistance between two CNTs with a being the contact area of CNTs; $\lambda = 5.0$ eV is the potential barrier height for CNTs dispersed in most polymers (Shiraishi and Ata, 2001; Takeda et al., 2011), $m = 9.10938291 \times 10^{-31}$ kg and $e = -1.602176565 \times 10^{-19}$ C are mass and electric charge of an electron, respectively, and $\hbar = 6.626068 \times 10^{-34}$ m^2 ·kg/s is the Planck constant.

With the nanoscale modeling of electron hopping as discussed above, the CNTs are modeled equivalently as solid cylinder fillers. These solid fillers are assumed as straight and uniformly dispersed in the polymer matrix. For such a two-phase composite, the overall effective electrical conductivity of the composite can be theoretically predicted by applying the micromechanics model as performed in the literature (Taya, 1995; Entchev and Lagoudas, 2002; Odegard et al., 2003; Taya, 2005; Seidel and Lagoudas 2009; Feng and Jiang, 2013). The routine procedure in these studies is that the overall electrical conductivity of the composite is derived by averaging the contribution of the fillers from all possible orientations in the representative volume element (RVE) as shown in Figure 4.1, i.e.,

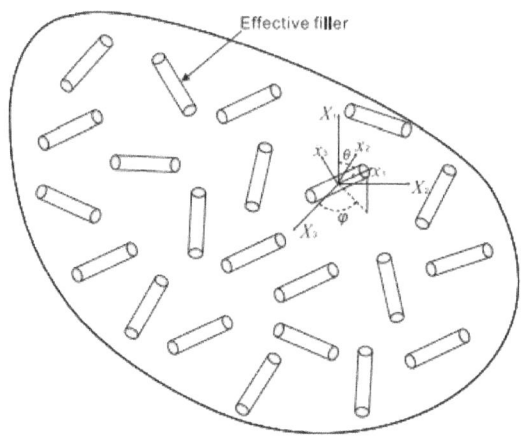

Figure 4.1: RVE containing effective fillers.

$$\sigma_{\text{eff}} = \sigma_{\text{m}} + \frac{\int_0^{2\pi} \int_0^{\pi} \rho(\varphi,\theta) f_{\text{eff}} (\sigma - \sigma_{\text{m}}) A \sin\theta \mathrm{d}\theta \mathrm{d}\varphi}{\int_0^{2\pi} \int_0^{\pi} \rho(\varphi,\theta) \sin\theta \mathrm{d}\theta \mathrm{d}\varphi},$$

(4.5)

where ϕ and θ are the Euler angles identifying the orientation of the fillers; $\rho(\phi,\theta)$ is the orientation distribution function (ODF) of the fillers; A is the electric field concentration tensor; σ_{m} is the electrical conductivity tensor of the polymer and σ is the electrical conductivity tensor of the effective filler in the global coordinates system (X^1, X^2, X^3) of the composites, which can be determined from the conductivity tensor $\tilde{\sigma}$ of the filler in its local coordinates system $(\tilde{x}_1, \tilde{x}_2, \tilde{x}_3)$ as

$$\sigma = Q^{\text{T}} \tilde{\sigma} Q,$$

(4.6)

with Q being the transformation matrix given as (Entchev and Lagoudas, 2002):

$$Q = \begin{bmatrix} \sin\theta\cos\varphi & -\cos\theta\cos\varphi & \sin\varphi \\ \sin\theta\sin\varphi & -\cos\theta\sin\varphi & -\cos\varphi \\ \cos\theta & \sin\theta & 0 \end{bmatrix}.$$

(4.7)

Following the percolation process proposed by Deng and Zheng (2008), it is understood that the overall electrical conductivity of the composite attributes to the electron hopping and the conductive networks depending on the CNT concentration. For example, only electron hopping contributes to the electrical behavior of the composite prior to the percolation. However, after CNT concentration reaches the percolation threshold f_c, certain percentage ξ of CNTs start to form conductive networks, which was estimated by Deng and Zheng (2008) as:

$$\xi = \frac{f_{CNT}^{1/3} - f_c^{1/3}}{1 - f_c^{1/3}} \ (f_c \le f_{CNT} < 1).$$

(4.8)

It is obvious from this expression that with the increase of the CNT concentration from fc to 1, the percentage of the CNTs forming conductive networks increases from 0 to 1. With the consideration of this percolation process, it is natural to incorporate both the electron hopping and the conductive networking mechanisms in the micromechanics modeling. Therefore, a mixed micromechanics model was developed in our previous work (Feng and Jiang, 2013) in which the effective electrical conductivity of the composite was derived from Eq. (2.5) in the following form:

$$\sigma_{eff} = \begin{cases} \sigma_m + \dfrac{\int_0^{2\pi}\int_0^{\pi} \{ f_{eff}(\sigma_{EH} - \sigma_m)A_{EH} \} \rho(\varphi,\theta)\sin\theta \, d\theta \, d\varphi}{\int_0^{2\pi}\int_0^{\pi} \rho(\varphi,\theta)\sin\theta \, d\theta \, d\varphi}, \ f_{CNT} < f_c \\[2em] \sigma_m + (1-\xi)\dfrac{\int_0^{2\pi}\int_0^{\pi} \{ f_{eff}(\sigma_{EH} - \sigma_m)A_{EH} \} \rho(\varphi,\theta)\sin\theta \, d\theta \, d\varphi}{\int_0^{2\pi}\int_0^{\pi} \rho(\varphi,\theta)\sin\theta \, d\theta \, d\varphi} \\[2em] + \xi\dfrac{\int_0^{2\pi}\int_0^{\pi} \{ f_{eff}(\sigma_{CN} - \sigma_m)A_{CN} \} \rho(\varphi,\theta)\sin\theta \, d\theta \, d\varphi}{\int_0^{2\pi}\int_0^{\pi} \rho(\varphi,\theta)\sin\theta \, d\theta \, d\varphi}, \ f_{CNT} \ge f_c \end{cases}$$

(4.9)

where the subscripts "EH" and "CN" denote terms contributed by electron hopping and conductive networks, respectively.

Here we adopt the Mori-Tanaka micromechanics model (Mori and Tanaka, 1973; Taya, 2005), for which the concentration tensor in Eq. (2.5) can be determined as (Taya, 1995; Entchev and Lagoudas, 2002; Odegard et al., 2003)

$$A = QA^{dil}Q^T \left\{ (1-f_{eff})\delta + f_{eff}\frac{\int_0^{2\pi}\int_0^{\pi} f_{eff} \{ QA^{dil}Q^T \} \rho(\varphi,\theta)\sin\theta \, d\theta \, d\varphi}{\int_0^{2\pi}\int_0^{\pi} \rho(\varphi,\theta)\sin\theta \, d\theta \, d\varphi} \right\}^{-1}$$

(4.10)

Where $A^{dil} = \left\{ \delta + S(\sigma_m)^{-1}(\tilde{\sigma} - \sigma_m) \right\}^{-1}$ with δ being the Kronecker delta tensor. The parameter S in the expression of A ^{dil} is the Eshelby tensor of the effective filler, which is a function of the effective aspect ratio of the filler (Taya, 2005; Seidel and Lagoudas, 2009; Feng and Jiang, 2013). For CNTs in conductive networks, the effective aspect ratio is taken as infinite due to the formed electrically conductive chains across the sample. In contrast, for CNTs associated with electron hopping, the effective aspect ratio is the real effective aspect ratio of single filler. Correspondingly, A_{EH} and A_{CN} in Eq. (2.9) can be calculated from Eq. (2.10) by using the associated aspect ratios.

Stretching induced changes

When the composite with effective fillers is subjected to a stretching, it is suggested in the literature (Taya et al., 1998; Taya 2005; Feng and Jiang, 2014) that three major changes are expected to occur: volume expansion of the composite, re-orientation of the fillers and the change in conductive networks. In the following, we will characterize these stretching induced changes and illustrate how such stretching effects are incorporated into the micromechanics model developed in the previous Section.

Volume Expansion And Re-Orientation

When subjected to a stretching, the fillers in the composites tend to re-orientate along the stretching direction. Following the fiber re-orientation model in references (Kuhn and Grün, 1942; Taya et al., 1998; Feng and Jiang, 2014), Figure 4.2 shows the orientation change of a filler in a cell before and after a bi-axial stretching. The original lengths of the cell in the X_1, X_2 and X_3 directions are l_0, w_0 and h_0, respectively. After the bi-axial stretching, the lengths of the cell become:

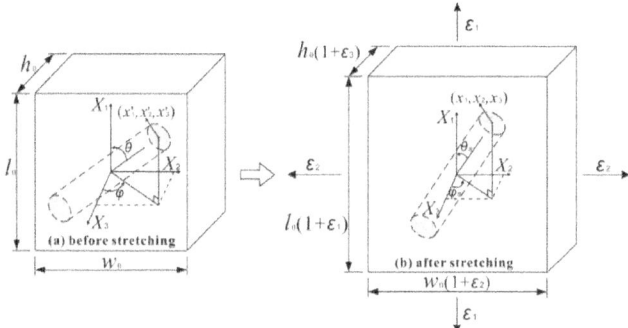

Figure 4.2: Sketch of orientation change of a filler due to bi-axial stretching.

$$l = l_0 (1 + \varepsilon_1), \quad w = w_0 (1 + \varepsilon_2) \quad \text{and} \quad h = h_0 (1 + \varepsilon_3),$$

(4.11)

where ε_1, ε_2 and ε_3 are the three principle strains along the X_1, X_2 and X_3 directions in the global coordinate system, respectively, which are assumed to satisfy $\varepsilon_1 \geq \varepsilon_2 \geq \varepsilon_3$. For any bi-axial stretching strains ε_1 and ε_2 under deformation, the stretching strain in the X_3 direction is derived as:

$$\varepsilon_3 = \left[(1 + \varepsilon_1)(1 + \varepsilon_2) \right]^{-\frac{v}{1-v}} - 1.$$

(4.12)

Due to the fact that the stiffness of the polymer is usually much lower than that of the CNT filler, the change of the filler's volume can be neglected since the deformation is mainly sustained by the polymer. Accordingly, the effective volume fraction of the fillers in the composite after the stretching can be approximated as:

$$f = \frac{V_0 f_{\text{eff}}}{V} == \frac{f_{\text{eff}}}{\left[(1 + \varepsilon_1)(1 + \varepsilon_2) \right]^{\frac{1-2v}{1-v}}},$$

(4.13)

where V_0 and V are the volumes of the cell before and after stretching, respectively. Obviously, for compressible materials, the bi-axial stretching induces a volume expansion, resulting in the decrease of the filler concentration. Therefore, the filler volume fraction in Eq. (2.9) needs to be replaced by the volume fraction defined in Eq. (3.3) when considering the bi-axial stretching.

As demonstrated in Figure 4.2, the re-orientation of the filler in the cell results in the variation of the two Euler angles, i.e., the polar angle changes from θ to θ_s and the azimuth angle changes from ϕ to φ_s. Correspondingly, the coordinates of the upper end of the filler in the cell before the bi-axial stretching are determined in terms of the Euler angles as (Taya et al., 1998)

$$x_1' = \frac{u'}{2} \cos \theta, \quad x_2' = \frac{u'}{2} \sin \theta \sin \varphi, \quad x_3' = \frac{u'}{2} \sin \theta \cos \varphi,$$

(4.14)

and after stretching as:

$$x_1 = \frac{u}{2} \cos \theta_s, \quad x_2 = \frac{u}{2} \sin \theta_s \sin \varphi_s, \quad x_3 = \frac{u}{2} \sin \theta_s \cos \varphi_s,$$

(4.15)

where u' and u are the lengths of the filler before and after stretching. Ignoring the change of the filler length, u' = u. From Eqs. (3.4) and (3.5), we have the following expressions:

$$\begin{cases} \dfrac{x_1}{x_1'} = \dfrac{u\cos\theta_s}{u'\cos\theta} = 1+\varepsilon_1 \\[2ex] \dfrac{x_2}{x_2'} = \dfrac{u\sin\theta_s\sin\varphi_s}{u'\sin\theta\sin\varphi} = 1+\varepsilon_2 \\[2ex] \dfrac{x_3}{x_3'} = \dfrac{u\sin\theta_s\cos\varphi_s}{u'\sin\theta\cos\varphi} = \left[(1+\varepsilon_1)(1+\varepsilon_2)\right]^{-\frac{v}{1-v}} \end{cases}.$$

$$(4.16)$$

Based on Eq. (3.6), the two Euler angles of the filler after the bi-axial stretching, φ_s and θ_s, can be determined in terms of the initial Euler angles φ and θ as:

$$\begin{cases} \tan\varphi_s = (1+\varepsilon_1)^{\frac{v}{1-v}}(1+\varepsilon_2)^{\frac{1}{1-v}}\tan\varphi \\[2ex] \tan\theta_s = \dfrac{1+\varepsilon_2}{1+\varepsilon_1}\cdot\dfrac{\sin\varphi}{\sin\varphi_s}\tan\theta \end{cases}.$$

$$(4.17)$$

From Eq. (3.7), it is revealed that the two Euler angles denoting the orientation of the filler vary with the stretching strains, which disrupts the randomness of the filler distribution in the composite. To quantify the stretching effects upon the re-orientation of the fillers, the ODF, $\rho(\varphi, \theta)$, is introduced to characterize the distribution of the fillers, which satisfies the following conditions (Gurp, 1995; Pérez et al., 2008):

$$\rho(\varphi,\theta)\geq 0 \text{ and } \frac{1}{4\pi}\int_0^{2\pi}\int_0^{\pi}\rho(\varphi,\theta)\sin\theta\,d\theta\,d\varphi = 1.$$

$$(4.18)$$

When fillers are randomly distributed in the polymer matrix before the stretching, the ODF equals unity, indicating a uniform distribution of fillers along any orientation. However, after a stretching, the two Euler angles vary with the stretching strains indicating the re-orientation of the fillers and thus the variation of the ODF. In order to determine the new ODF after the stretching, we assume there are a number of G fillers randomly distributed in the RVE of the micromechanics model before the stretching (as shown in Figure 4.1). Following Eq. (3.8), the total number of the fillers falling in the ranges of (θ, $\theta+d\theta$) and (φ, $\varphi+d\varphi$) in the RVE can be determined as (Kuhn and Grün, 1942):

$$dN_{\substack{\theta,\theta+d\theta \\ \varphi,\varphi+d\varphi}} = \frac{1}{4\pi}G\rho(\varphi,\theta)\sin\theta\,d\theta\,d\varphi.$$

$$(4.19)$$

After a bi-axial stretching, the G fillers will be re-oriented within the ranges of $(\phi_s, \varphi_s+d\varphi_s)$ and $(\theta_s, \theta_s+d\theta_s)$, i.e.,

$$dN_{\substack{\theta_s,\theta_s+d\theta_s \\ \varphi_s,\varphi_s+d\varphi_s}} = \frac{1}{4\pi} G\rho\left(\varphi_s,\theta_s\right)\sin\theta_s d\theta_s d\varphi_s = dN_{\substack{\theta,\theta+d\theta \\ \varphi,\varphi+d\varphi}},$$

(4.20)

where $\rho(\phi_s, \theta_s)$ is the new ODF after the stretching. Combining Eq. (3.7) and Eq. (3.10), the new ODF after the bi-axial stretching is derived as:

$$\rho\left(\varphi_s,\theta_s\right) = \frac{\left(\dfrac{1+\varepsilon_1}{1+\varepsilon_2}\cdot\dfrac{\sin\varphi_s}{\sin\varphi}\right)^{\frac{1}{2}}}{\left(1+\varepsilon_1\right)^{\frac{v}{1-v}}\left(1+\varepsilon_2\right)^{\frac{1}{1-v}}\cos^2\varphi_s + \dfrac{1}{\left(1+\varepsilon_1\right)^{\frac{v}{1-v}}\left(1+\varepsilon_2\right)^{\frac{1}{1-v}}}\sin^2\varphi_s}{\left[\left(\dfrac{1+\varepsilon_1}{1+\varepsilon_2}\right)^{-1}\cdot\left(\dfrac{\sin\varphi_s}{\sin\varphi}\right)^{-1}\cos^2\theta_s + \dfrac{1+\varepsilon_1}{1+\varepsilon_2}\cdot\dfrac{\sin\varphi_s}{\sin\varphi}\sin^2\theta_s\right]^{\frac{3}{2}}}.$$

(4.21)

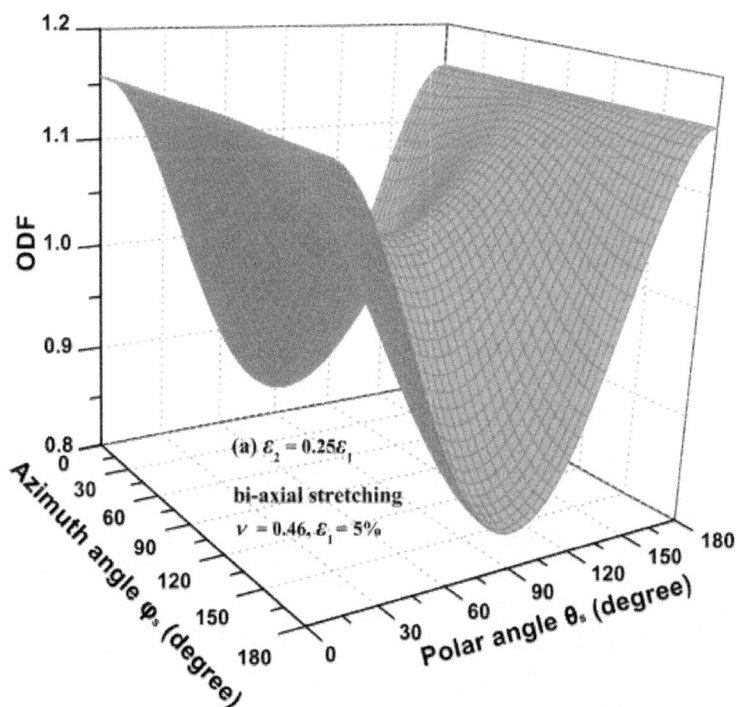

Figure 4.3: Variation of ODF with Euler angles (a) $\varepsilon_2 = 0.25\varepsilon_1$; (b) $\varepsilon_2 = \varepsilon_1$.

It should be noted here that with the consideration of the bi-axial stretching, the Euler angles and the ODF in Eq. (2.9) should be replaced by the new Euler angles and the ODF determined by Eqs. (3.7) and (3.11), respectively. Without any stretching, i.e., $\varepsilon_1 = \varepsilon_2 = \varepsilon_3 = 0$, the ODF defined in Eq. (3.11) reduces to unity for a random distribution as expected. Figure 4.3 plots the ODF profile with two Euler angles. From these figures, it can be seen that the stretching tends to re-align the fillers along the stretching directions ($\phi_s = 90°$ and $\theta s = 0°$ and $180°$), which was also commented by Mayoral et al. (2013). Comparing the ODF in Figure 4.3(a) and Figure 4.3(b), it is indicated that increasing the stretching strain in a particular direction, the X_2 direction for example, enhances the re-alignment of fillers along that direction.

Change in conductive networks

As discussed in the literature (Deng and Zheng; 2008; Seidel et al., 2008), when the CNT concentration reaches a certain value (the percolation threshold), the electrical conductivity of the CNT-polymer composite increases abruptly denoting the onset of the formation of conductive networks. According to the fitting from Monte Carlo simulation by Tallman and Wang (2013), it was found that the stretching increases the percolation threshold which influences the conductive networks in the composite. Here we adopt the excluded volume method (Balberg et al., 1984; Celzard et al., 1996; Tallman and Wang, 2013) to incorporate the stretching effects into the percolation threshold, which was given as:

$$f_c = 1 - \exp\left(\frac{-\langle V_{ex} \rangle V_{CNT}}{\langle V_e \rangle} \right),$$

(4.22)

Where

$$\langle V_e \rangle = \frac{4\pi}{3} D^3 + 2\pi D^2 L + 2 \cdot D \cdot L^2 \langle \sin \gamma \rangle_\mu$$

(4.23)

is the average excluded volume of the CNT, $\langle V_{ex} \rangle$ is the total average excluded volume of the CNT and V_{CNT} is the volume of the CNT. The term $\langle \sin\gamma \rangle_\mu$ in Eq. (3.13) is an averaging term accounting for the CNT orientation, with γ being the angle between the ith and the jth CNTs. As indicated in Figure 4.3, stretching induces the re-alignment of CNTs favoring the stretching directions. Such a re-alignment disrupts the randomness of the CNT distribution. For example, CNTs uniformly distribute in the ranges of the polar angle ($0 \leq \theta \leq 2\pi$) and the azimuth angle ($0 \leq \varphi \leq \pi$) before the stretching. However, it was understood from the work of Tallman and Wang (2013) that after the stretching more CNTs will concentrate within certain ranges, i.e., within the polar angle ranges of $[0, \theta\mu]$ and $[\pi-\theta_{\mu}, \pi]$, in which θ_{μ} was determined as:

$$\theta_{\mu} = \arcsin\left[(1+\varepsilon_3)/(1+\varepsilon_1)\right],$$

(4.24)

where ε_1 and ε_3 are the first and the third principle strains. Based on the fitting data from Monte Carlo simulation, the term $\langle \sin\gamma \rangle_{\mu}$ in Eq. (3.13) was determined in terms of $\theta\mu$ as (Tallman and Wang, 2013):

$$\langle \sin\gamma \rangle_{\mu} = 0.018\theta_{\mu}^5 + 0.021\theta_{\mu}^4 - 0.234\theta_{\mu}^3 - 0.015\theta_{\mu}^2 + 0.909\theta_{\mu}.$$

(4.25)

In addition, a linear relation between $\langle V_{ex} \rangle$ and $\langle \sin\gamma \rangle_{\mu}$ was also given by Tallman and Wang (2013] as:

$$\langle V_{ex} \rangle = 2.8 - 5.6\frac{\langle \sin\gamma \rangle_{\mu}}{\pi}.$$

(4.26)

Combining Eqs. (3.12)–(3.16), the percolation threshold with the consideration of the stretching effects was derived as (Feng and Jiang, 2014):

$$f_c = 1 - \exp\left[-\frac{\left(2.1\pi - 4.2\langle\sin\gamma\rangle_{\mu}\right)p}{4\pi + 6\pi p + 6p^2\langle\sin\gamma\rangle_{\mu}}\right],$$

(4.27)

where p = L/D is the aspect ratio of the CNTs. Obviously, the stretching induced reorientation of the CNTs will influence the percolation threshold, which results in the change of the average separation distance among CNTs and the percentage of the percolated CNTs as indicated by Eqs. (2.3) and (2.8), respectively

In order to see the stretching effects on the conductive networks, Figure 4.4 plots the variation of the normalized percentage $\xi_N = \xi_s/\xi_0$ of the percolated CNTs for a composite under bi-axial stretching, where ξ_0 and ξ_s denote the percentage of the percolated CNTs before and after stretching, respectively. From this figure, it is observed that with a fixed stretching strain in the X_1 direction, the stretching strain in the X_2 direction decreases the percentage of percolated CNTs as compared to the uni-axial stretching case. It is thus concluded that bi-axial stretching induces more breakdown of conductive networks. It is also found that the decreasing rate of the percolated CNTs is more sensitive to the stretching strain in the X_2 direction for the composite with lower CNT concentration than that with higher CNT concentration. Furthermore, the percentage of percolated CNTs of the composite with the same CNT concentration but different aspect ratio (fixed diameter but different length) is also provided in this figure for comparison. It is concluded that the composites with shorter CNTs are more susceptible to the breakdown of conductive networks.

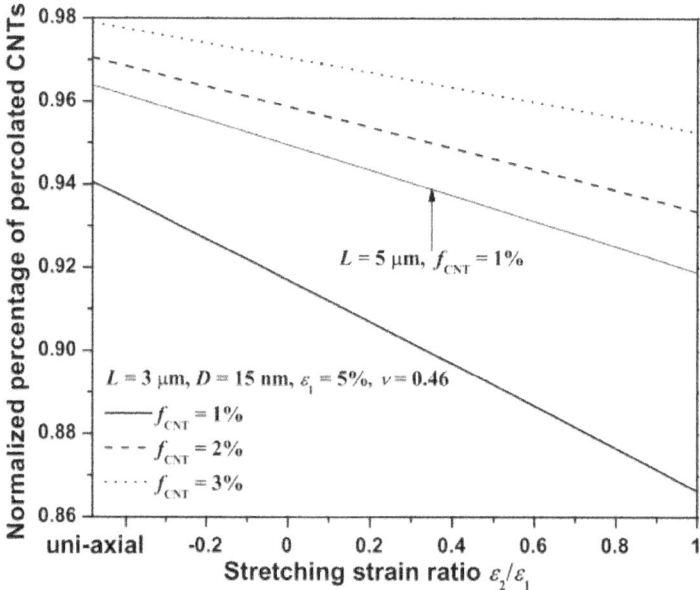

Figure 4.4: Variation of normalized percentage of percolated CNTs with stretching strain ratio.

Results and discussions

To investigate the bi-axial stretching effects on the electrical conductivity of the CNTpolymer composites, a MWCNT/PEO (Polyethylene Oxide) composite (Park and Kim,2006; Park et al., 2008] is chosen as the example material. The MWCNTs were functionalized and well dispersed in the PEO matrix by stirring and sonication before the stretching. Therefore, we can assume that the CNTs are uniformly and well dispersed in the polymer matrix before the stretching. The length and the diameter of the CNTs range from 1 μm to 5 μm and 10 nm to 20 nm, respectively. The electrical conductivity of the polymer and the CNTs is taken as $\sigma_m = 1 \times 10^{-13}$ S/m and $\sigma_{CNT} = 100$ S/m (Deng and Zheng, 2008; Feng and Jiang, 2013). For the value of the Poisson's ratio of the composites, it should be noted that the true value naturally depends on both the CNT concentration and the CNT orientation, and it varies with the stretching. However, according to the evaluation method provided by Pan (1996), the effect of the orientation of fillers upon the Poisson's ratio of the composites is negligible. For example, for a CNT composite with volume fraction 3%, ν is approximately determined as 0.46 for a random CNT distribution while 0.45 for a perfectly aligned distribution. In addition, our previous work (Feng and Jiang, 2014) shows that the Poisson's ratio has limited effects on the electrical conductivity of the composites. Therefore, we choose the Poisson's ratio of the polymer, i.e., $v = 0.46$, as the value of the composite.

Figure 4.5 plots the variation of the electrical conductivity of the CNT-polymer composite in the three principal directions with the CNT concentration for different stretching strain ratios when the CNT size is fixed. From this figure, it can be seen that the electrical conductivity increases with the CNT volume fraction as expected. It is also observed that both uni-axial stretching and bi-axial stretching decrease the electrical conductivity of the composite, which is attributed to the stretching induced change in conductive networks, i.e., the increase in separation distance among CNTs and the decrease in percentage of percolated CNTs. However, it is found that for particularly low CNT volume fraction (below percolation threshold) and high enough CNT volume fraction (far away from percolation threshold), small discrepancy is observed between different scenarios. It can be explained by the fact that for the composite with low CNT volume fraction, no conductive networks are formed and the electrical conductivity is mainly attributed to the electron hopping, which slightly depends on the stretching. In contrast, for the composites with high enough CNT volume fraction, there exist a large number of conductive networks in the composite.

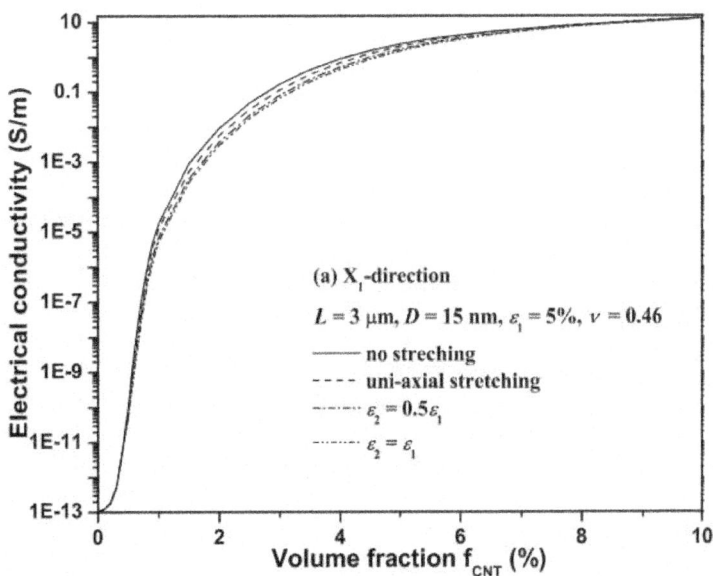

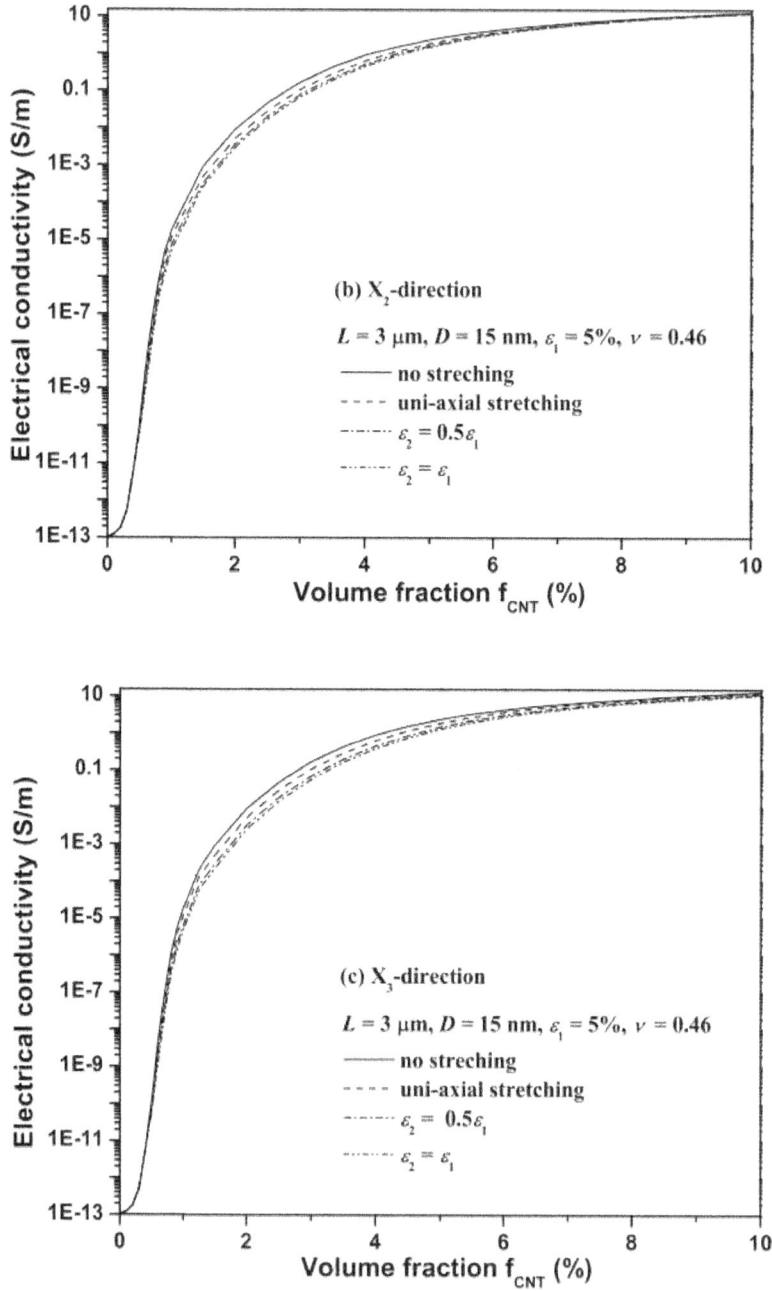

Figure 4.5: Variation of electrical conductivity with CNT volume fraction (a) X_1- direction; (b) X_2-direction; (c) X_3-direction.

Thus any change in the conductive networks, stretching induced breakdown for example, will have relatively limited effect on the overall electrical conductivity of the composite as argued by Jiang et al. (2007). Furthermore, it is found that compared to the uni-axial stretching, bi-axial stretching decreases the electrical conductivity more significantly since the stretching in the X2 direction enhances the re-orientation of CNTs in the stretching direction, resulting in more breakdowns of conductive networks. Such a bi-axial stretching enhanced decrease in the electrical conductivity was also experimentally observed by Mayoral et al.(2013).

As mentioned in Section 3, there are three major changes induced by the stretching, including volume expansion, re-orientation of fillers and breakdown of conductive networks (change in conductive networks). When the composite ($v = 0.46$) is under a maximum bi-axial stretching strain considered in the current work, i.e., $\varepsilon_1 = \varepsilon_2 = 5\%$, the volume expansion induces a relative decrease in the CNT concentration, which is calculated as less than 1.5% of the original CNT volume fraction. Therefore, such a small change in the CNT concentration is negligible under small deformation.

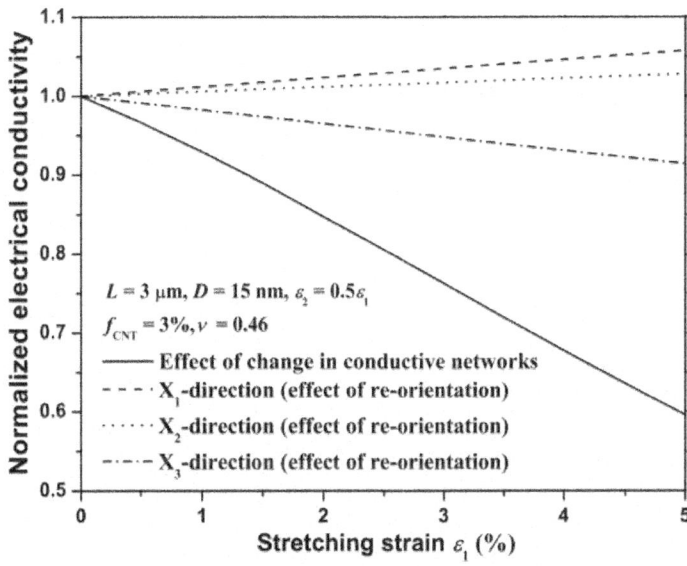

Figure 4.6: Effect of stretching induced re-orientation and change in conductive networks.

It should be mentioned that although stretching overall decreases the electrical conductivity of the composite (Figure 4.5), the re-orientation itself is expected to increase the electrical conductivity along the stretching directions.

To evaluate the relative effect of the other two stretching induced changes, Figure 4.6 demonstrates the variation of the normalized electrical conductivity with considering the individual effect of re-orientation and change in the conductive networks separately. From this figure, it can be seen that under the considered range of the volume fraction and the stretching strain, the decrease of the electrical conductivity induced by the change in the conductive networks is much more prominent than the re-orientation induced variation of the electrical conductivity, which indicates the dominate role of the change in the conductive networks.

Figure 4.7 shows the variation of the normalized electrical conductivity with stretching strain ratio for the three principle directions, in which the electrical conductivity was normalized by the value before the stretching. As expected, under uni-axial stretching the electrical conductivity in the first principle stretching direction (X_1 direction) is higher than that in the other two transverse directions (X_2 and X_3 directions) due to the realignment of CNTs along the stretching direction X_1. With the increase of the stretching ratio, the electrical conductivity decreases in all the three directions while the electrical conductivity in the X_2 direction is approaching that in the X_1 direction until the equal biaxial stretching condition is reached, indicating an increasing randomness of the CNT distribution in the bi-axial stretching plane (X_1-X_2 plane). This scenario was commented by Shen et al. (2012) that the second stretching enables CNTs to have less chance to be parallel, enhancing the randomness of the CNTs distribution in the bi-axial stretching plane.

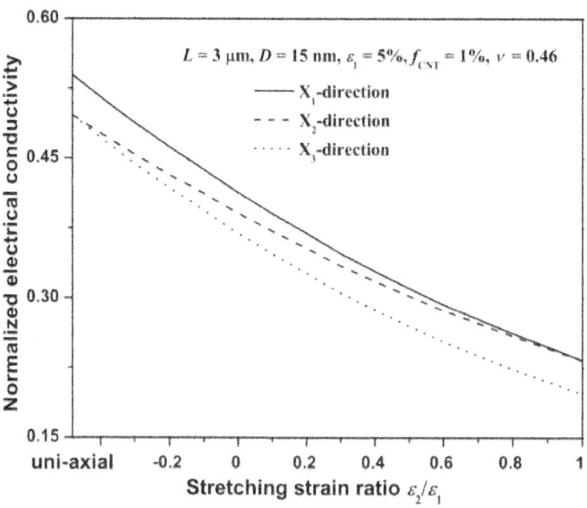

Figure 4.7: Variation of normalized electrical conductivity with stretching strain ratio.

Figure 4.8 investigates the variation of the electrical conductivity with stretching strain for different CNT volume fractions. It is observed that with a fixed stretching strain ratio ($\varepsilon_2/\varepsilon_1 = 0.5$), the electrical conductivity decreases with the stretching strain in the X_1 direction and the decreasing rate is enhanced for the composite with lower CNT volume fraction. This suggests that the electrical conductivity of the composite with lower CNT concentration is more sensitive to stretching. The reason behind such sensitivity relies on the combined dependency of the separation distance among CNTs and the percentage of percolated CNTs upon the CNT volume fraction as indicated by Eqs. (2.3) and (2.8). As argued by Jiang et al. (2007) that since only a few conductive networks exist in composites with lower CNT concentration, any small change in the networks, including stretching induced separation distance between CNTs, the percolation threshold and the percentage of percolated CNTs, will influence the electrical conductivity significantly. In contrast, any change in the conductive networks for composites with higher CNT concentration will have relatively limited effects on the electrical conductivity due to the existence of large amount of conductive networks.

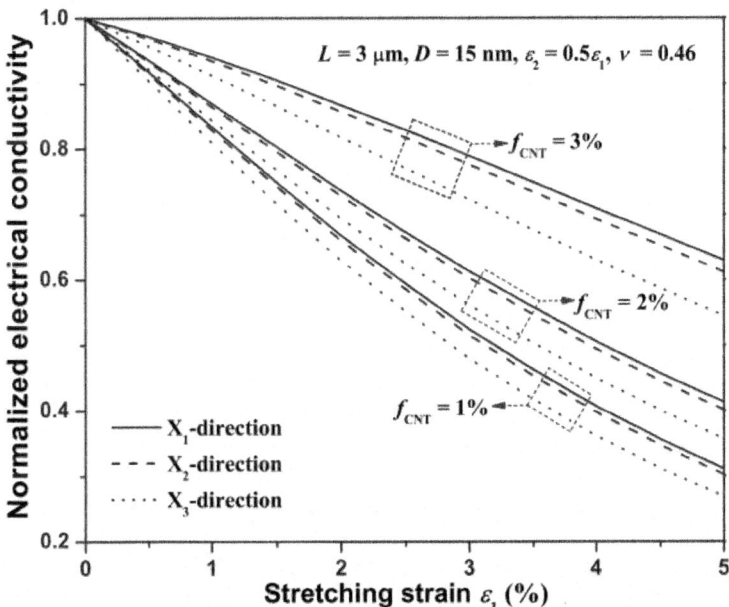

Figure 4.8: Variation of normalized electrical conductivity with stretching strain (a) X_1- direction; (b) X_2-direction; (c) X_3-direction.

Figure 4.9 demonstrates the effects of CNT size on the electrical conductivity of the composite under a bi-axial stretching. It is obvious that for the composite with a fixed CNT concentration (f_{CNT} = 1%) and fixed diameter of CNTs, the electrical conductivity of the composite with shorter CNTs decreases more significantly. It can be understood from the fact that the percolation threshold is highly dependent on the filler's aspect ratio, which results in the change of the separation distance among CNTs and the percentage of the percolated CNTs. As demonstrated in Figure 4.4, for the same CNT volume fraction the percentage of the percolated CNTs is reduced more significantly for the composite with shorter CNTs than that with longer CNTs. Therefore, it is concluded that the composite with shorter CNTs is more susceptible to decrease its electrical conductivity under stretching.

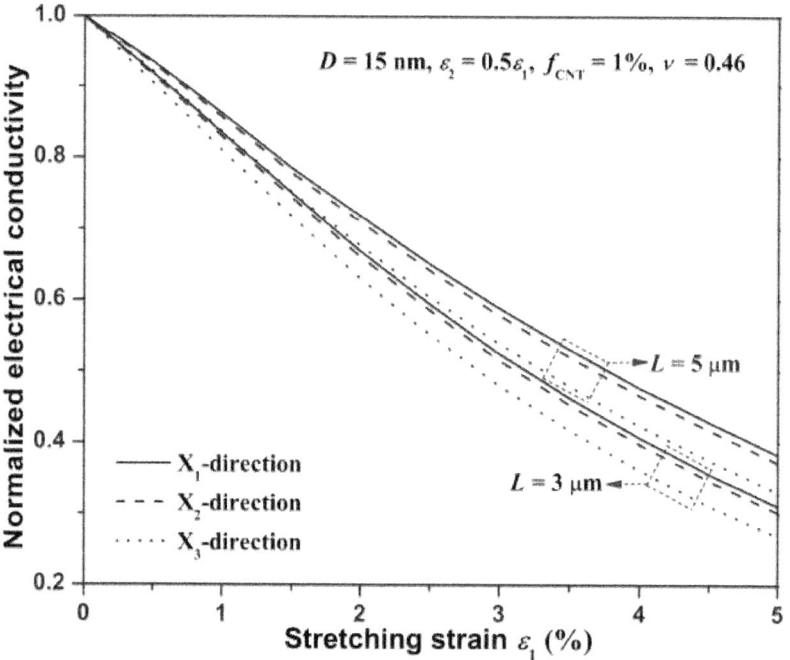

Figure 4.9: Variation of normalized electrical conductivity with stretching strain for different CNT lengths (a) X_1-direction; (b) X_2-direction; (c) X_3-direction.

It should be noted that in this current work we assume the CNTs dispersed in polymers are straight and randomly distributed in polymers before stretching. However, due to van der Waals forces among CNTs and low CNT bending stiffness, CNTs usually agglomerate and exist in a wavy state. Therefore, a more comprehensive model with the incorporation of the effects of agglomeration and waviness needs to be further developed.

Conclusions

In this work, stretching induced volume expansion, re-orientation of CNTs and change of conductive networks in the composite are incorporated into a mixed micromechanics model to study the bi-axial stretching effects upon the overall electrical conductivity of a CNT-polymer composite. Compared to a uni-axial stretching in which CNTs tend to realign along the uni-stretching direction, bi-axial stretching increases the randomness of CNTs' distribution in the bi-axial stretching plane, while causes more breakdown of conductive networks. Therefore, the simulation results conclude that the bi-axial stretching decreases the electrical conductivity of the composite and the decreasing rate is enhanced by the increasing strain in the second principle stretching direction due to the dominant role of the change of the conductive networks. It is also observed that the overall electrical conductivity of the CNT-polymer composite is more sensitive to stretching for the composites with lower CNT concentration and smaller CNT aspect ratio. The modeling in the current work is expected to provide an increased understanding on the stretching effects upon the electrical conductivity of CNT-polymer composites.

INFLUENCE OF CNT WAVINESS UPON THE ELECTRICAL CONDUCTIVITY OF CNT-POLYMER COMPOSITES UNDER UNI-AXIAL STRETCHING

Since the discovery of carbon nanotubes (CNTs) in 1991 (Iijima, 1991), their extraordinary electrical properties as well as high aspect ratio have made CNTs one of the most preferred conductive fillers to develop multi-functional conductive polymer composites. Compared to traditional conductive fillers, such as carbon black and metals, a very small amount of CNTs added into polymers, which are usually insulators, can significantly improve the electrical properties of the composites while still keep the beneficial features of the polymers, including flexibility, large deformation, easy processability, good chemical and biological compatibilities, etc (Yu et al., 2009; Nambiar and Yeow, 2011; Shang et al., 2011).

It has been experimentally and theoretically demonstrated that the electrical conductivity of CNT-polymer composites display a percolation-like behavior (Ounaies et al., 2003; Kim et al., 2005; Gojny et al., 2006; Yan et al., 2007; Feng and Jiang, 2013), i.e., the electrical conductivity increases abruptly when the CNT concentration is greater than a critical volume fraction, which is usually referred as percolation threshold. Such percolation-like behavior is attributed to two conductivity mechanisms: nanoscale electron hopping and microscale conductive networks (Deng and Zheng, 2008; Chang et al., 2009;

Zhang et al., 2009; Lu et al., 2010; Feng and Jiang, 2013). It is well accepted that the contribution of these two mechanisms is highly dependent on CNT volume fraction. For example, when the CNT concentration is below the percolation threshold, CNTs in the polymer are electrically independent and only the nanoscale electron hopping contributes to the electrical conductivity. However, when the CNT concentration is above the percolation threshold, certain percent of CNTs will form microscale conductive networks, which will significantly contribute to the electrical conductivity in addition to the nanoscale electron hopping. The dependency of the overall electrical conductivity of the CNT-polymer composites upon the CNT concentration and the conductivity mechanisms has been extensively explored (Kim et al., 2005; Gojny et al., 2006; Li et al., 2007; Chang et al., 2009; Seidel and Lagoudas, 2009; Takeda et al., 2011; Feng ang Jiang, 2013).

It should be mentioned that most of the existing theoretical studies of the electrical properties were focused on the composites with the assumption of straight conductive fillers. However, it is experimentally observed that CNTs dispersed in polymers are usually not straight but rather have a degree of waviness due to their large aspect ratio and low bending stiffness (Li et al., 2008). The waviness of the CNTs was suggested to have considerable effects on the overall electrical properties of the composites (Yi et al., 2004; Li et al., 2008; Takeda et al., 2011; Yu et al., 2013). Therefore, the consideration of CNT waviness effects is necessary and essential for predicting the electrical conductivity of the CNT-polymer composites in realistic cases. In the literature, people have made some efforts in incorporating the CNT waviness effects to investigate the electrical properties of the composites. For example, assuming wavy CNTs with a simple sinusoidal shape, Yi et al. (2004), Berhan and Sastry (2007) and Fisher et al. (2003) investigated the effect of the waviness on the percolation onset of CNT-polymer composites. It was observed that the CNT waviness increases the percolation threshold of the composites. Approximating wavy CNTs as elongated polygons, Li et al.'s (2008) computational simulation on the CNT waviness effects showed that the waviness increases the percolation threshold and decreases the electrical conductivity and elastic stiffness of the composites. By introducing a length ratio into a simplified micromechanics model, Deng and Zheng (2008) and Takeda et al. (2011) investigated the effect of the non-straightness of the CNTs. It was found that the non-straightness of CNTs could significantly decrease the electrical conductivity of the composites. Nevertheless, these existing studies focused on the percolation behavior of the composites while did not combine any stretching effect, which is an important consideration for the application of the composites as stretchable electronics. Therefore, the objective of the current work will focus on investigating the CNT waviness effect on the electrical conductivity of CNT-polymer composites under a uni-axial stretching through applying our previously developed micromechanics model.

Modeling and formulation

In this section, CNT waviness effects will be incorporated into the developed mixed micromechanics model (Feng and Jiang, 2013) by using straight CNTs with equivalent conductive performance. The iterations of the mixed micromechanics model with both nanoscale and microscale conductivity mechanisms, the stretching effects and the equivalence of wavy CNTs to straight CNTs are listed in the follows.

Nanoscale composite cylinder model

The nanoscale conductive mechanism of the CNT-polymer composites attributes to the electron hopping, which was captured by the introduction of an interphase outside the CNT to form an effective composite cylinder. Such a effective composite cylinder model has been widely adopted by researchers (Hashin, 1990; Yan et al., 2007; Seidel et al., 2009) as shown in Figure 5.1 in which the straight CNT is identified with length L_{CNT}^{str} and diameter D_{CNT}^{str} and the surrounding interphase is described with thickness t. By applying the law-of-mixture rule (Taya, 2005; Yan et al., 2007), the composite cylinder can be homogenized into an equivalent solid filler with effective longitudinal and transverse electrical conductivity determined as,

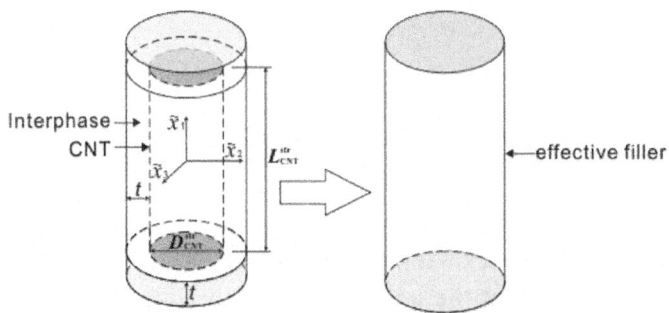

Figure 5.1: Sketch of a composite cylinder.

$$\bar{\sigma}^{L} = \frac{\left(L_{CNT}^{str}+2t\right)\sigma_{Int}\left[\sigma_{CNT}^{str}D_{CNT}^{str\,2}+4\sigma_{Int}\left(D_{CNT}^{str}t+t^{2}\right)\right]}{2\sigma_{CNT}^{str}D_{CNT}^{str\,2}t+8\sigma_{Int}\left(D_{CNT}^{str}t+t^{2}\right)t+\sigma_{Int}L_{CNT}^{str}\left(D_{CNT}^{str}+2t\right)^{2}}$$

$$\bar{\sigma}^{T} = \frac{\sigma_{Int}}{L_{CNT}^{str}+2t}\left[L_{CNT}^{str}\frac{D_{CNT}^{str\,2}\sigma_{CNT}^{str}+2\left(\sigma_{CNT}^{str}+\sigma_{Int}\right)\left(t^{2}+D_{CNT}^{str}t\right)}{D_{CNT}^{str\,2}\sigma_{Int}+2\left(\sigma_{CNT}^{str}+\sigma_{Int}\right)\left(t^{2}+D_{CNT}^{str}t\right)}+2t\right]$$

$$(5.1)$$

where the subscripts "L" and "T" denote the longitudinal and transverse directions, respectively; and σ_{CNT}^{str} and σ_{Int} are the electrical conductivity of the straight CNT and the interphase, respectively. Correspondingly, we can have the electrical conductivity tensor of the effective filler as $\bar{\sigma} = \mathrm{diag}\left(\bar{\sigma}^L, \bar{\sigma}^T, \bar{\sigma}^T\right)$. . With the equivalence of the CNTs to the effective solid fillers, the volume fraction of the fillers in the composite was expressed in terms of the original volume fraction of straight CNTs f_{CNT}^{str} (Yan et al., 2007; Seidel et al., 2009), i.e.

$$f_{eff} = \frac{\left(D_{CNT}^{str} + 2t\right)^2 \left(L_{CNT}^{str} + 2t\right)}{D_{CNT}^{str\ 2} L_{CNT}^{str}} L_{CNT}^{str}.$$

(5.2)

It is obvious from Eq. (5.1) that the interphase properties, i.e., the interphase electrical conductivity and the interphase thickness, are essential to determine the electrical properties of the composite cylinder. These interphase properties are naturally believed to correlate to the electrical conductivity mechanisms and vary with the CNT volume fraction. Some researchers have indicated that the average separation distance da between the adjacent CNTs forming conductive networks follows a power law relation (Allaoui et al., 2008; Deng et al., 2008; Takeda et al., 2011), i.e.,

$$d_a = \left(\frac{f_c^{str}}{f_{CNT}^{str}}\right)^{1/3} d_c$$

(5.3)

where f_c^{str} is the percolation threshold of the composites with straight CNTs. Obviously, this separate distance decreases with the increase of the CNT volume fraction f_c^{str} after the percolation threshold. Existing experiments (Li et al., 2007; Takeda et al., 2011) and simulations suggested that the critical separation distance for the formation of conductive networks is $d_c = 1.8$ nm. From the previous works (Simmons, 1963; Takeda et al., 2011; Feng and Jiang, 2013), the thickness and the conductivity of the interphase for conductive networks were expressed as:

$$t = \frac{1}{2}\left(\frac{f_c^{str}}{f_{CNT}^{str}}\right)^{1/3} d_c$$

(5.4)

and

$$\sigma_{\text{Int}} = \frac{d_a}{aR_{\text{Int}}(d_a)},$$

(5.5)

Where $R_{\text{Int}}(d_a) = \frac{d_a \hbar^2}{ae^2(2m\lambda)^{1/2}} \exp\left(\frac{4\pi d_a}{\hbar}(2m\lambda)^{1/2}\right)$ is the tunneling-type
contact resistance between two CNTs. $\lambda = 5.0$ eV is the potential barrier
height for CNTs dispersed in most polymers. $m = 9.10938291 \times 10^{-31}$ kg
and $e = -1.602176565 \times 10^{-19}$ C are mass and electric charge of an electron,
respectively. a is the contact area of CNTs and $\hbar = 6.626068 \times 10^{-34}$ m^2 ·kg·s^{-1}
is the Planck constant. In contrast, for the CNTs without forming conductive
network, the separation distance is larger than the critical distance 1.8 nm.
However, this distance was assumed as a constant for the CNTs without
forming conductive networks. Therefore, the thickness and the electrical
conductivity of the interphase for CNTs without forming conductive networks
can be approximated as (Feng and Jiang, 2013):

$$t = \frac{1}{2}d_c \text{ and } \sigma_{\text{Int}} = \frac{d_c}{aR_{\text{Int}}(d_c)}.$$

(5.6)

Micromechanics model

Based on the composite cylinder model, the CNT-polymer composite becomes a
composite filled with effective solid fillers with effective electrical conductivity
$\tilde{\sigma}$ and effective volume fraction f. For the two-phase composite with uniformly
and randomly distributed fillers, the overall electrical conductivity can be
predicted by applying a micromechanics model with selecting a representative
volume element (RVE) containing enough fillers (shown in Figure 5.2). In
the RVE, $(\tilde{x}_1, \tilde{x}_2, \tilde{x}_3)$ and (X_1, X_2, X_3) are the local and global coordinate systems
to describe the position of the fibers, respectively. The orientation of any
individual filler is identified by two Euler angles, ϕ and θ. By averaging the
contribution of the fillers from all possible orientations in the RVE, the overall
electrical conductivity of the two-phase composite can be determined as (Taya,
1995; Entchev and Lagoudas, 2002; Odegard et al., 2003; Taya, 2005; Seidel
et al., 2009):

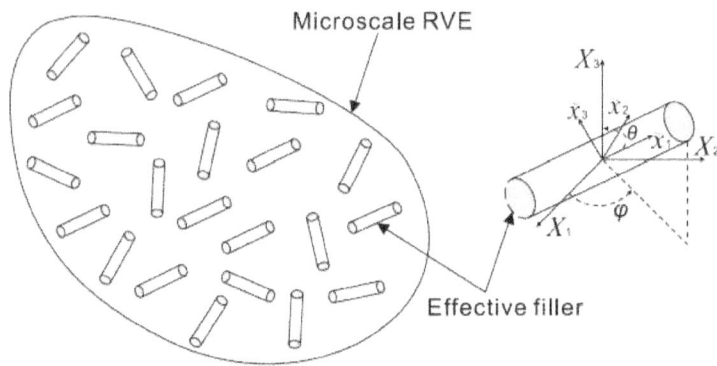

Figure 5.2: Sketch of a microscale RVE containing conductive fillers.

$$\sigma_{\text{eff}} = \sigma_{\text{m}} + \frac{\int_0^{2\pi} \int_0^{\pi} \rho(\varphi,\theta) f_{\text{eff}} (\sigma - \sigma_{\text{m}}) A \sin\theta \, d\theta \, d\varphi}{\int_0^{2\pi} \int_0^{\pi} \rho(\varphi,\theta) \sin\theta \, d\theta \, d\varphi}$$

(5.7)

where $\rho(\ ,\theta)$ is the orientation distribution function (ODF), which denotes the probability of the distribution of the fillers with a given orientation; A is the electric field concentration tensor; σ_{m} is the electrical conductivity tensor of the polymer and σ is the electrical conductivity tensor of the effective filler in the global coordinate system, which can be transformed from the electrical conductivity tensor in the local coordinate system as:

$$\sigma = Q^T \tilde{\sigma} Q$$

(5.8)

where Q is the transformation matrix given as (Entchev et al., 2002):

$$Q = \begin{vmatrix} \sin\theta\cos\varphi & -\cos\theta\cos\varphi & \sin\varphi \\ \sin\theta\sin\varphi & -\cos\theta\sin\varphi & -\cos\varphi \\ \cos\theta & \sin\theta & 0 \end{vmatrix}$$

(5.9)

Applying Mori-Tanaka method (Mori and Tanaka, 1973), the concentration tensor A in Eq. (5.7) can be defined as (Taya, 2005; Seidel and Lagoudas, 2009; Feng and Jiang, 2013):

$$A = QA^{\text{dil}}Q^T \left\{ (1 - f_{\text{eff}})\delta + f_{\text{eff}} \frac{\int_0^{2\pi} \int_0^{\pi} f_{\text{eff}} \{QA^{\text{dil}}Q^T\} \rho(\varphi,\theta) \sin\theta \, d\theta \, d\varphi}{\int_0^{2\pi} \int_0^{\pi} \rho(\varphi,\theta) \sin\theta \, d\theta \, d\varphi} \right\}^{-1},$$

(5.10)

where A ^{dil} is defined as:

$$A^{dil} = \left\{ \delta + S\left(\sigma_m\right)^{-1}\left(\tilde{\sigma} - \sigma_m\right)\right\}^{-1}$$

(5.11)

with δ being the Kronecker delta tensor. S is the Eshelby tensor of the effective filler, which is given by:

$$S = \begin{bmatrix} S_{11} & 0 & 0 \\ 0 & S_{22} & 0 \\ 0 & 0 & S_{33} \end{bmatrix},$$

(5.12)

Where

$$S_{22} = S_{33} = \frac{A_{re}}{2\left(A_{re}^2 - 1\right)^{3/2}} \left[A_{re}\left(A_{re}^2 - 1\right)^{1/2} - \cosh^{-1} A_{re}\right]$$

(5.13)

with A_{ro} being the aspect ratio of the effective filler, i.e $A_{re} = \left(L_{CNT}^{str} + 2t\right)\left(D_{CNT}^{str} + 2t\right)$, and $S_{11} = 1 - 2S_{22}$.

According to Deng and Zheng's (2008) argument, conductive networks start to contribute to the electrical conductivity after the percolation and the percentage of CNTs forming conductive networks can be estimated as:

$$\xi = \frac{\left(f_{CNT}^{str}\right)^{1/3} - \left(f_c^{str}\right)^{1/3}}{1 - \left(f_c^{str}\right)^{1/3}} \quad \left(f_{CNT}^{str} \geq f_c^{str}\right)$$

(5.14)

Considering the percolation process of CNT-polymer composites, i.e., both electron hopping and conductive networks contribute to the electrical conductivity of the composite after the percolation, while only electron hopping is responsible for the electrical conductivity before the percolation, the overall electrical conductivity can be determined by a mixed micromechanics model as (Feng andg Jiang, 2013):

$$\sigma_{eff} = \begin{cases} \sigma_m + \dfrac{\int_0^{2\pi}\int_0^{\pi}\left\{f_{eff}\left(\sigma_{EH} - \sigma_m\right)A_{EH}\right\}\rho(\varphi,\theta)\sin\theta d\theta d\varphi}{\int_0^{2\pi}\int_0^{\pi}\rho(\varphi,\theta)\sin\theta d\theta d\varphi}, & f_{CNT}^{str} < f_c^{str} \\[3em] \sigma_m + \left(1-\xi\right)\dfrac{\int_0^{2\pi}\int_0^{\pi}\left\{f_{eff}\left(\sigma_{EH} - \sigma_m\right)A_{EH}\right\}\rho(\varphi,\theta)\sin\theta d\theta d\varphi}{\int_0^{2\pi}\int_0^{\pi}\rho(\varphi,\theta)\sin\theta d\theta d\varphi} \\[3em] +\xi\dfrac{\int_0^{2\pi}\int_0^{\pi}\left\{f_{eff}\left(\sigma_{CN} - \sigma_m\right)A_{CN}\right\}\rho(\varphi,\theta)\sin\theta d\theta d\varphi}{\int_0^{2\pi}\int_0^{\pi}\rho(\varphi,\theta)\sin\theta d\theta d\varphi}, & f_{CNT}^{str} \geq f_c^{str} \end{cases}$$

(5.15)

where the superscripts "EH" and "CN" denote the terms contributed by the electron hopping and the conductive networks, respectively. When conductive networks are formed, several CNTs will be electrically connected to each other while not in physical contact due to van der Waals forces between CNTs. Thus, the effective aspect ratio of the formed networks can be taken as infinite due to the large aspect ratio of CNTs. However, quantities associated with the electron hopping correspond to the real effective filler aspect ratio as defined. Correspondingly, AEH and ACN in Eq. (5.15) can be determined from Eq. (5.10) by using different aspect ratios for CNTs.

Equivalence of wavy carbon nanotubes

As mentioned, CNTs dispersed in polymers usually exist in a wavy state. The basic idea of considering waviness effects is to convert wavy CNTs into equivalent straight fillers as adopted by previous studies (Deng and Zheng, 2008; Takeda et al., 2011). In the current work, we assume the wavy CNT with length of wavy L_{CNT}^{wavy} and diameter of D_{CNT}^{wavy} has a sinusoidal shape as illustrated in Figure 5.3. The sinusoidal shape is described by two parameters, amplitude a_w and wavelength λ_w, as:

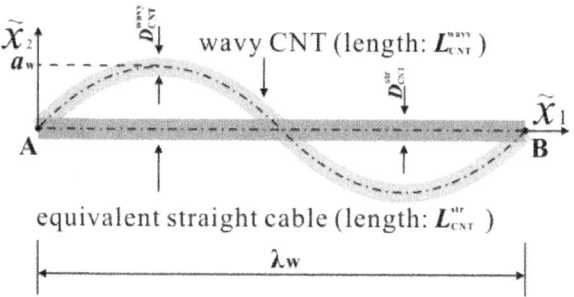

Figure 5.3: Sketch of a wavy CNT and its equivalent straight counterpart.

$$\tilde{x}_2 = a_w \sin(\frac{2\pi\tilde{x}_1}{\lambda_w}).$$

(5.16)

To characterize the waviness of the CNT, a parameter referred as waviness ratio is defined as following (Fisher et al., 2003)

$$\chi = \frac{a_w}{\lambda_w}.$$

(5.17)

Combining Eqs. (5.16)–(5.17), the length of the wavy CNT can be correlated to the amplitude and the wavelength as:

$$L_{CNT}^{wavy} = \int_0^{\lambda_w} \sqrt{1 + \left(\frac{d\tilde{x}_2}{d\tilde{x}_1}\right)^2}\, d\tilde{x}_1 = \lambda_w \int_0^1 \sqrt{1 + \left(2\pi\chi\cos(2\pi u)\right)^2}\, du \,,$$

(5.18)

Where

$$u = \tilde{x}_1 / \lambda_w \,.$$

The equivalence of the wavy CNTs into straight fillers is carried out according to the rules from the perspective of electrical conduction mechanisms. When the wavy CNT as shown in Figure 5.3 is subjected to a two-ends potential difference, $\Delta V = V_2 - V_1$, the CNT can be taken as an electrical CNT cable and the electrical flux J through the CNT can be approximated (Deng and Zheng, 2008)

$$J = \sigma_{CNT}^{wavy} \frac{\Delta V}{L_{CNT}^{wavy}}\,.$$

(5.19)

On the other side, the wavy CNT can be regarded as an equivalent straight CNT with the capability of conducting the same electrical flux J between the two ends of the CNT with a minimum distance L_{CNT}^{str} which equals to the wavelength λ_w of the wavy CNT. Such an analysis can produce the effective electrical conductivity of the equivalent straight CNT as:

$$\sigma_{CNT}^{str} = \alpha \sigma_{CNT}^{wavy} \,,$$

(5.20)

Where $\alpha = L_{CNT}^{str} / L_{CNT}^{wavy}$ is the length ratio. It is suggested that the length ratio α usually ranges from 0.5 to 1.0 (Fisher et al., 2003; Takeda et al., 2011) for composites with wavy CNTs. Particularly, α equals unity for straight CNTs. To fully convert the wavy CNT into an equivalent straight CNT, in addition to conducting the same electrical flux, the equivalent straight CNT should have the capability of transporting the same amount of electrical charges from end A to end B as that for the wavy CNT within a certain time interval, from which we can have the following relationship

$$R_{CNT}^{str} = R_{CNT}^{wavy} \,,$$

(5.21)

Where R_{CNT}^{wavy} and R_{CNT}^{str} are the resistance of the wavy CNT and the equivalent straight CNT respectively. Combining Eqs. (5.20) and (5.21), it can be easily obtained that the diameter of the equivalent CNT is the same as that of the wavy CNT, i.e., str wavy $D_{CNT}^{str} = D_{CNT}^{wavy}$, which was also adopted by Deng and Zheng (2008). Due to the reduced equivalent length, the volume fraction of the equivalent straight CNT in the composites reduces to str wavy $f_{CNT}^{str} = \alpha f_{CNT}^{wavy}$, where f_{CNT}^{wavy} is the volume fraction of the wavy CNTs.

Replacing wavy CNTs with equivalent straight CNTs, the electrical conductivity and the geometric property of the CNTs in the formulation of the previous sections take the values for the equivalent straight CNTs.

Uni-axial stretching induced changes

When composites with fillers are under stretching, it has been accepted that stretching induces three changes in the composites, including volume expansion of the composites, re-orientation of the fillers and change in conductive networks (Taya et al., 1998; Taya, 2005; Feng and Jiang, 2014). In the following, the stretching effects on the electrical conductivity of the composites will be investigated by characterizing these three changes and incorporating them into the micromechanics model. Figure 5.4 shows a cell containing an effective filler under a uni-axial stretching with stretching strain ε in the X_3 direction. After stretching, the dimensions of the beam become (Feng and Jiang, 2014):

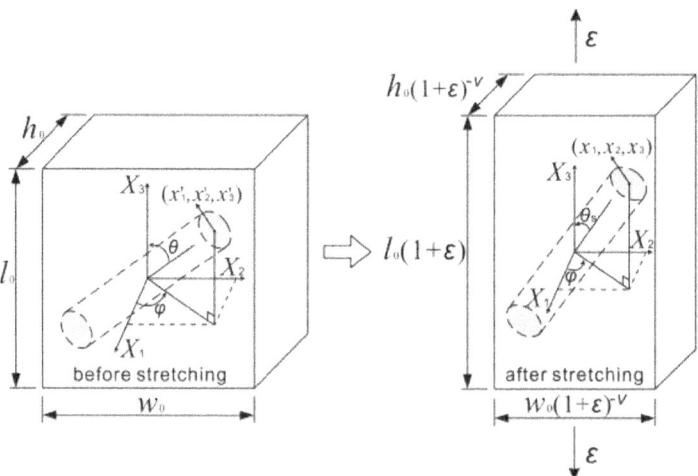

Figure 5.4: Orientation description of a conductive filler in a cell.

$$l = l_0(1+\varepsilon), \quad w = w_0(1+\varepsilon)^{-\nu} \text{ and } h = h_0(1+\varepsilon)^{-\nu}, \tag{5.22}$$

where l_0, w_0 and h_0 are the initial length, width and height of the cell, respectively, and v is the Poisson's ratio of the composites. Due to the expansion of the composites, the volume fraction of the effective filler after stretching reduces to (Feng and Jiang, 2014)

$$f_{\text{update}} = f_{\text{eff}} / (1+\varepsilon)^{1-2v} .$$

(5.23)

In addition to volume expansion, the fillers in the cell are also re-oriented, resulting in the change of the polar angel, i.e.,

$$\theta_s = \arctan\left[(1+\varepsilon_3)^{-1-v} \cdot \tan\theta \right].$$

(5.24)

The change of the other Euler angle, φ, can be neglected here under the uni-axial stretching condition (Kuhn and Grün, 1942; Feng et al., 2014). Such a re-orientation will disrupt the random distribution of fillers before the stretching, i.e., the ODF in Eq. (5.7) changes from unity to the following expression (Feng and Jiang, 2014)

$$\rho(\varphi,\theta_s) = \frac{(1+\varepsilon)^{\frac{1+v}{2}}}{\left[(1+\varepsilon)^{-(1+v)} \cos^2\theta_s + (1+\varepsilon)^{1+v} \sin^2\theta_s \right]^{3/2}} .$$

(5.25)

Our previous work demonstrated that after the uni-axial stretching, fillers in the polymer tend to re-align along the stretching direction and the increase of the Poisson's ratio can enhance such a re-alignment (Feng and Jiang, 2014a and 2014b). In addition, stretching also induces change in conductive networks, including the increase of the percolation threshold and the separation distance between CNTs (Taya et al., 1998; Lin et al., 2010; Tallman and Wang, 2013). Based on the excluded volume method (Balberg et al., 1984; Celzard et al., 1996; Tallman et al., 2013), the percolation threshold can be determined as:

$$f_c^{\text{str}} = 1 - \exp\left(-\frac{\langle V_{\text{ex}} \rangle V_{\text{CNT}}^{\text{str}}}{\langle V_e \rangle} \right)$$

(5.26)

Where $\langle V_{\text{ex}} \rangle$ is the total average excluded volume of equivalent straight CNT, $V_{\text{CNT}}^{\text{str}}$ is the volume of the equivalent straight CNT and $\langle V_e \rangle = \frac{4\pi}{3} \left(D_{\text{CNT}}^{\text{str}} \right)^3 + 2\pi \left(D_{\text{CNT}}^{\text{str}} \right)^2 L_{\text{CNT}}^{\text{str}} + 2 D_{\text{CNT}}^{\text{str}} \left(L_{\text{CNT}}^{\text{str}} \right)^2 \langle \sin\gamma \rangle$ is the average excluded volume of the equivalent straight CNT with Dstr and Lstr being the diameter and the length, respectively, and γ being the angle between two CNTs.

Fitting from Monte Carlo simulation, Tallman and Wang (2013) proposed an approximate expression for $\langle \sin \gamma \rangle_\mu$, i.e.,

$$\langle \sin \gamma \rangle_\mu = 0.018 \theta_\mu^5 + 0.021 \theta_\mu^4 - 0.234 \theta_\mu^3 - 0.015 \theta_\mu^2 + 0.909 \theta_\mu \tag{5.27}$$

Where $\theta_\mu = \arcsin[(1+\varepsilon)^{-1-\nu}]$ is a critical polar angle representing the randomness of the filler distribution. From Eqs. (5.26) and (5.27), it can be seen that the stretching will change the percolation threshold due to the variation of the θ_μ with the stretching strain. To further consider the stretching effect on the percolation threshold, Tallman and his coworker (2013) proposed a linear expression for the total average excluded volume, i.e.,

$$\langle V_{ex} \rangle = 2.8 - 5.6 \frac{\langle \sin \gamma \rangle_\mu}{\pi} . \tag{5.28}$$

Combining Eqs. (5.26)-(5.28), the percolation threshold is cast as:

$$f_c^{str} = 1 - \exp\left[-\frac{\left(2.1\pi - 4.2\langle \sin \gamma \rangle_\mu\right) p}{4\pi + 6\pi p + 6p^2 \langle \sin \gamma \rangle_\mu} \right] \tag{5.29}$$

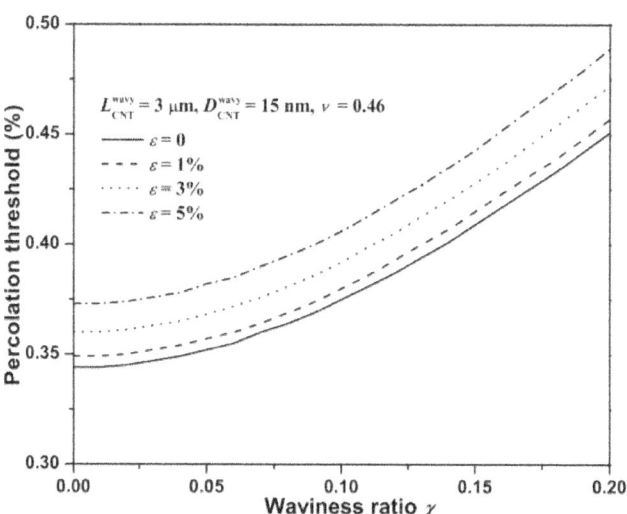

Where $p = L_{CNT}^{str} / D_{CNT}^{str}$ is the aspect ratio of the equivalent straight CNT. Figure 5.5 presents the variation of the percolation threshold with the waviness ratio. From this figure, it can be seen that the percolation threshold increases with the waviness ratio, which was also observed by researchers applying

Monte Carlo simulations (Yi et al., 2004; Li et al., 2008; Yu et al., 2013). In addition, the uni-axial stretching increases the percolation threshold as well.

Figure 5.5: Variation of percolation threshold with waviness ratio.

Results and discussion To study the CNT waviness effects on the electrical conductivity of the composites under a uni-axial stretching, we take the MWCNT/PEO (Polyethylene Oxide) composite in references (Park and Kim, 2006; Park et al., 2008) as the example material. The length and the diameter of the CNTs are taken as $L_{CNT}^{wavy} = 3$ µm and $D_{CNT}^{str} = 15$ nm, respectively. We adopt $\sigma_m = 1 \times 10^{-13}$ S/m and $\sigma_{CNT}^{wavy} = 100$ S/m as the electrical conductivity of the polymer and the wavy CNTs, respectively (Deng et al., 2008; Feng and Jiang, 2013). The Poisson's ratio of the composite is set as $v = 0.46$. In Park and co-workers' experiments (Park et al., 2006; Park et al., 2008), it was observed that MWCNTs with functionalized surface were well dispersed in the PEO matrix before the stretching. To investigate the waviness effects on the percentage of percolated CNTs, Figure 5.6 plots the variations of the normalized percentage of the percolated CNTs, $\zeta_N = \zeta_w/\zeta_0$, for a composite under a uni-axial stretching for different CNT volume fractions and different

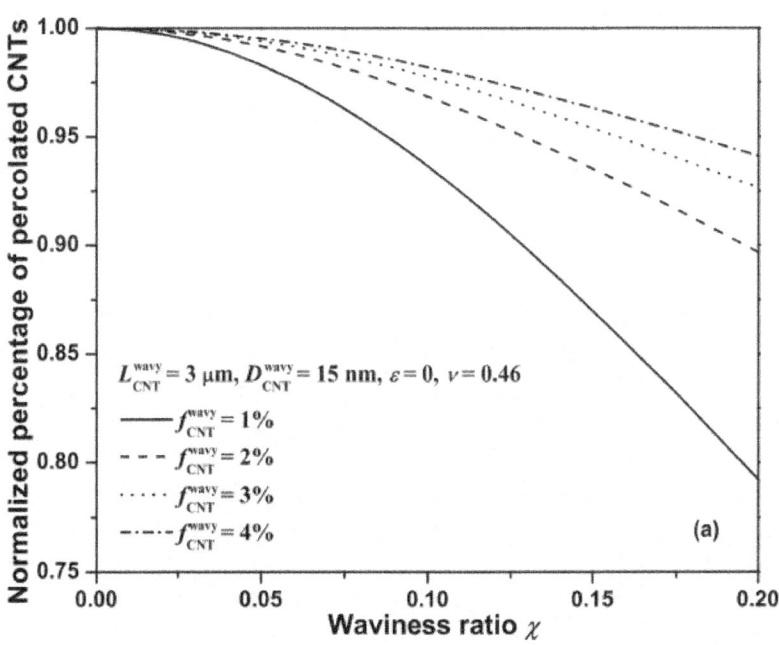

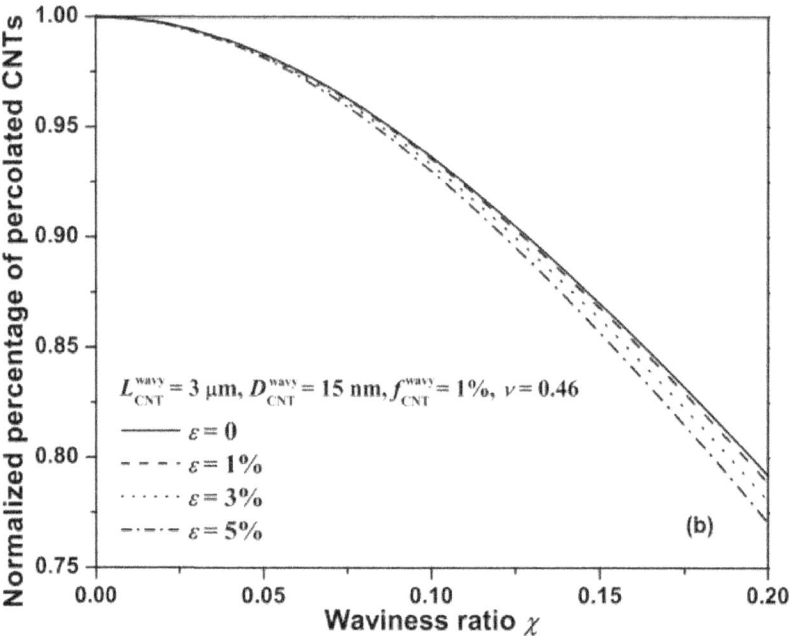

Figure 5.6: Variation of normalized percentage of percolated CNTs with waviness ratio (a) Different CNT volume fractions; (b) Different stretching strains.

stretching strains, where ξ_0 and ξ_w denote the percentages of percolated CNTs for the composite with straight and wavy CNTs, respectively. From the figure, it is found that the percentage decreases with the increase of the waviness ratio. It is also suggested in the figures that the percentage is more sensitive to the waviness for the composites with lower CNT volume fraction and larger stretching strain.

Figure 5.7 plots the variation of the electrical conductivity of the composite with CNT volume fraction for different waviness ratios. From this figure, it can be seen that the electrical conductivity of the composite increases with the increase of CNT volume fraction as expected. However, it is noticed that for the same CNT volume fraction, the electrical conductivity of the composite with wavy CNTs is much lower than that of the composite with straight CNTs. Such decrease in the electrical conductivity can be attributed to the waviness induced decrease in the effective electrical conductivity of the equivalent straight CNT and the change of the conductive networks.

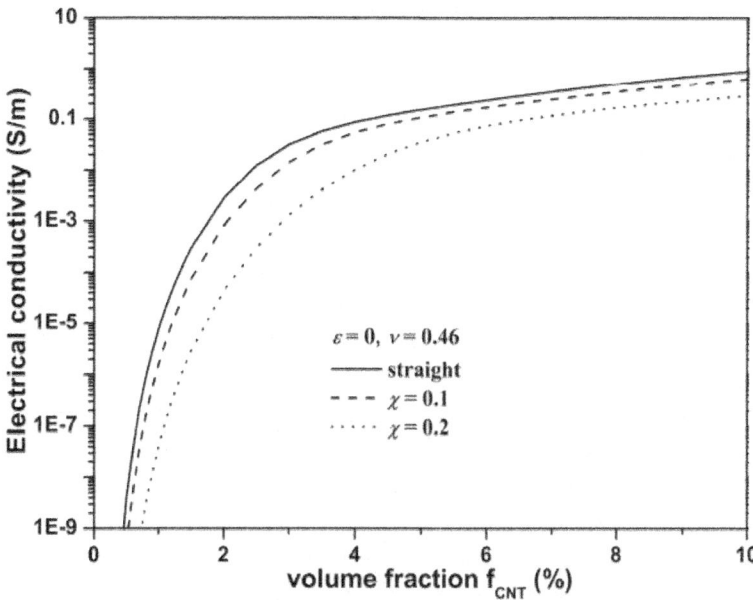

Figure 5.7: Variation of electrical conductivity with CNT volume fraction for different waviness ratios.

Figure 5.8 demonstrates the variation of the normalized electrical conductivity, $\sigma_N = \sigma_{eff}^{wavy} / \sigma_{eff}^{str}$ with the waviness ratio for different CNT volume fractions, where σ_{eff}^{wavy} and σ_{eff}^{str} denote the electrical conductivity of the composite with wavy CNTs and straight CNTs, respectively. It can be observed that the electrical conductivity decreases with the increase of the waviness ratio. In addition, it is found that with the same waviness ratio, the electrical conductivity decreases with the waviness ratio more significantly for the composite with lower CNT concentration. The sensitivity of the electrical conductivity to the CNT concentration can be explained by the variations of the normalized percentage with the waviness ratio as demonstrated in Figure 5.6a. For lower CNT volume fraction, since only a few conductive networks exist in the composite, any small change, such as CNT waviness, will have relatively more significant effect on the conductive networks, which results in more decrease in the overall electrical conductivity of the composites.

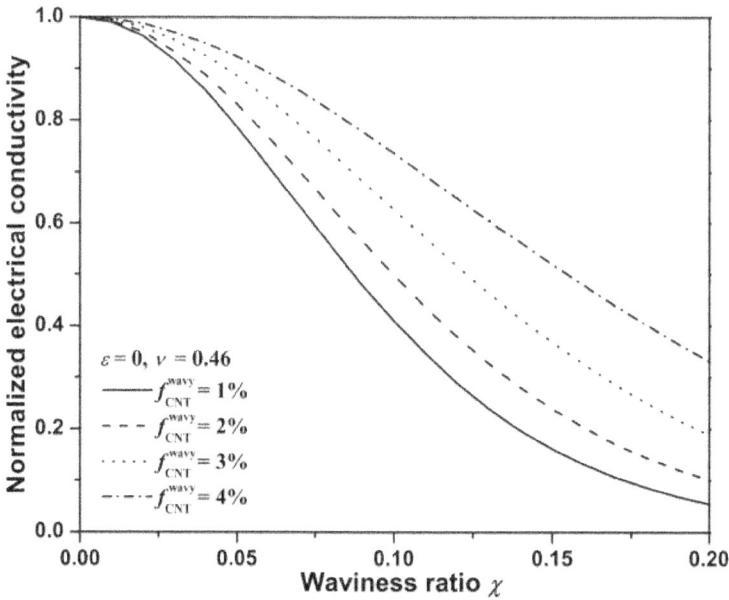

Figure 5.8: Variation of normalized electrical conductivity with waviness ratio for different volume fractions.

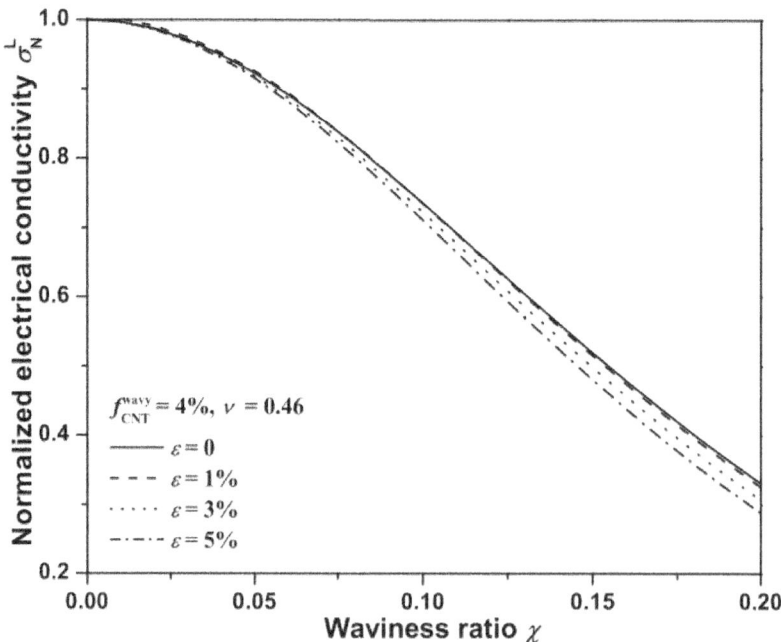

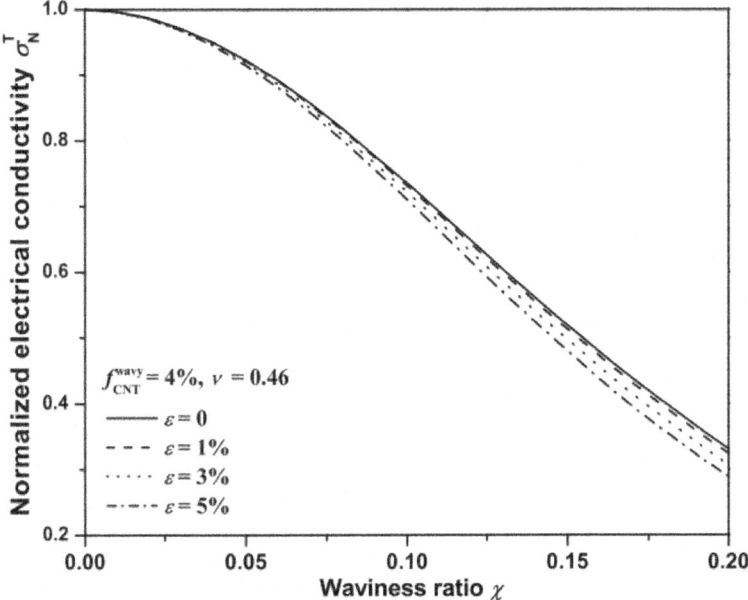

Figure 5.9: Variation of normalized electrical conductivity with waviness ratio for different stretching strains (a) Longitudinal direction; (b) Transverse direction.

Figure 5.9 shows the dependency of the normalized electrical conductivity on the waviness ratio for different stretching strains. From this figure, it can be seen that similar to Figure 5.7 and Figure 5.8, the electrical conductivity decreases with the CNT waviness ratio for both the longitudinal and the transverse directions. Furthermore, it is demonstrated that for a fixed CNT volume fraction the electrical conductivity is more sensitive to the waviness ratio for the composite under a larger uni-axial stretching strain. The stretching effects on the sensitivity of the electrical conductivity to the waviness can be interpreted by the demonstration in Figure 5.6b, in which the normalized percentage of percolated CNTs decreases more significantly with the waviness ratio for the composites with larger stretching strain.

CONCLUSIONS

Based on electrical conduction mechanisms, wavy CNTs characterized by a sinusoidal function are converted into equivalent straight CNTs with effective electrical conductivity and dimensions. Then CNT waviness effects on the electrical behaviors of CNT-polymer composites under a uni-axial stretching are studied through applying our previously developed mixed micromechanics model. The modeling results show that the electrical conductivity of the composites with wavy CNTs is much lower than that with straight CNTs. It is observed that the increasing waviness of CNTs enhances the stretching induced decrease of the electrical conductivity of the composites. It is also found that the electrical conductivity decreases with the increase of CNT waviness more significantly for the composites with lower CNT volume fraction and under larger stretching strain. The investigation in the current work is expected to provide qualitative predictions on the CNT waviness effects on the electrical conductivity of CNT-polymer composites.

CONCLUSIONS AND FUTURE WORK

Conclusions

Compared to conductive polymer composites with traditional fillers, such as carbon black and metals, the electrical conductivity of CNT-polymer composites demonstrates a significant percolation-like behavior. For engineering application, understanding of the mechanisms that underpin the macroscopic behaviors is essential for accurate prediction of the electrical behaviors of the composites. In addition, for the application of the composites as stretchable electronics, stretching and CNT waviness have significant effects upon the electrical conductivity of the composites. In this work, a mixed micromechanics model with the incorporation of the electrical conductivity mechanisms, stretching effects and CNT waviness is developed to settle the as-mentioned challenges. The contributions of the thesis include:

1. Based on percolation process of the CNT-polymer composites, the work is the first to develop a mixed form of micromechanics model, in which nanoscale electron hopping and microscale conductive networks are incorporated. Also instead of fixing the properties of the interphase surrounding the CNTs, such as the thickness and the electrical conductivity, the properties of the interphase vary with the CNT volume fraction.

2. We incorporate the stretching induced three changes, including volume expansion of composites, re-orientation of fillers and change in conductive networks, into our developed micromechanics model to investigate the uni-axial and bi-axial stretching effects on the electrical conductivity of the composites. Theoretical modeling work on the stretching effects upon the electrical properties of CNTpolymer composites through incorporating the stretching induced changes has not been reported thus far in the literature.

3. We investigate the CNT waviness effects on the electrical conductivity of the composites under uni-axial stretching through converting wavy

CNTs into equivalent straight fillers and the developed micromechanics model. There exists some work on considering CNT waviness effects on mechanical properties of the composites. However, limited theoretical work has been found on study the CNT waviness effects on the electrical conductivity of the composites.

Based on the developed model and the work done, we have some conclusions, which are listed in the following:

1. The developed micromechanics model is validated by comparing modeling results with experimental data. It was found that both electron hopping and conductive networks contribute to the electrical conductivity of the nanocomposite, while conductive networks become dominant after the percolation. Meanwhile, it was observed that both CNT length and diameter significantly affect the percolation concentration of nanocomposites, while having moderate effects on the overall electrical conductivity of the nanocomposites after the percolation.

2. Modeling results show that uni-axial stretching decreases the electrical conductivity in both the longitudinal and transverse directions. The electrical conductivity is more sensitive to the stretching for composites with lower CNT concentration than those with higher CNT concentration. It is also indicate that the stretching induced change in conductive networks plays a dominant role for the variation of the electrical conductivity along the stretching direction.

3. The investigation found that the bi-axial stretching decreases the electrical conductivity of the composites and the decreasing rate is enhanced by the increasing strain in the second principle stretching direction due to the dominant role of stretching induced change in conductive networks. Compared to uni-axial stretching, in which CNTs tend to be re-aligned along the one stretching direction, bi-axial stretching enables CNTs to be re-oriented in the two stretching directions and increases the randomness of CNTs' distribution in the stretching plane. It is also observed that the composites under bi-axial stretching are more sensitive to stretching with lower CNT concentration and smaller aspect ratio.

4. It concludes that the electrical conductivity of the composites increases with the CNT concentration while the electrical conductivity of the composites with wavy CNTs is much lower than that with straight CNTs. The waviness decreases the electrical conductivity of the composites in both the stretching and the transverse direction. It is also found that the electrical conductivity is more sensitive to CNT waviness for the composites with lower CNT volume fraction and larger stretching strain.

Future work

The current work developed a mix micromechanics model to investigate the overall electrical conductivity of the CNT-polymer composites. The work is expected to provide theoretical predictions and better understanding on the electrical conductivity of the composites and useful guidelines for the design and optimization of conductive polymer nanocomposites. However, there exist limitations for the developed model in the current work, which include:

1. The micromechanics model developed in this current work is based on small strain formulation, which used linear kinematics for the traditional micromechanics theory. Therefore, the model is not applicable for composites under finite deformation.

2. It is assumed that the intrinsic electrical conductivity of CNTs does not change under small stretching strain. However, it is argued that the intrinsic electrical conductivity of CNTs would vary due to CNT's elongation when the composite is under stretching. The assumption may cause inaccuracy on the model's prediction, particularly for the composites under large stretching.

Based on the existing limitations, the following work is proposed:

1. For certain CNT–polymer composites with the capability of undergoing large deformation, such as CNT–rubber composites, linear kinematics assumed for traditional micromechanics theory may be violated. Therefore, a more comprehensive micromechanics model based on finite deformation formulation needs to be developed.

2. When the composites are under large deformation, the elongation of the CNTs will become obvious resulting in the variation of the intrinsic electrical conductivity of the CNTs, which may become a significant contribution to the variation of the electrical conductivity of the composites. Such intrinsic variation of the electrical conductivity of the CNTs needs to be taken into account.

3. As materials are required to sustain more and more extreme loading conditions, in addition to the mechanisms of the translation of CNTs' extraordinary performance into the bulk composites, the effects of bonding/debonding between CNTs and the polymer molecules at the nanoscale and the failure of the composites under large deformation, are still unanswered due to the multiscale nature of the problems. Therefore, a multiscale model with the combination of continuum modeling and atomistic simulation needs to be developed in the future.

REFERENCES

1. Allaoui, A., Hoa, S. V. and Pugh, M. D., 2008. The electronic transport properties and microstructure of carbon nanofiber/epoxy composites. Compos. Sci. Technol. 68, 410–416.

2. Chang, L., Friedrich, K., Ye, L. and Toro, P., 2009. Evaluation and visualization of the percolating networks in multi-wall carbon nanotube/epoxy composites. J. Mater. Sci. 44, 4003–4012

3. Deng, F. and Zheng, Q. S., 2008. An analytical model of effective electrical conductivity of carbon nanotube composites. Appl. Phys. Lett. 92, 071902.

4. Du, F. M., Scogna, R. C., Zhou, W., Brand, S., Fischer, J. E. and Winey, K. I., 2004.

5. Nanotube networks in polymer nanocomposites: Rheology and electrical conductivity.

6. Macromolecules 37, 9048–9055.

7. Ebbesen, T. W., Lezec, H. J., Hiura, H., Bennett, J. W., Ghaemi, H. F. and Thio, T.,1996. Electrical conductivity of individual carbon nanotubes. Nature 382, 54–56.

8. Gao, L. and Li, Z., 2003. Effective medium approximation for two-component nonlinear composites with shape distribution. J. Phys.: Condens. Matter 15, 4397–4409.

9. Ghazavizadeh, A., Baniassadi, M., Safdari, M., Atai, A. A., Ahzi, S., Patlazhan, S. A., Gracio, J. and Ruch, D., 2011. Evaluating the effect of mechanical loading on the electrical percolation threshold of carbon nanotube reinforced polymers: a 3D MonteCarlo study. J. Comput. Theor. Nanosci. 8, 2087–2099.

10. Gojny, F. H., Wichmann, M. H. G., Fiedler, B., Kinloch, I. A., Bauhofer, W., Windle, A. H. et al., 2006. Evaluation and identification of electrical and thermal conduction mechanisms in carbon nanotube/epoxy composites. Polymer 47, 2036–2045.

11. Golosova, A. A., Adelsberger, J., Sepe, A., Niedermeier, M. A., Lindner, P., Funari, S. S. et al., 2012. Dispersions of Polymer-Modified Carbon Nanotubes: A Small-Angle Scattering Investigation. J. Phys. Chem. C 116,15765−15774

12. Grimmett, G., 1999. Percolation. Springer Verlag, Berlin. Hashin, Z., 1990. Thermoelastic properties and conductivity of carbon/carbon fiber composites. Mech. Mater. 8, 293–308.

13. Hatta, H. and Taya, M., 1985. Effective Thermal-Conductivity of a Misoriented Short Fiber Composite. J. Appl. Phys. 58, 2478–2486.

14. Hu, N., Masuda, Z., Yan, C., Yamamoto, G., Fukunaga, H. and Hashida, T., 2008. The electrical properties of polymer nanocomposites with carbon nanotube fillers. Nanotechnology 19, 215701.

15. Kim, Y. J., Shin, T. S., Choi, H. D., Kwon, J. H., Chung, Y. C. and Yoon, H. G., 2005.

16. Electrical conductivity of chemically modified multiwalled carbon nanotube/epoxy composites. Carbon 43, 23–30.

17. Kirkpatrick, S., 1973. Percolation and conduction. Rev. Mod. Phys. 45, 574–588.

18. Li, X., Rong, J. P. and Wei, B.Q., 2010. Electrochemical Behavior of Single-Walled

19. Carbon Nanotube Supercapacitors under Compressive Stress. Acs Nano 4, 6039–6049.

20. Li, C. Y., Thostenson, E. T. and Chou, T. W., 2007. Dominant role of tunneling resistance in the electrical conductivity of carbon nanotube-based composites. Appl. Phys. Lett. 91, 223114.

21. Li, J., Ma, P. C., Chow, W. S., To, C. K., Tang, B. Z. and Kim. J. K., 2007. Correlations between percolation threshold, dispersion state, and aspect ratio of carbon nanotubes. Adv. Funct. Mater. 17, 3207–3215.

22. Lu, W. B., Chou, T. W. and Thostenson, E. T., 2010. A three-dimensional model of electrical percolation thresholds in carbon nanotube-based composites. Appl. Phys. Lett. 96, 223106.

23. Ma, H. M. and Gao. X. L., 2008. A three-dimensional Monte Carlo model for electrically conductive polymer matrix composites filled with curved fibers. Polymer 49, 4230–4238.

24. Martin, C., Sandler, J. K. W., Shaffer, M. S. P., Schwarz, M-K., Bauhofer, W., Schulte, K. et al., 2004. Formation of percolating networks in multi-wall carbon-nanotube-epoxy composites. Compos. Sci. Technol. 64, 2309–2316.

25. McLachlan, D. S., Chiteme, C., Park, C., Wise, K. E., Lowther, S. E., Lillehei, P. T. et

26. al., 2005. AC and DC percolative conductivity of single wall carbon nanotube polymer composites. J. Polym. Sci. Pol. Phys. 43, 3273–3287.

27. Mori, T. and Tanaka, K., 1973. Average Stress in Matrix and Average Elastic Energy of Materials with Misfitting Inclusions. Acta Metall. Mater. 21, 571–574.

28. Odegard, G. M., Gates, T., Wise, K. E., Park, C. and Siochi, E., 2003. Constitutive modeling of nanotube-reinforced polymer composites. Compos. Sci. Technol. 63,1671– 1687.

29. Ounaies, Z., Park, C., Wise, K. E., Siochi, E. J. and Harrison, J. S., 2003. Electrical properties of single wall carbon nanotube reinforced polyimide composites. Compos. Sci. Technol. 63,1637–1646.

30. Ramasubramaniam, R., Chen, J. and Liu, H. Y., 2003. Homogeneous carbon nanotube/polymer composites for electrical applications. Appl. Phys. Lett. 83, 2928– 2930.

31. Seidel, G. D. and Lagoudas, D. C., 2009. A Micromechanics Model for the Electrical Conductivity of Nanotube-Polymer Nanocomposites. J. Compos. Mater. 43, 917–941.

32. Shiraishi, M. and Ata, M., 2001. Work function of carbon nanotubes. Carbon 39,1913– 1917.

33. Simmons, J. G., 1963. Generalized formula for the electric tunnel effect between similar electrodes separated by a thin insulating film. J. Appl. Phys. 34,1793–1803.

34. Song, Y. S. and Youn, J. R., 2005. Influence of dispersion states of carbon nanotubes on physical properties of epoxy nanocomposites. Carbon 43, 1378–1385

35. Subramanian, V., Zhu, H. and Wei, B., 2006. Synthesis and electrochemical characterizations of amorphous manganese oxide and single walled carbon nanotube composites as supercapacitor electrode materials. Electrochem. Commun. 8, 827–832.

36. Takeda, T., Shindo, Y., Kuronuma, Y. and Narita, F., 2011. Modeling and characterization of the electrical conductivity of carbon nanotube-based polymer composites. Polymer 52, 3852–3856.

37. Taya, M., 2005. Electronic composites. Cambridge University Press, Cambridge.

38. Yan, K. Y., Xue, Q. Z., Zheng, Q. B. and Hao, L. Z., 2007. The interface effect of the effective electrical conductivity of carbon nanotube composites. Nanotechnology 18, 255705.

39. Yu, C. J., Masarapu, C., Rong, J. P., Wei, B. Q. and Jiang, H. Q., 2009. Stretchable Supercapacitors Based on Buckled Single-Walled Carbon Nanotube Macrofilms. Adv. Mater. 21, 4793–1797.

40. Zhang, R., Dowden, A., Deng, H., Baxendale, M. and Peijs, T., 2009. Conductive network formation in the melt of carbon nanotube/ thermoplastic polyurethane composite. Compos. Sci. Technol. 69, 1499–1504.

41. Zhang, T. and Yi, Y. B., 2008. Monte Carlo simulations of effective electrical conductivity in short-fiber composites. J. Appl. Phys. 103, 014910.

42. Allaoui, A., Hoa, S. V. and Pugh, M. D., 2008. The electronic transport properties and microstructure of carbon nanofiber/epoxy composites. Compos. Sci. Technol. 68, 410–416.

43. Balberg, I., Anderson, C. H., Alexander, S. and Wagner, N., 1984. Excluded volume and its relation to the onset of percolation. Phys. Rev. B 30, 3933–3943.

44. Bao, S. P., Liang, G. D. and Tjong, S. C., 2011. Effect of mechanical stretching on electrical conductivity and positive temperature coefficient characteristics of poly(vinylidene fluoride)/carbon nanofiber composites prepared by non-solvent precipitation. Carbon 49, 1758–1768.

45. Berson, S., de Bettignies, R., Bailly, S., Guillerez, S. and Jousselme, B., 2007. Elaboration of P3HT/CNT/PCBM composites for organic photovoltaic cells. Adv. Funct. Mater. 17, 3363–3370.

46. Chang, L., Friedrich, K., Ye, L. and Toro, P., 2009. Evaluation and visualization of the percolating networks in multi-wall carbon nanotube/epoxy composites. J. Mater. Sci. 44 4003–4012.

47. Celzard, A., McRae, E., Deleuze, C., Dufort, M., Furdin, G. and Mareche, J. F., 1996.

48. Critical concentration in percolating systems containing a high-aspect-ratio filler Phys. Rev. B 53, 6209–6214. Cheng, Q. F., Bao, J. W., Park, J., Liang, Z. Y., Zhang, C. and Wang, B., 2009.

49. High mechanical performance composite conductor: multi-Walled carbon nanotube sheet/bismaleimide nanocomposites. Adv. Funct. Mater. 19, 3219–3225.

50. Dang, Z. M., Yao, S. H. And Xu, H. P., 2007. Effect of tensile strain on morphology and dielectric property in nanotube/polymer nanocomposites. Appl. Phys. Letter. 90, 012907.

51. Das, N. C., Chaki, T. K. And Khastgir, D., 2002. Effect of axial stretching on electrical resistitivty of short carbon fibre and carbon black filled conductive rubber composites. Polym. Int. 51, 156–163.

52. Deng, F. and Zheng, Q. S., 2008. An analytical model of effective electrical conductivity of carbon nanotube composites. Appl. Phys. Lett. 92, 071902.

53. Du, F. M., Fischer, J. E. and Winey, K. I., 2005. Effect of nanotube alignment on percolation conductivity in carbon nanotube/polymer composites. Phys. Rev. B 72, 121404.

54. Entchev, P. B. and Lagoudas, D. C., 2002. Modeling porous shape memory alloys using micromechanical averaging techniques. Mech. Mater. 34, 1–24.

55. Feng, C. and Jiang, L. Y., 2013. Micromechanics modeling of the electrical conductivity of carbon nanotube (CNT)–polymer nanocomposites. Compos. Pt. A-Appl. Sci. Manuf. 47, 143–149.

56. Gojny, F. H., Wichmann, M. H. G., Fiedler, B., Kinloch, I. A., Bauhofer, W., Windle, A. H. et al., 2006.

57. Evaluation and identification of electrical and thermal conduction mechanisms in carbon nanotube/epoxy composites. Polymer 47, 2036–2045.

58. Hu, N., Karube, Y., Arai, M., Watanabe, T., Yan, C., Li, Y., et al., 2010.

59. Investigation on sensitivity of a polymer/carbon nanotube composite strain sensor. Carbon 48, 680–687.

60. Jiang, M. J., Dang, Z. M. and Xu, H. P., 2007. Giant dielectric constant and resistancepressure sensitivity in carbon nanotubes/rubber nanocomposites with low percolation threshold. Appl. Phys. Lett. 90, 042914.

61. Kim, Y. J., Shin, T. S., Choi, H. D., Kwon, J. H., Chung, Y. C. and Yoon, H. G., 2005. Electrical conductivity of chemically modified multiwalled carbon nanotube/epoxy composites. Carbon 43, 23–30.

62. Kuhn, W. and Grün, F., 1942. Beziehungen zwischen elastischen Konstanten und Dehnungsdoppelbrechung hochelastischer Stoffe. Colloid Polym. Sci. 101, 248–271.

63. Li, C. Y., Thostenson, E. T., and Chou, T. W., 2007. Dominant role of tunneling resistance in the electrical conductivity of carbon nanotube-based composites. Appl. Phys. Lett. 91, 223114.

64. Lin, C. A., Wang, H. T. and Yang, W., 2010. Variable percolation threshold of composites with fiber fillers under compression. J. Appl. Phys. 108, 013509.

65. Lu, W. B., Chou, T. W. and Thostenson, E. T., 2010. A three-dimensional model of electrical percolation thresholds in carbon nanotube-based composites. Appl. Phys. Lett. 96, 223106.

66. Miao, Y., Chen, L., Lin, Y. Z., Sammynaiken, R., Zhang, W. J., 2011. On finding of high piezoresistive response of CNT films without surfactants for in-plane strain detection. J. Intell. Mater. Syst. Struct. 22, 2155–2159.

67. Miao, Y., Chen, L., Sammynaiken, R., Lin, Y. and Zhang, W. J., 2011. Optimization of Piezoresistive Response of Pure CNT Networks as In-plane Strain Sensors. Rev. Sci. Instru. 82, 126104.

68. Miao, Y., Yang, Q., Chen, L., Sammynaiken, R. and Zhang, W. J., 2012.

69. Modeling of piezoresistive response of carbon nanotube network based films under in-plane straining by percolation theory. Appl. Phys. Lett. 101, 063120.

70. Mori, T. and Tanaka, K., 1973. Average Stress in Matrix and Average Elastic Energy of Materials with Misfitting Inclusions. Acta. Metall. Mater. 21, 571–574.

71. Nambiar, S. and Yeow, J. T. W., 2011. Conductive polymer-based sensors for biomedical applications. Biosens. Bioelectron. 26, 1825–1832.

72. Odegard, G. M., Gates, T. S., Wise, K. E., Park, C. and Siochi, E. J., 2003.

73. Constitutive modeling of nanotube-reinforced polymer composites. Compos. Sci. Technol. 63, 1671– 1687.

74. Ounaies, Z., Park, C., Wise, K. E., Siochi, E. J. and Harrison, J. S., 2003. Electrical properties of single wall carbon nanotube reinforced polyimide composites. Compos. Sci. Technol. 63, 1637–1646.

75. Pan, N., 1996. The elastic constants of randomly oriented fiber composites: A new approach to prediction. Sci. Eng. Compos. Mater. 5, 63–72.

76. Park, M., Kim, H. and Youngblood, J. P., 2008. Strain-dependent electrical resistance of multi-walled carbon nanotube/polymer composite films. Nanotechnology 19, 055705.

77. Park, M. and Kim, H., 2006. Evaporation-based method for fabricating conductive MWCNT/PEO composite film and its application as strain sensor. Proc. 12th US-Japan Conf. Compos. Mater. Michigan: University of Michigan, pp 78–86.

78. Perez, R., Banda, S. and Ounaies, Z., 2008. Determination of the orientation distribution function in aligned single wall nanotube polymer nanocomposites by polarized Raman spectroscopy. J. Appl. Phys. 103, 074302.

79. Seidel, G. D. and Lagoudas, D. C., 2009. A micromechanics model for the electrical conductivity of nanotube-polymer nanocomposites. J. Compos. Mater. 43, 917–941.

80. Shang, S. M., Zeng, W. and Tao, X. M., 2011. High stretchable MWNTs/ polyurethane conductive nanocomposites. J. Mater. Chem. 21, 7274– 7280.

81. Shiraishi, M. and Ata, M., 2001. Work function of carbon nanotubes. Carbon 39, 1913– 1917. Simmons, J. G., 1963.

82. Generalized formula for the electric tunnel effect between similar electrodes separated by a thin insulating film. J. Appl. Phys. 34, 1793– 1803. Takeda, T., Shindo, Y., Kuronuma, Y. and Narita, F., 2011.

83. Modeling and characterization of the electrical conductivity of carbon nanotube-based polymer composites. Polymer 52, 3852–3856.

84. Tallman, T. and Wang, K. W., 2013. An arbitrary strains carbon nanotube composite piezoresistivity model for finite element integration. Appl. Phys. Lett. 102, 011909.

85. Taya, M., 1995. Micromechanics modeling of electronic composites. J. Eng. Mater-Tran. ASME 117, 462–469.

86. Taya, M. 2005. Electronic composites: modeling, characterization, processing, and MEMS applications. Cambridge: Cambridge University Press. Taya, M., Kim, W. J. and Ono, K., 1998.

87. Piezoresistivity of a short fiber/elastomer matrix composite. Mech. Mater. 28, 53–59.

88. Vangurp, M., 1995. The Use of Rotation Matrices in the Mathematical-Description of Molecular Orientations in Polymers. Colloid Polym. Sci. 273, 607–625.

89. Wang, X., Bradford, P. D., Liu, W., Zhao, H. B., Inoue, Y., Maria, J. P. et al., 2011.

90. Mechanical and electrical property improvement in CNT/Nylon composites through drawing and stretching. Compos. Sci. Technol. 71, 1677–1683.

91. Yan, K. Y., Xue, Q. Z., Zheng, Q. B. and Hao, L. Z., 2007. The interface effect of the effective electrical conductivity of carbon nanotube composites. Nanotechnology 18, 255705.

92. Yang, Y., Gupta, M. C., Dudley, K. L. and Lawrence, R. W., 2005.

93. Novel carbon nanotube-polystyrene foam composites for electromagnetic interference shielding. Nano Lett. 5, 2131–2134.

94. Yu, C. J., Masarapu, C., Rong, J. P., Wei, B. Q. and Jiang, H. Q., 2009. Stretchable supercapacitors based on buckled single-walled carbon nanotube macrofilms. Adv. Mater. 21, 4793–4797.

95. Zhang, R., Dowden, A., Deng, H., Baxendale, M. and Peijs, T., 2009. Conductive network formation in the melt of carbon nanotube/ thermoplastic polyurethane composite. Compos. Sci. Technol. 69, 1499–1504.

96. Allaoui, A., Hoa S. V. and Pugh M. D., 2008. The electronic transport properties and microstructure of carbon nanofiber/epoxy composites. Compos. Sci. Technol. 68, 410–416.

97. Balberg, I. Binenbaum, N. and Wagner, N., 1984. Excluded Volume and Its Relation to the Onset of Percolation. Physical Review B 30, 3933–3943.

98. Bao, S. P., Liang G. D. and Tjong S. C., 2011. Effect of mechanical stretching on electrical conductivity and positive temperature coefficient characteristics of poly(vinylidene fluoride)/carbon nanofiber composites prepared by non-solvent precipitation. Carbon 49, 1758–1768.

99. Berson, S., de Bettignies, R., Bailly, S., Guillerez, S. and Jousselme, B., 2007. Elaboration of P3HT/CNT/PCBM composites for organic photovoltaic cells. Adv. Funct. Mater. 17, 3363–3370.

100. Celzard, A., McRae, E., Deleuze, C., Dufort, M., Furdin, G. and Marêché, J. F., 1996. Critical concentration in percolating systems containing a high-aspect-ratio filler. Phys. Rev. B 53, 6209–6214.

101. Chang, L., Friedrich, K., Ye, L. and Toro, P., 2009. Evaluation and visualization of the percolating networks in multi-wall carbon nanotube/epoxy composites, J. Mater. Sci. 44, 4003–4012.

102. Deng, F. and Zheng, Q. S., 2008. An analytical model of effective electrical conductivity of carbon nanotube composites. Appl. Phys. Lett. 92, 071902.

103. Entchev, P. B. and Lagoudas, D. C., 2002. Modeling porous shape memory alloys using micromechanical averaging techniques. Mech. Mater. 34, 1–24.

104. Feng, C. and Jiang, L., 2013. Micromechanics modeling of the electrical conductivity of carbon nanotube (CNT)–polymer nanocomposites. Compos. Pt. A-Appl. Sci. Manuf. 47, 143–149.

105. Feng, C. and Jiang, L., Investigation of uni-axial stretching effects on the electrical conductivity of CNT-polymer nanocomposites. Journal of Physics D: Applied Physics. Accepted. Gojny, F. H., Wichmann, M. H. G., Fiedler, B. and Schulte, K., 2006.

106. Evaluation and identification of electrical and thermal conduction mechanisms in carbon nanotube/epoxy composites. Polymer 47, 2036–2045.

107. Gurp, M., 1995. The use of rotation matrices in the mathematical description of molecular orientations in polymers. Colloid Polym. Sci. 273, 607–625.

108. Hu, N., Karube, Y., Arai, M., Watanabe, T., Yan, C., Li, Y., Liu, Y. L. and Fukunaga, H., 2010. Investigation on sensitivity of a polymer/carbon nanotube composite strain sensor. Carbon 48, 680–687.

109. Jiang, M. J., Dang Z. M. and Xu, H. P., 2007. Giant dielectric constant and resistancepressure sensitivity in carbon nanotubes/rubber nanocomposites with low percolation threshold. Appl. Phys. Lett. 90, 042914.

110. Kim, Y. J., Shin, T. S., Choi, H. D., Kwon, J. H., Chung, Y. C. and Yoon, H. G., 2005.

111. Electrical conductivity of chemically modified multiwalled carbon nanotube/epoxy composites. Carbon 43, 23–30. Kuhn, W. and Grün, F., 1942.

112. Beziehungen   zwischen   elastischen   Konstanten   und Dehnungsdoppelbrechung hochelastischer Stoffe. Colloid Polym. Sci. 101, 248–271.

113. Li, C. Y., Thostenson, E. T. and Chou, T. W., 2007. Dominant role of tunneling resistance in the electrical conductivity of carbon nanotube-based composites. Appl. Phys. Lett. 91, 223114.

114. Lin, C. A., Wang, H. T. and Yang, Wei., 2010. Variable percolation threshold of composites with fiber fillers under compression. J. Appl. Phys. 108, 013509.

115. Lu, W. B., Chou, T. W. and Thostenson, E. T., 2010. A three-dimensional model of electrical percolation thresholds in carbon nanotube-based composites. Appl. Phys. Lett. 96, 223106.

116. Mayoral, B., Hornsby, P. R., McNally, T., Schiller T. L., Jack, K. and Marti, D. J., 2013.

117. Quasi-solid state uniaxial and biaxial deformation of PET/MWCNT composites: structural evolution, electrical and mechanical properties. Rsc Adv. 3, 5162–5183.

118. Mori, T. and Tanaka, K., 1973. Average Stress in Matrix and Average Elastic Energy of Materials with Misfitting Inclusions. Acta Metall. Sin. 21, 571–574.

119. Nambiar, S. and Yeow, J. T. W., 2011. Conductive polymer-based sensors for biomedical applications. Biosens. Bioelectron. 26, 1825–1832.

120. Odegard, G. M., Gates, T. S., Wise, K. E., Park, C. and Siochi, E. J., 2003. Constitutive modeling of nanotube-reinforced polymer composites. Compos. Sci. Technol. 63, 1671– 1687. Ounaies, Z., Park, C., Wise, K. E., Siochi, E. J. and Harrison, J. S., 2003.

121. Electrical properties of single wall carbon nanotube reinforced polyimide composites. Compos. Sci.Technol. 63, 1637–1646. Park, M. and Kim, H., 2006.

122. Evaporation-based method for fabricating conductive MWCNT/PEO composite film and its application as strain sensor. Proc. 12th US-Japan Conf. Compos. Mater. Michigan, USA, pp. 78–86.

123. Park, M., Kim, H. and Youngblood, J. P., 2008. Strain-dependent electrical resistance of multi-walled carbon nanotube/polymer composite films. Nanotechnology 19, 055705.

124. Pérez, R., Banda, S. and Qunaies, Z., 2008. Determination of the orientation distribution function in aligned single wall nanotube polymer nanocomposites by polarized Raman spectroscopy. J. Appl. Phys. 103, 074302.

125. Seidel, G. D. and Lagoudas, D. C., 2009. A Micromechanics Model for the Electrical Conductivity of Nanotube-Polymer Nanocomposites. J. Compos. Mater. 43, 917–941.

126. Shang, S. M., Zeng, W. and Tao, X. M., 2011. High stretchable MWNTs/ polyurethane conductive nanocomposites. J. Mater. Chem. 21, 7274– 7280

127. Shen, J. B., Champagne, M. F., Yang, Z., Yu, Q., Gendron, R. and Guo, S. Y., 2012.

128. The development of a conductive carbon nanotube (CNT) network in CNT/polypropylene composite films during biaxial stretching. Compos. Pt. A- Appl. Sci. Manuf. 43, 1448– 1453.

129. Shiraishi, M. and Ata, M., 2001. Work function of carbon nanotubes. Carbon 39, 1913– 1917.

130. Simmons, J. G., 1963. Generalized Formula for the Electric Tunnel Effect between Similar Electrodes Separated by a Thin Insulating Film. J. Appl. Phys. 34, 1793–1803.

131. Takeda, T., Shindo, Y., Kuronuma, Y. and Narita, F., 2011. Modeling and characterization of the electrical conductivity of carbon nanotube-based polymer composites. Polymer 52, 3852–3856.

132. Tallman, T. and Wang, K. W., 2013. An arbitrary strains carbon nanotube composite piezoresistivity model for finite element integration. Appl. Phys. Lett. 102, 011909.

133. Taya, M., 1995. Micromechanics modeling of electronic composites. J. Engineering Mater. Technol.-Trans. ASME 117, 462–469. Taya, M., 2005.

134. Electronic composites: modeling, characterization, processing, and MEMS applications. Cambridge University Press, Cambridge. Taya, M., Kim, W. J. and Ono, K., 1998.

135. Piezoresistivity of a short fiber/elastomer matrix composite. Mech. Mater. 28, 53–59. Yan, K. Y., Xue, Q. Z., Zheng, Q. B. and Hao, L. Z., 2007.

136. The interface effect of the effective electrical conductivity of carbon nanotube composites. Nanotechnology 18, 255705.

137. Yang, Y., Gupta, M. C., Dudley, K. L. and Lawrence, W., 2005. Novel carbon nanotubepolystyrene foam composites for electromagnetic interference shielding. Nano Lett. 5, 2131–2134.

138. Yu, C. J., Masarapu, C., Rong, J. P., Wei, B. Q. and Jiang, H. P., 2009. Stretchable Supercapacitors Based on Buckled Single-Walled Carbon Nanotube Macrofilms. Adv. Mater. 21, 4793–4397.

139. Zhang, R., Dowden, A., Deng, H., Baxendale, M. and Peijs, T., 2009. Conductive network formation in the melt of carbon nanotube/thermoplastic polyurethane composite. Compos. Sci. Technol. 69, 1499–1504.

140.

141. Allaoui, A., Hoa, S. V. and Pugh, M. D., 2008. The electronic transport properties and microstructure of carbon nanofiber/epoxy composites. Compos. Sci. Technol. 68, 410–416.

142. Balberg, I., Binenbaum, N. and Wagner, N., 1984. Percolation thresholds in the threedimensional sticks system. Phys. Rev. Lett. 52, 1465–1468. Berhan, L. and Sastry, A. M., 2007.

143. Modeling percolation in high-aspect-ratio fiber systems. II. The effect of waviness on the percolation onset. Phys. Rev. E 75, 041121.

144. Celzard, A., McRae, E., Deleuze, C., Dufort, M., Furdin, G. and Marêché, J., 1996. Critical concentration in percolating systems containing a high-aspect-ratio filler. Phys. Rev. B 53, 6209–6214.

145. Chang, L., Friedrich, K., Ye, L. and Toro, P., 2009. Evaluation and visualization of the percolating networks in multi-wall carbon nanotube/epoxy composites. J. Mater. Sci. 44, 4003–4012.

146. Dastgerdi, J. N., Marquis, G. and Salimi, M., 2013. The effect of nanotubes waviness on mechanical properties of CNT/SMP composites. Compos. Sci. Technol. 86, 164–169.

147. Deng, F. and Zheng, Q. S., 2008. An analytical model of effective electrical conductivity of carbon nanotube composites. Appl. Phys. Lett. 92, 071902.

148. Entchev, P. B. and Lagoudas, D. C. 2002. Modeling porous shape memory alloys using micromechanical averaging techniques. Mech. Mater. 34, 1–24.

149. Feng, C. and Jiang, L., 2013. Micromechanics modeling of the electrical conductivity of carbon nanotube (CNT)–polymer nanocomposites. Compos. Pt. A-Appl. Sci. Manuf. 47, 143–149.

150. Feng, C. and Jiang, L., 2014. Investigation of uni-axial stretching effects on the electrical conductivity of CNT-polymer nanocomposites. Submitted. Feng, C. and Jiang, L., 2014.

151. Micromechanics modeling of bi-axial stretching effects on the electrical conductivity of CNT-polymer composites. Submitted. Fisher, F. T., Bradshaw, R. D. and Brinson, L. C., 2003.

152. Fiber waviness in nanotubereinforced polymer composites-1: Modulus predictions using effective nanotube properties. Compos. Sci. Technol. 63, 1689–1703.

153. Gojny, F. H., Wichmann, M. H. G., Fiedler, B., Kinloch, I. A., Bauhofer, W., Windle, A. H. and Schulte, K. 2006. Evaluation and identification of electrical and thermal conduction mechanisms in carbon nanotube/epoxy composites. Polymer 47, 2036–2045.

154. Hashin, Z., 1990.

155. Thermoelastic Properties and Conductivity of Carbon Carbon-Fiber Composites. Mech. Mater. 8, 293–308. Iijima, S., 1991. Helical microtubules of graphitic carbon. Nature 354, 56–58.

156. Kim, Y. J., Shin, T. S., Choi, H. D., Kwon, J. H., Chung, Y. C. and Yoon, H. G., 2005. Electrical conductivity of chemically modified multiwalled carbon nanotube/epoxy composites. Carbon 43, 23–30.

157. Kuhn, W. and Grün, F., 1942.

158. Beziehungen zwischen elastischen Konstanten und Dehnungsdoppelbrechung hochelastischer Stoffe. Kolloid-Zeitschrift 101, 248–271.

159. Li, C., Thostenson, E. T. and Chou, T. W., 2008. Effect of nanotube waviness on the electrical conductivity of carbon nanotube-based composites. Compos. Sci. Technol. 68, 1445–1452.

160. Li, C. Y., Thostenson, E. T. and Chou, T. W., 2007.

161. Dominant role of tunneling resistance in the electrical conductivity of carbon nanotube-based composites. Appl. Phys. Lett. 91, 223114.

162. Li, J., Ma, P. C., Chow, W. S., To, C. K., Tang, B. Z. and Kim, J. K., 2007. Correlations between Percolation Threshold, Dispersion State, and Aspect Ratio of Carbon Nanotubes. Adv. Funct. Mater. 17, 3207–3215.

163. Lin, C., Wang, H. and Yang, W., 2010. Variable percolation threshold of composites with fiber fillers under compression. J. Appl. Phys. 108, 013509.

164. Lu, W. B., Chou, T. W. and Thostenson, E. T., 2010. A three-dimensional model of electrical percolation thresholds in carbon nanotube-based composites. Appl. Phys. Lett. 96, 223106.

165. Mori, T. and Tanaka, K., 1973. Average Stress in Matrix and Average Elastic Energy of Materials with Misfitting Inclusions. Acta Metall. Mater. 21, 571–574.

166. Nambiar, S. and Yeow, J. T. W., 2011. Conductive polymer-based sensors for biomedical applications. Biosens. Bioelectron. 26, 1825–1832.

167. Odegard, G. M., Gates, T. S., Wise, K. E., Park, C. and Siochi, E. J., 2003. Constitutive modeling of nanotube-reinforced polymer composites. Compos. Sci. Technol. 63, 1671– 1687.

168. Ounaies, Z., Park, C., Wise, K. E., Siochi, E. J. and Harrison, J. S., 2003. Electrical properties of single wall carbon nanotube reinforced polyimide composites. Compos. Sci. Technol. 63, 1637–1646.

169. Park, M. and Kim, H., 2006. Evaporation-based method for fabricating conductive MWCNT/PEO composite film and its application as strain sensor. Proc. 12th US-Japan Conf. Compos. Mater. Michigan: University of Michigan, pp 78–86.

170. Park, M., Kim, H. and Youngblood, J. P., 2008. Strain-dependent electrical resistance of multi-walled carbon nanotube/polymer composite films. Nanotechnology 19, 055705.

171. Seidel, G. D. and Lagoudas, D. C., 2009. A Micromechanics Model for the Electrical Conductivity of Nanotube-Polymer Nanocomposites. J. Compos. Mater. 43, 917–941.

172. Shang, S. M., Zeng, W. and Tao, X. M., 2011. High stretchable MWNTs/ polyurethane conductive nanocomposites. J. Mater. Chem. 21, 7274–7280.

173. Shi, D. L., Feng, X. Q., Huang, Y. G. Y., Hwang, K. C. and Gao, H. J., 2004. The effect of nanotube waviness and agglomeration on the elastic property of carbon nanotubereinforced composites. J. Eng. Mater.-Tran. ASME 126, 250–257.

174. Takeda, T., Shindo, Y., Kuronuma, Y. and Narita, F., 2011. Modeling and characterization of the electrical conductivity of carbon nanotube-based polymer composites. Polymer 52, 3852–3856.

175. Tallman, T. and Wang, K. W., 2013. An arbitrary strains carbon nanotube composite piezoresistivity model for finite element integration. Appl. Phys. Lett. 102, 011909.

176. Taya, M., 1995. Micromechanics modeling of electronic composites. J. Eng. Mater.-Tran. ASME 117, 462–469. Taya, M., 2005.

177. Electronic composites: modeling, characterization, processing, and MEMS applications, Cambridge University Press, Cambirdge. Taya, M., Kim, W. and Ono, K., 1998. Piezoresistivity of a short fiber/elastomer matrix composite. Mech. Mater. 28, 53–59.

178. Yan, K. Y., Xue, Q. Z., Zheng, Q. B. and Hao, L. Z., 2007. The interface effect of the effective electrical conductivity of carbon nanotube composites. Nanotechnology 18, 255705.

179. Yanase, K., Moriyama, S. and Ju, J. W., 2013. Effects of CNT waviness on the effective elastic responses of CNT-reinforced polymer composites. Acta Mech. 224, 1351–1364.

180. Yi, Y., Berhan, L. and Sastry, A., 2004. Statistical geometry of random fibrous networks, revisited: Waviness, dimensionality, and percolation. J. Appl. Phys. 96, 1318–1327.

181. Yi, Y. B., Berhan, L. and Sastry, A. M., 2004. Statistical geometry of random fibrous networks, revisited: Waviness, dimensionality, and percolation. J. Appl. Phys. 96, 1318– 1327.

182. Yu, C. J., Masarapu, C., Rong, J. P., Wei, B. Q. and Jiang, H. Q., 2009. Stretchable Supercapacitors Based on Buckled Single-Walled Carbon Nanotube Macrofilms. Adv. Mater. 21, 4793–4797

183. Yu, Y., Song, S., Bu, Z., Gu, X., Song, G. and Sun, L., 2013. Influence of filler waviness and aspect ratio on the percolation threshold of carbon nanomaterials reinforced polymer nanocomposites. J. Mater. Sci. 48, 5727–5732.

184. Yu, Y., Song, S. Q., Bu, Z. X., Gu, X. F., Song, G. B. and Sun, L., 2013. Influence of filler waviness and aspect ratio on the percolation threshold of carbon nanomaterials reinforced polymer nanocomposites. J. Mater. Sci. 48, 5727–5732.

185. Zhang, R., Dowden, A., Deng, H., Baxendale, M. and Peijs, T., 2009. Conductive network formation in the melt of carbon nanotube/ thermoplastic polyurethane composite. Compos. Sci. Technol. 69, 1499– 1504.

Chapter 9

EFFECT OF THE CURING PROCESS ON THE TRANSVERSE TENSILE STRENGTH OF FIBER-REINFORCED POLYMER MATRIX LAMINA USING MICROMECHANICS COMPUTATIONS

Royan J D'Mello[1,2], Marianna Maiarù[1,2] and Anthony M Waas [1,2]

[1]Composite Structures Laboratory, Department of Aerospace Engineering, University of Michigan

[2]William E. Boeing Department of Aeronautics and Astronautics, University of Washington

ABSTRACT

The effect of the curing process on the mechanical response of fiber-reinforced polymer matrix composites is studied using a computational model.

Computations are performed using the finite element (FE) method at the microscale where representative volume elements (RVEs) are analyzed with periodic boundary conditions (PBCs). The commercially available finite element (FE) package ABAQUS is used as the solver, supplemented by user-written subroutines. The transition from a continuum to damage/failure is effected by using the Bažant-Oh crack band model, which preserves mesh objectivity. Results are presented for a hexagonally packed RVE whose matrix portion is first subjected to curing and subsequently to mechanical loading. The effect of the fiber packing randomness on the microstructure is analyzed by considering multi-fiber RVEs where fiber volume fraction is held constant but with random packing of fibers. The possibility of failure is accommodated throughout the analysis—failure can take place during the curing process prior to the application of in-service mechanical loads. The analysis shows the differences in both the cured RVE strength and stiffness, when cure-induced damage has and has not been taken into account.

BACKGROUND

Fiber-reinforced polymer matrix composites (FRPCs) are high-strength and lightweight advanced materials widely used in the aerospace and automotive industries. Since FRPCs are manufactured by curing the matrix that surrounds the interspersed fibers, good understanding of the matrix state during the curing process is necessary to have sufficient control over the quality of the cured product. The mechanical properties of the matrix during curing can be altered by the presence of fibers and also by details of the curing cycle. The curing matrix undergoes shrinkage due to chemical processes, which gives rise to self-equilibrating internal stresses. Plepys and Farris [1] and Plepys et al. [2] have used finite element calculations using incremental elasticity to show tensile residual stress buildup of up to 28 MPa post cure in a three-dimensionally constrained Epon 828 epoxy resin. Merzlyakov et al. [3] reported the development of tensile stresses in a constrained thermosetting resin system undergoing cure and also quantified the variation of these tensile stresses during subsequent thermal cycling. Depending on the constituent chemistry of the matrix, the thermal cycle prescribed, and the fracture and strength properties of a curing matrix, a fiber-reinforced composite can and may undergo damage and cracking in the matrix during the cure cycle. Chekanov et al. [4] have reported various types of defects that may form in a constrained epoxy resin system undergoing curing. Rabearison et al. [5] studied the curing of a thick epoxy tube using a finite element model and concluded that high stress gradients developed during differential curing can cause cracking. Therefore, the state of the matrix within a cured FRPC structure exhibits in situ matrix properties, which are effective properties of the matrix that take into account imperfections caused in the matrix due to the cure process, including the presence of residual stresses. That is, the in situ matrix properties, where the matrix is treated as a 'new' material with a reference configuration that corresponds to the post-cured state, deviate from idealized or 'virgin' matrix properties of the bulk matrix. The in situ matrix properties can be extracted from an inverse analysis [6] through the uniaxial tensile response of a ±45° laminate, and this is convenient in engineering analysis of cured composites. Song and Waas [7] have shown that the use of bulk matrix properties in numerical predictions of compression response of a 2D triaxially braided composite RVE can lead to erroneous results - the computed compressive strength being noticeably higher than the experimentally measured strength. They observed that the tow kinking failure mode, which controls the compression strength was found to be sensitive to the nonlinear shear response of the matrix. Cure shrinkage in the matrix surrounded by randomly dispersed fibers can also influence the final shape of the structure [8]. Therefore, it is necessary to have good knowledge of the influence of the cure cycle on the subsequent mechanical response of the laminate. For a particular fiber-matrix laminate system, the optimal cure cycle can be identified such that the cured product has the highest strength and stiffness. Efforts to optimize various aspects of the cure cycle for mitigating the residual stresses generated during cure can be found in the studies of Li et al. [9], Gopal et al. [10], and White and Hahn [11].

In the present investigation, the effects of the cure cycle on possible damage accumulation during cure and subsequent in-service performance at the microstructural level are studied. A hexagonally packed representative volume element (RVE) having a total of two fibers (one full center fiber and quarter fibers at four corners) with different volume fractions, and a randomly packed RVE having multiple fibers are studied. First, the influence of fiber volume fraction on the strength of the cured RVE using the hexagonally packed RVE with two fibers is studied. Next, the effect of the randomness of the packing for RVEs having fixed volume fraction is investigated. For illustrating the findings of this study, the strength investigated is the transverse tensile strength (S+22S22+), which is obtained by mechanically loading each of the virtually cured RVEs along the transverse direction under tension. Then, the initial slope and peak stress value of the nominal stress-strain response are the transverse stiffness E_{22} and transverse tensile strength S+22S22+, respectively. For low to moderate fiber volume fractions, the transverse stiffness is controlled by matrix stiffness (see [12]). The transverse tensile strength associated with transverse matrix cracking is controlled by a combination of factors such as matrix tensile strength, matrix fracture toughness, fiber packing, and adhesion strength between fibers and the matrix. Hence, it is expected that both E_{22} and S+22S22+ are influenced by the details of the cure process.

METHODS

Cure process

The curing process of a thermoset polymer can be divided into two parts: The first part consists of the chemical reaction, heat generation, and conduction. The second is the generation of self-equilibrating stresses and development of the structural integrity via the evolution of matrix stiffness. The stress generation has been modeled by Mei [13], Mei et al. [14], and Heinrich et al. [15]. The degree of cure (φ) of the matrix is defined as $\phi = H(t)/H_r$, where $H(t)$ is the heat generated up to time t, and H_r is the total heat of reaction at the end of the cure cycle. Mathematically, the rate of cure $\left(\frac{d\phi}{dt}\right)$ can be expressed as,

$$\frac{d\phi}{dt} = f(T, \phi)$$

(1)

where $f(T,\phi) \geq 0$ is a function. The evolution of temperature (T) and degree of cure (ϕ) for the matrix material system is determined through a coupled system that considers the heat equation and an empirical curing law or can be supplied from the output of a simulation that takes into account a cure kinetics model. Kamal [16] has proposed a semi-empirical expression for the function $f(T,\phi)$ in terms of Arrhenius terms that depend on temperature

$$f(T, \phi) = \left[A_1 \exp\left(\frac{\Delta E_1}{TR}\right) + A_2 \exp\left(\frac{\Delta E_2}{TR}\right) \phi^m \right] (1 - \phi)^n$$

(2)

where T is temperature, R is the gas constant, and ΔE_1 and ΔE_2 are activation energies. The frequency-like constants A_1, A_2 and exponents m and n, in theory, have to be determined by fitting the above equation to the experimental data. However, due to the complexity of the function $f(T,\phi)$, a general closed formed solution to Equation 1 is elusive, and often times, this differential equation has to be solved using some numerical method. Assuming the form for $f(T,\phi)$ in Equation 2 by setting m=A_2=ΔE_2=0,n=1 and under isothermal conditions, an explicit relation between the degree of cure and time can be found as a solution to the differential equation 1, which is

$$\phi(t) = 1 - \exp(-\lambda t)$$

(3)

where the Arrhenius parameter $\lambda = A_1 \exp\left(\frac{-\Delta E_1}{TR}\right)$. Cure data as a function of time for Epon 862/Epikure 9553 resin under isothermal conditions are chosen for the present work and are available in [15]. The constants obtained by curve fitting with experimental data at various temperatures are as follows: A_1=3.62×10^{11} s^{-1} and ΔE_1=8.854×10^4 J.

During curing, the matrix heats up due to an exothermic chemical reaction and due to conduction from the heating source at the boundary. This process can be modeled using the equation

$$\rho c \frac{\partial T}{\partial t} = \frac{\partial}{\partial x_i} \left(\kappa(T, \phi) \frac{\partial T}{\partial x_i} \right) + \rho H_r \frac{\partial \phi}{\partial t}$$

(4)

where ρ is the mass density, c_p is the specific heat, and κ is the thermal conductivity. The evolution of self-equilibrating stresses $\sigma_{ij}(t)$ during curing is included in the analysis by using a model proposed by Heinrich et al. [15]:

$$\underline{\underline{\sigma}}(t) = \int_0^t \frac{d\phi}{ds} \underline{1} \Bigg[K(s) tr \Big(\underline{\underline{\varepsilon}}(t) - \underline{\underline{\varepsilon}}(s) + \underline{\underline{\varepsilon}}_c(s) - \underline{1}\alpha(s)\Delta T(t,s) \Big)$$

$$+ 2\mu(s) \Big(\underline{\underline{\varepsilon}}(t) - \underline{\underline{\varepsilon}}(s) + \underline{\underline{\varepsilon}}_c(s) - \underline{1}\frac{1}{3} tr\{\underline{\underline{\varepsilon}}(t) - \underline{\underline{\varepsilon}}(s) + \underline{\underline{\varepsilon}}_c(s)\} \Big) \Bigg] ds$$

$$+ (1 - \phi(t))K(0) tr(\underline{\underline{\varepsilon}}(t) - \underline{1}\alpha(0)\Delta T(t))\underline{1} \tag{5}$$

where K, μ, α, and ε_c are the per-network bulk modulus, shear modulus, coefficient of thermal expansion, and cure shrinkage, respectively. The first term having the integral is the contribution to stress evolution due to the curing matrix, whereas the second term captures the contribution of the uncured liquid resin. The constants K(0) and α(0) correspond to the bulk modulus and coefficient of thermal expansion of the liquid resin, respectively. The coefficient of thermal expansion $\alpha(\varphi)$ of the curing matrix is assumed to have a constant value of 61×10^{-6}m/mK. As shown by Heinrich et al. [15], the per-network properties can be obtained from experimentally measured values of the plane wave modulus (M_{exp}) and shear modulus (μ_{exp}) for the curing matrix as

$$M(\phi) = \frac{dM_{exp}}{d\phi} + K_{exp}(0)$$

$$\mu(\phi) = \frac{d\mu_{exp}}{d\phi} \tag{6}$$

The moduli values M_{exp} and μ_{exp} are measured as a function of time by concurrent Raman and Brillouin light scattering for the pure resin, that is, for a resin curing in the absence of fibers. These moduli are assumed to correspond to the virgin matrix as a function of degree of cure. The effect of the presence of fibers around the matrix on matrix degradation during cure will be demonstrated later in this paper. Once M(φ) and μ(φ) are known, the per-network bulk modulus K(φ) can be obtained from the isotropic material relation $K = M - \frac{4}{3}\mu.$ The per-network shrinkage strain $\varepsilon_c(\Phi)$ up to a certain degree of cure $\varphi = \Phi$ is given by

$$\varepsilon_c = \frac{1}{3K(\Phi)} \left[\left(\varepsilon(\Phi) - (1 - \Phi)\frac{d\varepsilon(\Phi)}{d\Phi} \right) K_{exp} - \frac{d\varepsilon(\Phi)}{d\Phi} \int_0^\Phi M(\phi)d\phi \right] \tag{7}$$

A gravimetric test method (see [17]) can be used to obtain shrinkage of all networks $\varepsilon(\Phi)$. A 2 % per-network cure shrinkage has been chosen for the present investigation.

Damage during cure

During curing, the matrix gradually solidifies (stiffness increases) and simultaneously contracts (cure shrinkage) due to network formation. Residual stresses develop in the matrix owing to cure shrinkage and thermal strains. Depending on the magnitude of tensile stresses developed, the degree of cure (φ), and the rate of cure $\left(\dfrac{d\phi}{dt}\right)$ the material may crack locally during curing. A crack band model is used to simulate the possibility of tensile cracking during the curing of the matrix. The critical tensile stress for cracking typically increases with the degree of cure. If certain matrix regions crack locally, it would result in a reduction in the matrix stiffness in that local region along with some energy dissipated due to cracking. Such a reduction in local matrix stiffness can control the mechanical properties of the cured RVE. Two assumptions are enforced because the degree of cure and the coefficient of thermal expansion of a partially cured local volume of material with microcracks are unknown and physically this local volume does not represent a continuum in the strictest sense. First, if a certain local volume of material cracks, it is assumed that no further curing can take place in that local volume. Second, it is assumed that if cracking occurs locally, the local cracked volume cannot expand or contract under temperature variations. In the context of the finite element framework that is used to numerically simulate cure-induced damage, the local volume is a single finite element. Therefore and because the crack band method is used, mesh objectivity is included in the formulation.

At the end of step II, the curing process is complete. In step III, the cured RVE (containing cracks or not, as the case maybe) is subjected to transverse tension loading along the 2-direction. The objective here is to compute the strength (S+22S22+) and stiffness (E$_{22}$) of the virtually cured RVE. Based on the temperature and cure parameters, computation of the stress evolution during cure (step II) and strength calculation based on mechanical loading (step III) is done in a unified step in the commercial software ABAQUS/Standard [18]. In this study, it is assumed that cracking in the curing matrix can occur only for $\varphi > 0.2$ and only under tensile stresses.

The crack band model of Bažant and Oh [19] is used to model failure in the matrix. This model assumes that once the critical fracture stress σ_{cr} has been reached, microcracks are formed and the additional opening due to cracking is smeared over a band of material. Here the width of that band is taken to be that which lies within an element and perpendicualr to the crack plane. The maximum principal stress criterion is used to determine the failure initiation. In the post-peak regime, the traction-separation law controls the behavior of the damaging material as shown in Figure 1 and the stiffness of the material (matrix) is reduced using the secant value. In the present investigation, σ_{cr} is assumed to be independent of φ. However, in reality, it is expected that the strength would vary with φ. Under mode I cracking, the energy dissipated during the fracturing process is the critical mode I energy release rate (G_{IC}) given by

$$G_{IC} = \int_0^{\delta_f} \sigma_{11}(\delta)\,d\delta = h \int_0^{\varepsilon_f} \sigma_{11}(\varepsilon_{11})\,d\varepsilon$$

(8)

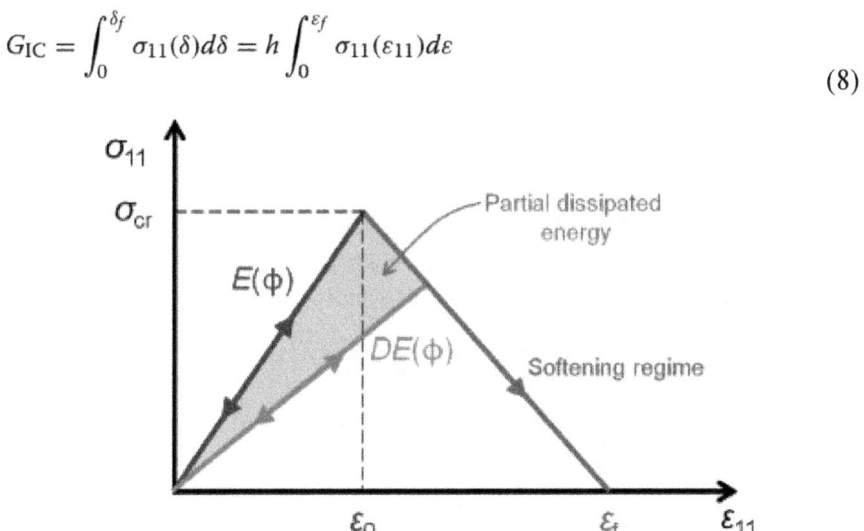

Figure 1: Crack band law in terms of maximum principal stress σ_{11} and maximum principal strain ε_{11}.

where stress σ_{11} and ε_{11} are the maximum principal stress and strain values, respectively, and the maximum separation $\delta_f = h\,\varepsilon_f$ where ε_f corresponds to the critical failure strain of the material (accompanied by complete loss of stiffness). Here, h is the characteristic element length that preserves mesh objectivity (see [20]), defined by prescribing a normalized value of G_{IC} for each element such that $g_{IC} = \dfrac{G_{IC}}{h}$. Consequently, the value of g_{IC} equals the area under the σ_{11}–ε_{11} law shown in Figure 1. The value of G_{IC} is chosen to be 0.6 N/mm in all the computations. For a given epoxy system, the values of G_{IC} and σ_{cr} have to be obtained from an experiment, each as a function of the degree of cure φ.

From the crack band model formulation, the stiffness reduction factor D with ($0 \leq D \leq 1$) for a material with initial stiffness $E = E(\varphi)$ which is now in the softening region of the traction-separation law is computed as

$$D = \frac{\sigma_{cr}}{E(\varepsilon_f - \varepsilon_{cr})}\left(\frac{\varepsilon_f}{\varepsilon_{11}} - 1\right)$$

(9)

where ε_{11} is the current maximum principal strain value. Thus, D=1 corresponds to no damage, 0<D<1 corresponds to damage but no two-piece failure, while D=0 would indicate complete failure. This D parameter will be used to quantify the extent of stiffness reduction after cure has been completed (i.e., at the end of step II).

Boundary conditions

During curing and mechanical loading, the RVE is subjected to periodic boundary conditions, in concert with the assumption that the RVE is a small volume within an infinite medium. The use of periodic boundary conditions for fiber-reinforced RVEs can be found in the studies of Gonzalez and Llorca [21] and Xia et al. [22], among others. During the cure process (step II), the RVE boundaries are allowed to contract or expand. The RVE can contract or expand depending on temperature change and can contract due to cure shrinkage.

Consider an arbitrary cuboid RVE in the undeformed configuration having lengths L_1, L_2, and L_3 along the x_1, x_2, and x_3 directions with one corner point placed at the origin (0,0,0). Then, the equations corresponding to the 3D periodic boundary conditions are

$$u_1(L_1, x_2, x_3) - u_1(0, x_2, x_3) = \epsilon_{11}L_1$$
$$u_2(L_1, x_2, x_3) - u_2(0, x_2, x_3) = 2\epsilon_{12}L_1$$
$$u_3(L_1, x_2, x_3) - u_3(0, x_2, x_3) = 2\epsilon_{13}L_1$$
$$u_1(x_1, L_2, x_3) - u_1(x_1, 0, x_3) = 2\epsilon_{21}L_2$$
$$u_2(x_1, L_2, x_3) - u_2(x_1, 0, x_3) = \epsilon_{22}L_2$$
$$u_3(x_1, L_2, x_3) - u_3(x_1, 0, x_3) = 2\epsilon_{23}L_2$$
$$u_1(x_1, x_2, L_3) - u_1(x_1, x_2, 0) = 2\epsilon_{31}L_3$$
$$u_2(x_1, x_2, L_3) - u_2(x_1, x_2, 0) = 2\epsilon_{32}L_3$$
$$u_3(x_1, x_2, L_3) - u_3(x_1, x_2, 0) = \epsilon_{33}L_3$$

(10)

u_1, u_2, and u_3 are the displacements of the RVE boundary along the x_1, x_2, and x_3 directions, respectively, and ε_{ij} are the tensorial strains.

Analysis procedure

In summary, the analysis procedure is divided into three steps as shown in Figure 2.

1. Step I: A thermochemical analysis is performed using the cure parameters described earlier. Temperature cycle, the degree of cure, and the cure rate in the matrix are provided. Since the RVE dimensions are on the micron scale, there is little to no variation in the temperature field across the RVE. The temperature profile, the degree of cure (φ), and the rate of cure (dφdt)(dφdt) used in the present study are shown in Figure 3.

2. Step II: The stress evolution calculations are preformed as described in Equation 5. Shrinkage during cure is modeled using Equation 7. At the end of this step, we have a virtually cured solid. Possibility of damage during curing is taken into account using a crack band model. Periodic boundary conditions are enforced throughout this step.

3. Step III: The virtually cured RVE is subjected to transverse tensile loading (with periodic boundary conditions in place) to back out the stiffness and strength. Again, the crack band model is used to simulate tensile failure, and periodic boundary conditions are enforced during this step.

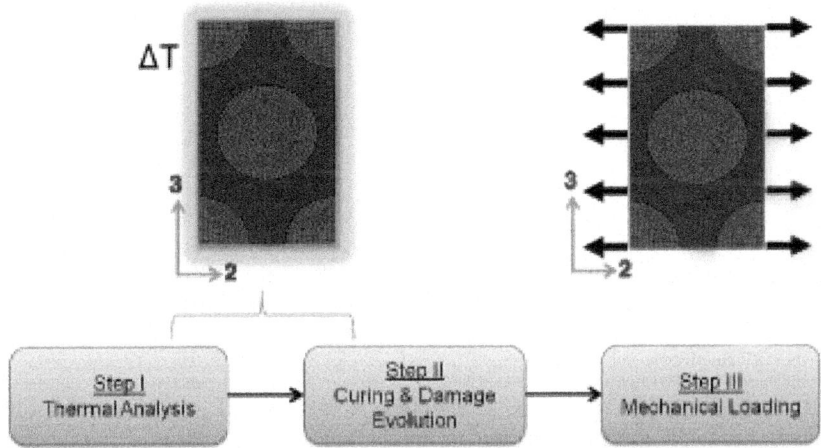

Figure 2: Schematic showing major steps in the analysis. Curing under the prescribed temperature field which includes step I and step II (left). Mechanical loading in the transverse direction to assess the mechanical properties of the virtually cured RVE in step III (right).

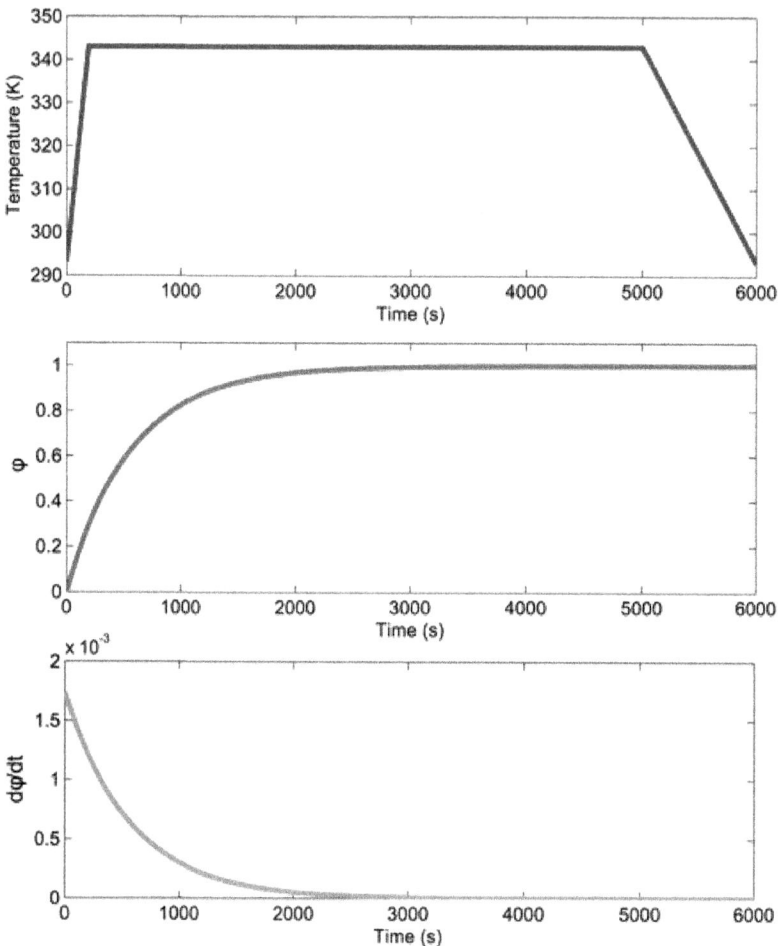

Figure 3: Temperature profile (top), degree of cure φ (middle), and rate of cure dφdtdφdt (bottom) as functions of time.

RESULTS AND DISCUSSION

Hexagonally packed fiber RVEs

Three 3D hexagonally packed RVEs with fiber volume fractions (V_f) of 0.5, 0.6, and 0.7 are studied. These RVEs are first subjected to the curing cycle (steps I and II) and then to tensile loading (step III) in the transverse direction. The latter step leads to the determination of the transverse stiffness E_{22} and tensile strength S+22S22+ of the virtually cured RVEs. The analysis is done using the finite element software ABAQUS/Standard. The stress evolution expression along with the crack band model is implemented using ABAQUS/ Standard's user subroutine UMAT. In each of the RVEs shown in Figure 4, the thickness t along the fiber direction is chosen to be 0.30 μm and carbon fibers are 6 μm in diameter. Both the fiber and the matrix are modeled as isotropic solids. Young's modulus and Poisson's ratio of the fibers are taken to be 200 MPa and 0.3, respectively.

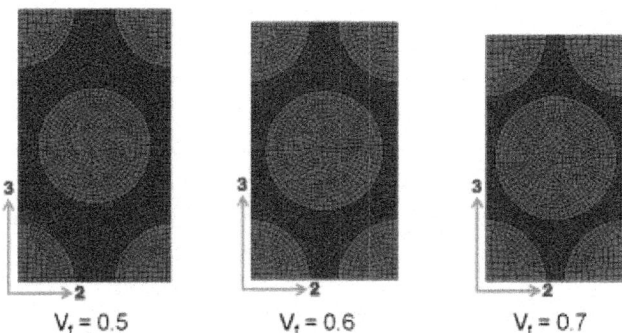

$V_f = 0.5$ $V_f = 0.6$ $V_f = 0.7$

Figure 4: Hexagonally packed fiber-matrix RVEs with fiber volume fractions of 0.5, 0.6, and 0.7. Notice that the dimensions of the fiber are held fixed across these three RVEs.

For each of the RVEs shown in Figure 4, three critical fracture strength values (σ_{cr}) of 20, 30, and 45 MPa are chosen which are independent of the degree of cure φ, while the critical mode I energy release rate G_{IC} is chosen to be 0.6 N/mm. The strength and toughness are assumed here to be independent of the degree of cure (φ). The objective of this portion of the study is to understand how the strength and stiffness of the cured product change with changes in fiber volume fraction and changes in the imposed critical fracture strength during cure. For a given RVE in step II, the matrix tensile stresses can exceed σ_{cr} and microcracks appear leading to a reduction in stiffness. To assess the amount of cure-induced damage, we can keep track of the stiffness reduction factor D at various times during the curing process. Figure 5 shows the average D value for each of the RVEs undergoing cure. Recall that D=1 corresponds to no loss of instantaneous stiffness, whereas D=0 corresponds to the complete loss of stiffness. Here, for each of the RVEs, the D values drop first for the case with σ_{cr}=20 MPa followed by the case with σ_{cr}=30 MPa and lastly by the case with σ_{cr}=45 MPa. This is expected as damage would occur first in the RVE that has the lowest critical strength. Consequently, the RVEs with σ_{cr}=20 MPa are also the first to stop curing (locally, at those locations where cracking has occured), on account of microcrack formation. The simulations with σ_{cr} values of 30 and 45 MPa first exhibit microcracks during the cooling phase (5,000 s≤t≤6,000 s) of the cure cycle where additional shrinkage occurs due to cooling. It is interesting to note that although microcracks appear last for the case with σ_{cr}=45 MPa, the drop in D is more drastic when compared to drops corresponding to the other two cases. Hence, at the end of the cure cycle, for each RVE, the extent of damage varies inversely with the critical fracture strength σ_{cr} of the curing matrix. The spatial variation of D at the end of cure is shown in Figure 6. Even though the RVE is symmetric about a vertical and horizontal line passing through its center, there is nonhomogeneity in the contour of D across the matrix. The nonhomogeneity in D arises because the stress distribution in the RVE does not strictly follow the symmetry present in the hexagonal packing during cure on account of small numerical differences. Hence, once cracking starts at locations where stresses are highest, this breaks the symmetry in stress distribution, thus leading to subsequent nonhomogeneity in D as the curing progresses. In these curing simulations, two-piece failure (corresponding to D=0) was not observed in the cured matrix.

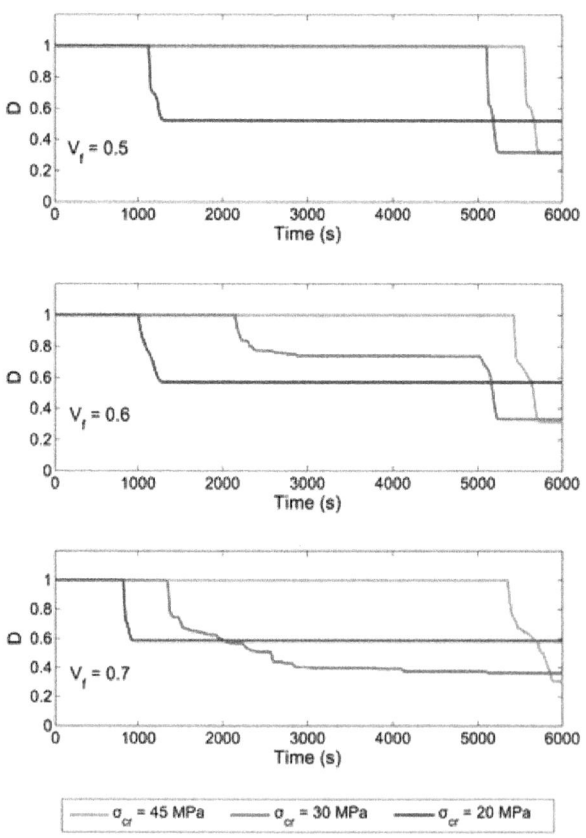

Figure 5: Plots of 'average' stiffness reduction factor D for virtually cured hexagonally packed RVEs during the cure cycle.

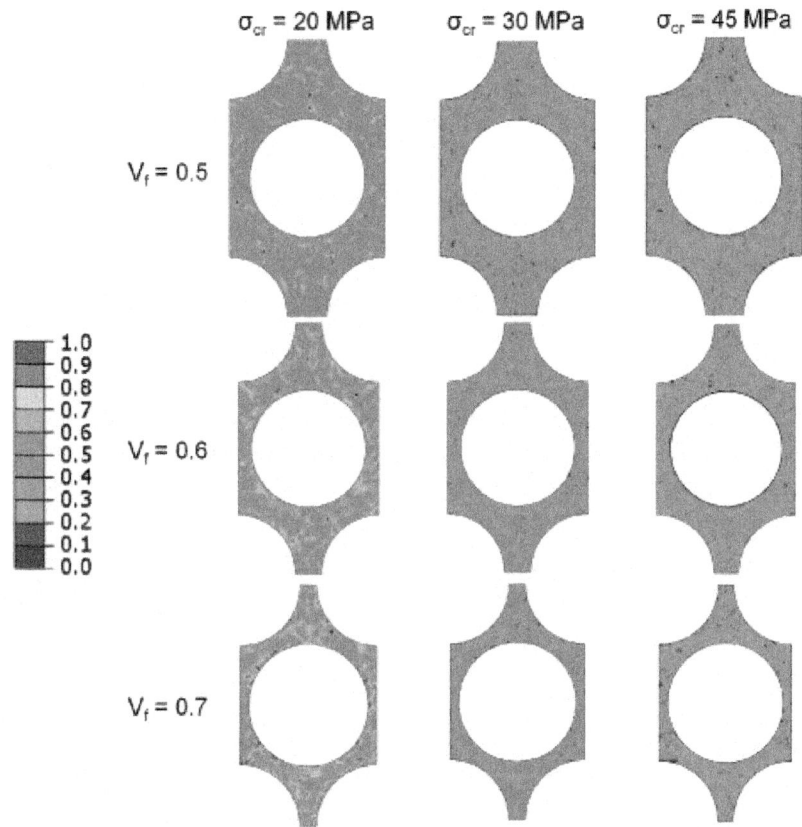

Figure 6: Stiffness reduction factor D for the hexagonally packed RVEs at the end of cure.

The virtually cured RVEs are now loaded in tension along the transverse direction in step III. As in the previous step, periodic boundary conditions are enforced. The nominal stress-strain (σ_{22}–ε_{22}) response is shown in Figure 7. Each of the cured RVEs exhibits a fairly linear response during the initial stages of loading (up to nominal strain), followed by a nonlinear softening response before attaining the peak. Past the peak, a rapid drop in stress is observed. The peak stress values correspond to the transverse strength S_{22}^{+} of the virtually cured RVEs. In the case where cure-induced damage is ignored, and when the RVEs are loaded under tension in the transverse direction, the resulting stress-strain response is shown in Figure 8. These RVEs also exhibit a fairly linear response during the initial stages of loading. However, the extent of nonlinearity present before the peak is much lesser than the case when cure-induced damage is accounted for (see Figure 7). The RVEs with no cure-induced damage exhibit higher global stiffness compared to those when cure-induced damage is taken into account. This is as expected. In the post-peak region, the crack paths for simulations with σ_{cr}=20 MPa for virtually cured RVEs and for RVEs where cure-induced damage has not been taken into account are shown in Figure 9 and in Figure 10, respectively. Figure 11 shows the variation of the initial stiffness (E_{22}) of RVEs under mechanical loading. For a given RVE with volume fraction held fixed, the lowest stiffness is exhibited by the RVE having the highest σ_{cr} value of 45 MPa in step II. Recall that from Figure 5, this case with σ_{cr}=45 MPa had the lowest value of D at the end of the cure cycle. Thus, the stiffness reduction factor D is seen to have a positive correlation with global transverse stiffness E_{22} under mechanical loading. Figure 12 shows the comparison of transverse tensile strength S_{22}^{+} values of the virtually cured RVEs and for those RVEs where cure-induced damaged has not been taken into account. It can be seen that for all the volume fractions and all σ_{cr} values considered, the RVEs that have cure-induced damage have noticeably lower strength values.

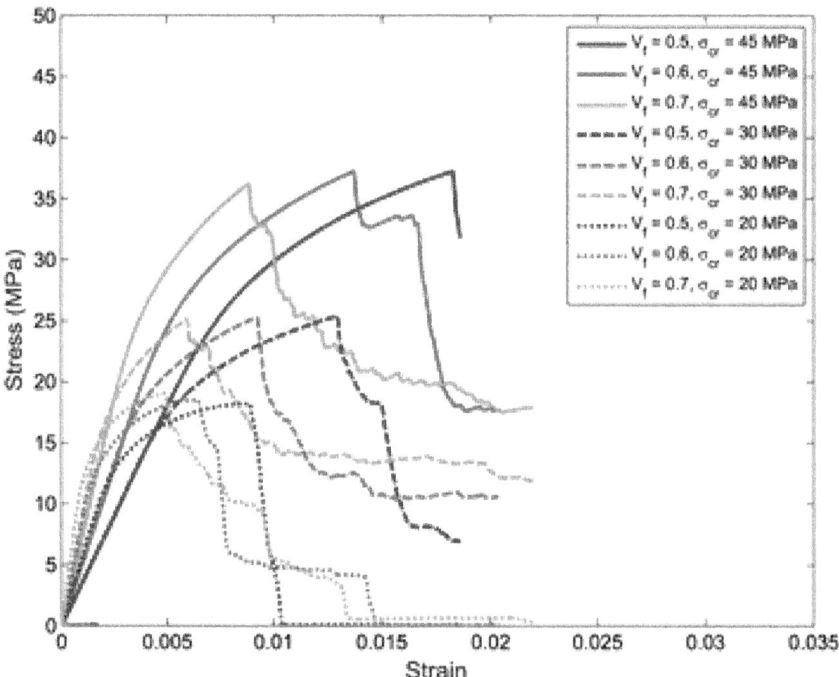

Figure 7: Nominal stress-strain (σ_{22}–ε_{22}) response of the hexagonally packed RVEs during step III under transverse tension.

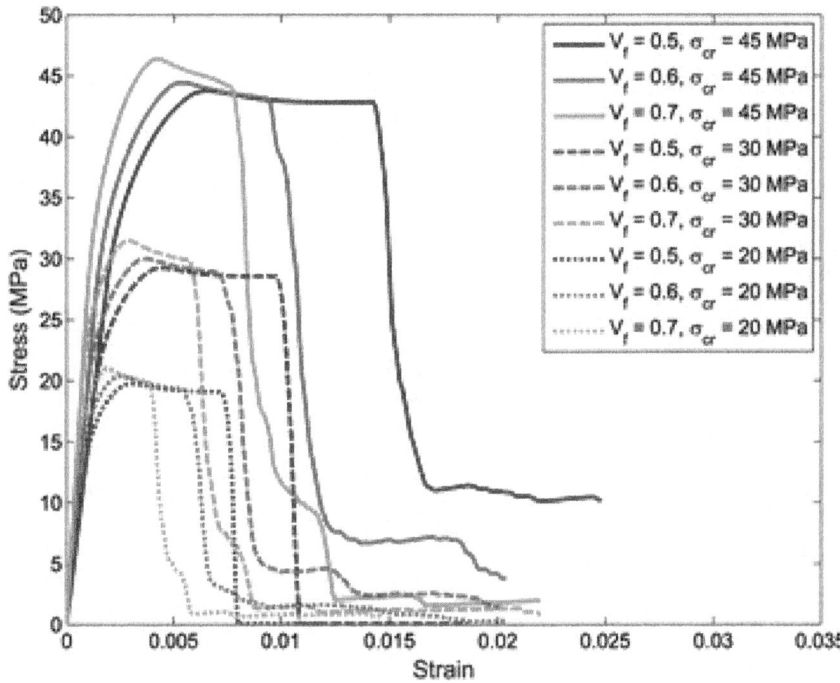

Figure 8: Nominal stress-strain (σ_{22}–ε_{22}) response of the hexagonally packed RVEs under transverse tension loading when no damage during cure is assumed.

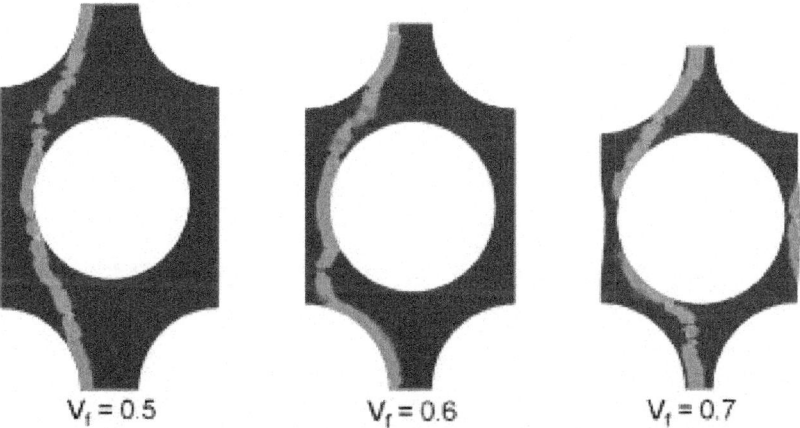

Figure 9: Crack paths under transverse tension loading in hexagonally packed RVEs with σ_{cr}=20 MPa.

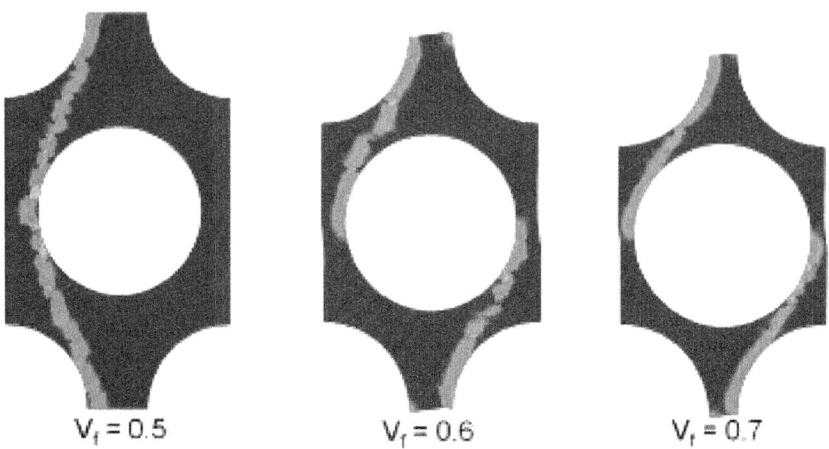

$V_f = 0.5$ $V_f = 0.6$ $V_f = 0.7$

Figure 10: Crack paths under transverse tension loading in hexagonally packed RVEs with σ_{cr}=20 MPa, ignoring cure-induced damage.

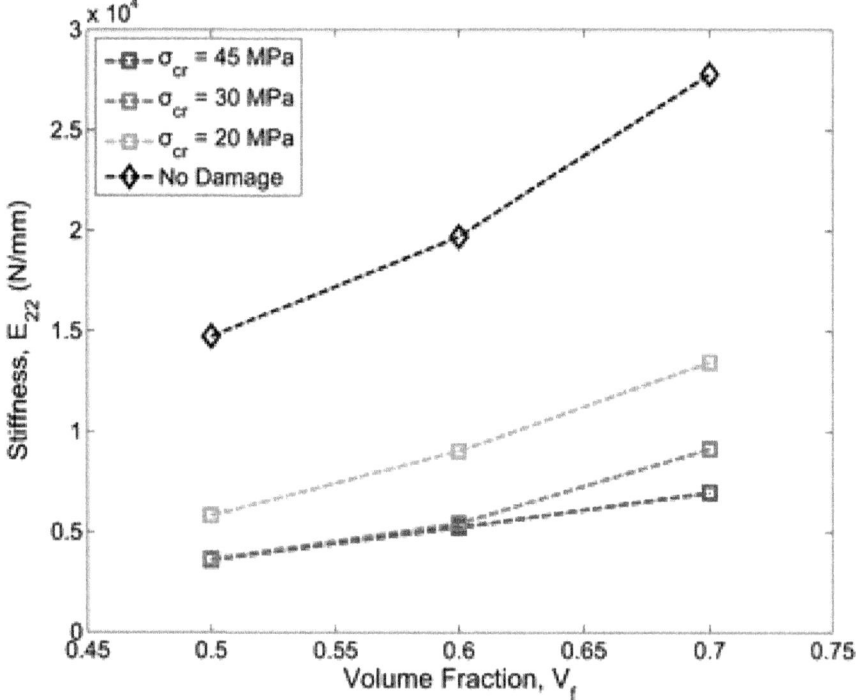

Figure 11: Transverse stiffness E_{22} in hexagonally packed RVEs compared to the case with no cure-induced damage.

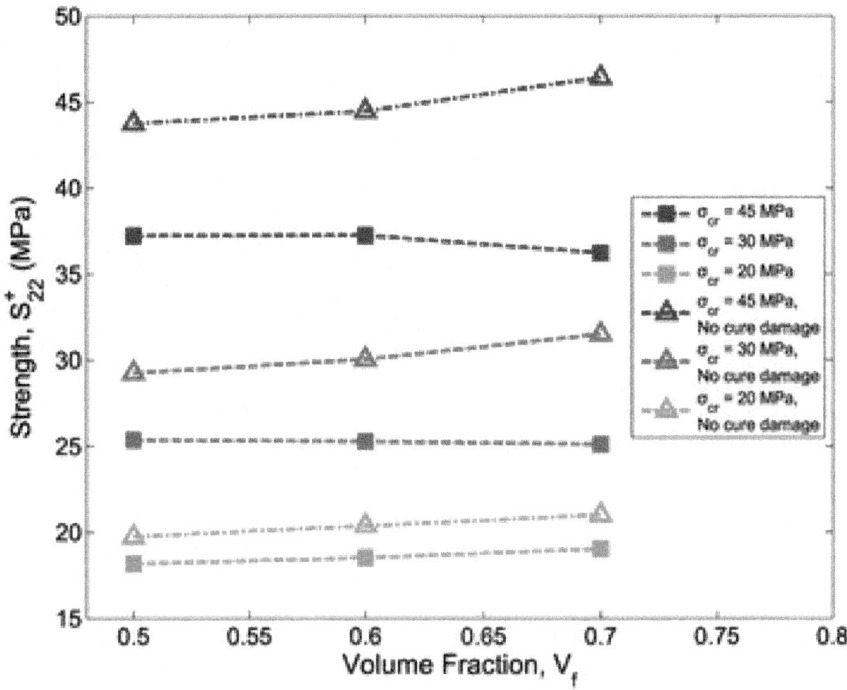

Figure 12: Transverse tensile strength S+22S22+ in hexagonally packed RVEs: with and without cure-induced damage.

Randomly packed fiber RVEs

Although the RVEs discussed thus far are idealized hexagonally packed geometries, they do not represent a RVE of a realistic FRPC sample. In realistic FRPCs, the fibers are randomly distributed which give rise to several matrix-rich pockets. It would be instructive to understand the severity of the cure-induced damage on the mechanical response, as a function of the randomness in fiber position in an RVE. Eight renditions of square FRPC RVEs with randomly distributed fibers are analyzed in this section. The distribution of fibers within the RVEs was done manually, in that the fibers were arbitrarily placed within the square RVE boundary. The fiber volume fraction (V_f) in all these renditions is chosen to be 0.55. These RVEs are shown in Figure 13. Few strategies to generate random RVEs may be found in the studies of Melro et al. [23], Yang et al. [24], and Vaughan and McCarthy [25]. Recently, using a heuristic random microstructure algorithm, Romanov et al. [26] have generated RVEs that are statistically well correlated with real FRPC RVEs.

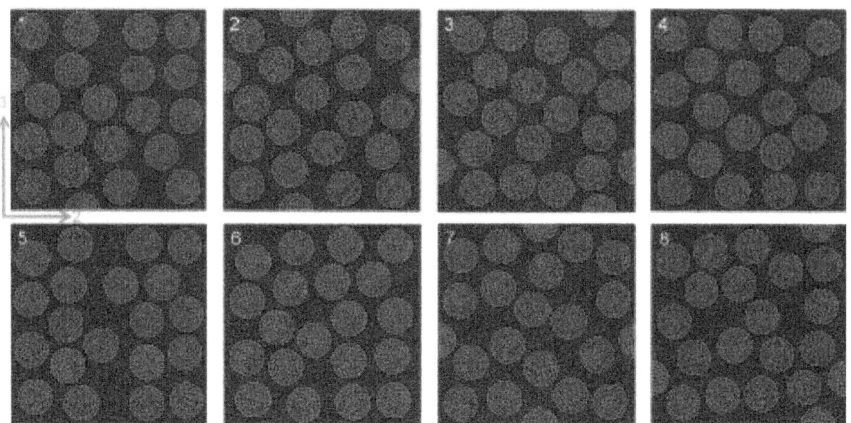

Figure 13: Eight renditions of random 20-fiber RVEs with volume fraction V_f=0.55.

The cure cycle, fiber, and matrix properties are similar to those used in the aforementioned study with hexagonally packed RVEs. A preliminary analysis on the mesh size has been conducted on a random RVE to establish that important features such as the stiffness reduction factor D at the end of step II and crack path at the end of step III are both mesh insensitive for the range of element sizes analyzed in this study. Figure 14 shows three different levels of refinement for a random packed case study. Results in terms of the factor D and crack path are shown in the top and bottom images of Figure 15, respectively. It can be seen that the spatial distribution of D and the two-piece failure paths are fairly consistent between the three meshes considered, thus establishing mesh objectivity.

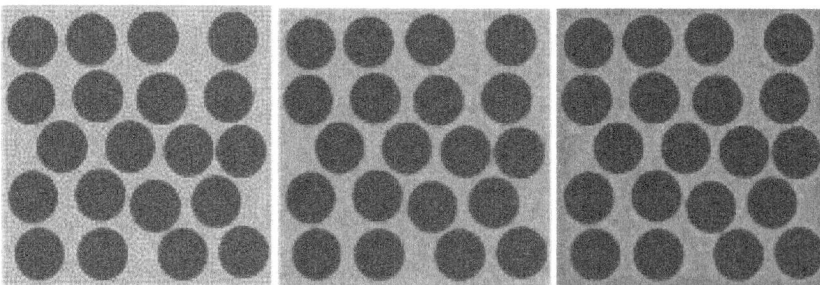

Figure 14: Three mesh sizes chosen for the mesh convergence study.

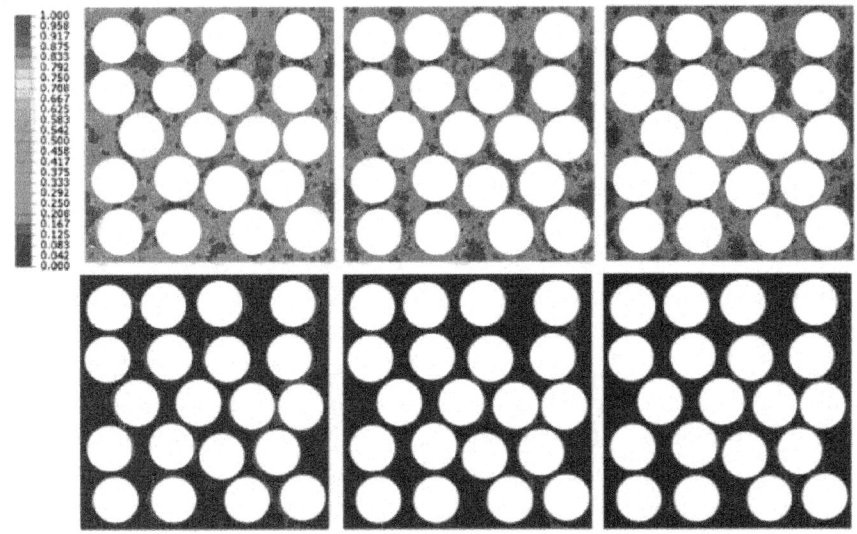

Figure 15: Stiffness reduction factor D at the end of cure (top) and crack path (bottom) for three mesh sizes.

Next, the eight random fiber RVEs shown in Figure 13 are cured with the crack band model (with critical fracture stress σ_{cr}=30 MPa during cure) prescribed to capture any local matrix damage during cure. Periodic boundary conditions are enforced. The stiffness reduction factor D for these renditions is shown in Figure 16. Each of the RVEs exhibits nonhomogeneity in the contour for D. Note that on account of inherent randomness in fiber packing, there is no symmetry in the RVE at the start of cure. Matrix region areas that are surrounded by closely packed fibers are seen to exhibit higher stresses. This introduces stress gradients in different parts of the RVE during cure. Then, damage initiates at locations where the tensile stresses attain the critical fracture strength σ_{cr}. Therefore, different regions in the virtually cured matrix end up having nonhomogeneous stiffness values owing to different regions damaging differently during cure. These RVEs are next subjected to transverse tension loading along the 2-direction. The nominal stress-strain response is shown in Figure 17. It can be seen that the response is fairly linear during initial stages of loading (for nominal strain $0 \leq \varepsilon \leq 0.0025$). Beyond $\varepsilon > 0.0025$, there is nonlinear response followed by a peak value between strains of about $\varepsilon \approx 0.008$. Past the peak, the response is like that of a brittle solid, i.e., there is a drastic drop in stress due to two-piece matrix tensile failure. The two-piece crack paths for each of the random RVEs are shown in Figure 18. In some of the RVEs, the crack path is more tortuous than others. It is interesting to look at RVE #5, where there is a prominent and continuous matrix-rich regiontransverse to the loading direction. The two-piece crack path in this RVE at the end of step III is seen to propagate along a zone that has fibers that are more closely packed, which is away from the matrix-rich region. Similar observation holds for RVEs #1, #7, and #9 which have prominent but isolated matrix-rich regions. The matrix which is in a region where fibers are closely packed encounters higher stresses during curing as well as during mechanical loading and is more susceptible to cracking. Thus, cracks tend to initiate and propagate from such sites. The global stress-strain response of the random RVEs when cure-induced damage is not considered is shown in Figure 19. Here, the initial region is fairly linear (for nominal strain $0 \leq \varepsilon \leq 0.002$). However, the stiffness values in the initial region are identical compared to the RVEs with cure-induced damage, where there is a larger spread of the initial stiffness value owing to nonuniform stiffness distribution in the damaged matrix. There is some nonlinearity in the response which is much lesser than that seen in the RVEs when cure-induced damage was considered. A peak is attained beyond which the stress value plateaus momentarily followed by a drastic drop in stress. Finally, the strength values of the two cases (with and without cure-induced damage) are shown in Figure 20. For each of the RVEs with no cure-induced damage, the transverse tensile strengths are higher, i.e., mean S_{22}^{+} =31.2 MPa compared to mean S_{22}^{+} =25.75 MPa for RVEs with cure-induced damage. Moreover, the scatter in strength values in the RVEs with cure-induced damage is much larger than when no cure-induced damage is taken into account.

Figure 16: Stiffness reduction factor D in the random fiber RVEs at the end of cure.

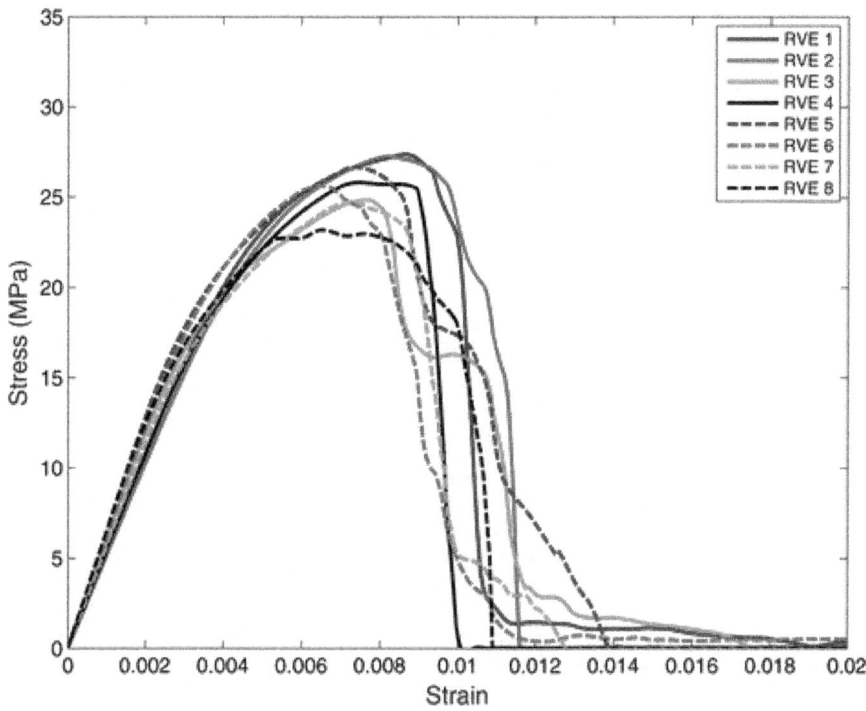

Figure 17: Nominal stress-strain (σ_{22}–ε_{22}) response of virtually cured random fiber RVEs having cure-induced damage.

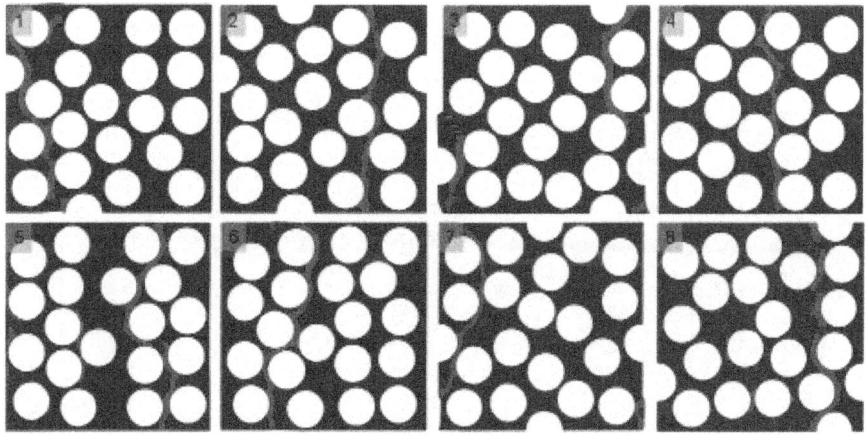

Figure 18: Two-piece crack paths in the randomly packed RVEs under transverse tension loading.

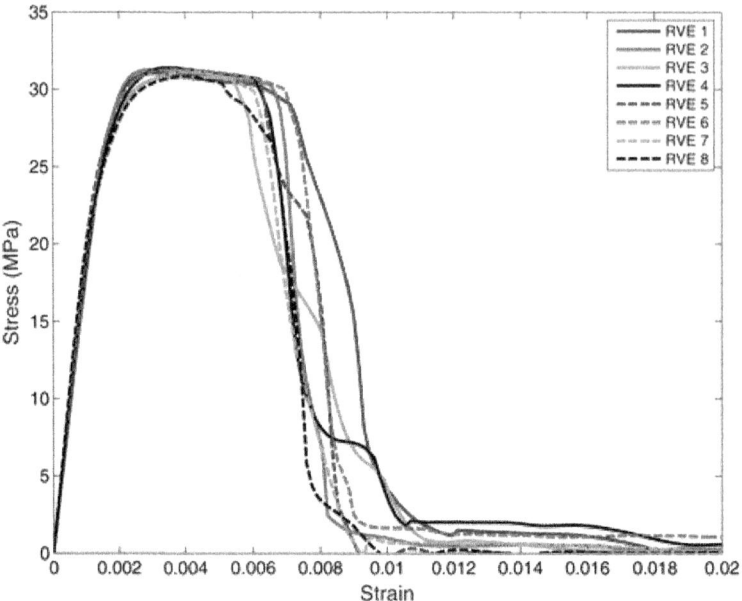

Figure 19: Nominal stress-strain ($\sigma_{22}-\varepsilon_{22}$) response of virtually cured random fiber RVEs with no cure-induced damage.

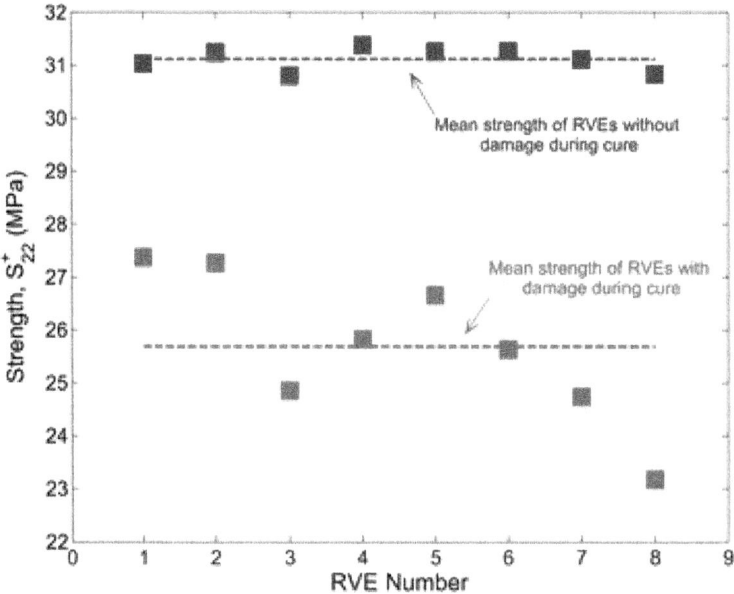

Figure 20: Transverse tensile strength of each random fiber RVE plotted against RVE number.

In the foregoing sections, we have established that damage in the matrix during cure results in a lower transverse strength value in the virtually cured hexagonally packed RVEs as well as virtually cured randomly packed RVEs. Thus, a comparison of the transverse tensile responses of these two types of RVEs is in order. In hexagonally packed RVEs having a fixed value of critical tensile fracture stress σ_{cr}, the transverse strength S_{22}^{+} did not seem to vary with volume fractions ranging between 0.5 and 0.7 (see Figure 12). However, in randomly packed RVEs, the S_{22}^{+} values exhibited significant scatter (see Figure 20). When compared to the case with no damage during cure, the mean transverse tensile strength reduction for hexagonally packed RVEs (15 % reduction) and that for randomly packed RVEs (17 %reduction) are similar. The worst strength reduction in randomly packed RVE is 25 % for RVE #8. This shows that fiber packing within an RVE has an effect on the strength of the cured RVE with volume fraction held fixed. The effect of microstructural randomness on mechanical response is detailed in the monograph by Ostoja-Starzewski [27].

CONCLUSIONS

The influence of cure on the mechanical response of virtually cured fiber-reinforced polymer matrix RVEs has been studied using a previously reported network model proposed by Heinrich et al. [15] in conjunction with the Bažant-Oh crack band model. Transverse tensile strength (S_{22}^{+} and transverse stiffness (E_{22}) of these virtually cured RVEs were compared with those when no cure-induced damage was taken into account. These two quantities were calculated from the nominal stress-strain response of the virtually cured RVEs subjected to tensile loading along the transverse direction. Damage during cure is seen to reduce both stiffness and strength of the cured RVEs. Moreover, it is seen that even though fiber volume fraction is held fixed, the transverse strength of the virtually cured RVEs depend on the fiber packing. Also, the scatter in transverse strength values for RVEs with cure-induced damage is appreciably higher than in the case where cure-induced damage has been neglected. Since fiber packing is seen to influence strength of the cured RVE of constant fiber volume fraction, the study of correlating strength values with some metric associated with fiber packing is delegated to a future study. Using the approach described in this paper, several cure cycles can be considered and eventually tailored to arrive at an optimal cure cycle to reduce damage in the microstructure during the curing process, leading to superior mechanical strength and stiffness of the cured product.

ACKNOWLEDGEMENTS

The authors thank Dr. Pascal Meyer and Dr. Christian Heinrich, of the Aerospace Engineering Department at the University of Michigan, Ann Arbor, and Prof. Pavana Prabhakar, Mechanical Engineering Department, University of Texas, El-Paso, for support with the user-defined subroutines used in the present work. The support of the Department of Aerospace Engineering, University of Michigan, Ann Arbor and the William E. Boeing Department of Aeronautics and Astronautics at the University of Washington, Seattle, is gratefully acknowledged.

COMPETING INTERESTS

The authors declare that they have no competing interests.

REFERENCES

1. Plepys AR, Farris RJ (1990) Evolution of residual stresses in three-dimensionally constrained epoxy resins. Polymer 31(10): 1932–1936.

2. Plepys AR, Vratsanos MS, Farris RJ (1994) Determination of residual stresses using incremental linear elasticity. Composite Struct 27(1-2): 51–56.

3. Merzlyakov M, McKenna GB, Simon SL (2006) Cure-induced and thermal stresses in a constrained epoxy resin. Composites: Part A 37: 585–591.

4. Chekanov YA, Korotkov VN, Rozenberg BA, Dhzavadyan EA, Bogdanova LM (1995) Cure shrinkage defects in epoxy resins. Polymer 36: 2013–2017.

5. Rabearison N, Jochum C h, Grandidier JC (2009) A FEM coupling model for properties prediction during the curing of any epoxy matrix. Comput Mater Sci 45(3): 715–724.

6. Ahn J, Waas AM (2002) Prediction of compressive failure in laminated composites at room and elevated temperature. AIAA Journal 40(2): 346–358.

7. Song S, Waas AM, Shahwan KW, Xiao X, Faruque O (2007) Braided textile composites under compressive loads: modeling the response, strength and degradation. Composite Sci Technol No. 67: 3059–3070.

8. Kim K, Hahn H (1989) Residual stress development during processing of graphite/epoxy composites. Composites Sci Technol 36: 121–132.

9.   Li M, Zhu Q, Geubelle PH, Tucker III CL (2001) Optimal curing for thermoset matrix composites: thermochemical considerations. Polymer Composites 22: 118–131.

10. Gopal AK, Adali S, Verijenko VE (2000) Optimal temperature profiles for minimum residual stress in the cure process of polymer composites. Composite Struct 48: 99–106.

11. White S, Hahn H (1993) Cure cycle optimization for the reduction of processing-induced residual stresses in composite materials. J Composite Mater 27: 1352–1378.

12. Halpin JC, Kardos JL (1976) Halpin-Tsai equations: a review. Polymer Eng Sci 16(5): 344–352.

13. Mei Y (2000) Stress evolution in a conductive adhesive during curing and cooling. Ph.D Thesis, University of Michigan.

14. Mei Y, Yee AS, Wineman AS, Xiao C (1998) Stress evolution during thermoset cure. Mater Res Soc Symp Proc 515: 195–202.

15. Heinrich C, Alridge M, Wineman AS, Kieffer J, Waas AM, Shahwan KW (2012) Generation of heat and stress during the cure of polymers used in fiber composites. Int J Eng Sci 53: 85–111.

16. Kamal MR (1974) Thermoset characterization for moldability analysis. Polymer Eng Sci 14(3): 231–239.

17. Li C, Potter K, Wisnom MR, Stinger G (2004) In-situ measurement of chemical shrinkage of MY750 epoxy resin by a novel gravimetric method. Composites Sci Technol 64(1): 55–64.

18. Simulia (2012) Abaqus user manual, version 6.12. Dassault Systèmes, Providence, RI, USA.

19. Bažant ZP, Oh B (1983) Crack band theory for fracture of concrete. Mater Struct 16(3): 155–177.

20. Jirasek M, Bažant ZP (2002) Inelastic analysis of structures. John Wiley & Sons, London and New York.

21. Gonzalez C, Llorca J (2007) Mechanical behavior of unidirectional fiber-reinforced polymers under transverse compression: microscopic mechanisms and modeling. Composites Sci Technol 7: 2795–2806.

22. Xia Z, Zhang Y, Ellyin F (2003) A unified periodical boundary conditions for representative volume elements of composites and applications. Int J Solids Struct 40: 1907–1921.

23. Melro AR, Camanho PP, Pinho ST (2008) Generation of random distributions of fibers in long-fibre reinforced composites. Composites Sci Technol 68(9): 2092–2102.

24. Yang L, Ying Y, Ran Z, Liu Y (2013) A new method for generating random fiber distributions for fiber reinforced composites. Composites Sci Technol 76: 14–20.

25. Vaughan TJ, McCarthy CT (2010) A combined experimental-numerical approach for generating statistically equivalent fiber distributions for high strength laminated composite materials. Composite Sci Technol 70(2): 291–297.

26. Romanov V, Lomov SV, Swolfs Y, Orlova S, Gorbatikh L, Verpoest I (2013) Statistical analysis of real and simulated fibre arrangements in unidirectional composites. Composite Sci Technol 87: 126–134.

27. Ostoja-Starzewski M (2007) Microstructural randomness and scaling in mechanics of materials. Chapman and Hall-CRC 2007, Florida, USA.

Chapter 10

MICROMECHANICS MODELING OF THE ELECTRICAL CONDUCTIVITY OF CARBON NANOTUDE (CNT)- POLYMER NANOCOMPOSITES

ABSTRACT

The addition of carbon nanotubes (CNTs) in polymers to form conductive composites has been attracting great interest from research and industry communities due to their potential applications. Experiments and simulations have demonstrated that the addition of a very small amount of CNTs into polymers can significantly improve the electrical conductivity of the composites. Such significant improvement in the electrical conductivity is attributed to two conductivity mechanisms: nanoscale electron hopping and microscale conductive networks. Understanding and prediction of the overall electrical conductivity of the composites with the incorporation of the conductivity mechanisms that underpin the macroscopic electrical properties are essential for their engineering applications. One of the most promising applications of the conductive composites is for stretchable electronics. For such an application, it is naturally necessary to investigate the stretching effects upon the overall electrical conductivity of the composites. Furthermore, CNTs dispersed in polymers are usually not straight but rather have a certain degree of waviness due to the CNTs' large aspect ratio and low bending stiffness. It has been suggested that the waviness can have considerable effect on the electrical conductivity of the composites. Therefore, the investigation of CNT waviness effect is of great importance for the prediction of the overall electrical conductivity of the composites.

In this thesis, based on the micromechanics theory, a mixed micromechanics model with the incorporation of the nanoscale electron hopping and the microscale conductive networks is first developed to predict the electrical conductivity of the composites. The modeling results successfully predict the trend of existing experimental data. It is found that both the electron hopping and the conductive networks contribute to the electrical conductivity of the composites while conductive networks become dominant to the electrical conductivity of the composites after percolation. It was also indicated that the sizes of CNTs have significant effects on the percolation threshold and the overall electrical conductivity of the nanocomposites. Based on the developed micromechanics model, stretching effects are then investigated by incorporating the stretching induced changes into the micromechanics model. The investigation found that the stretching, including uni-axial and bi-axial stretching, decreases the electrical conductivity of the composites in the stretching direction and the decrease is more evident for the bi-axial stretching compared to uni-axial stretching. It is also observed that the electrical conductivity is more sensitive to stretching for the composites with lower CNT volume fraction. Finally, we studied the CNT waviness effects upon the electrical conductivity of the composites under a uni-axial stretching. It is demonstrated that the waviness significantly decreases the electrical conductivity of the composites and the electrical conductivity is more sensitive to the waviness for the composites with lower CNT volume fraction and larger stretching strain. Reasons for the observed variations and phenomena are interpreted. The work in this thesis is expected to obtain increased understanding on the overall electrical conductivity of CNT-polymer composites from the theoretical perspective and provide useful guidelines for the design and optimization of the composites.

CNT-POLYMER NANOCOMPOSITES

Electrically conductive polymer composites have been attracting interest from academic and industrial communities since 1950s due to their excellent combination of electrical conductivity and beneficial attributes of polymers. The basic idea of producing conductive polymer composites is adding conductive fillers into polymers, which are usually insulators (Narkis et al., 1978; Tchoudakov et al., 1997; Feller et al., 2002). Among conductive fillers, carbon black and metals are two typical conductive fillers which are traditionally used to produce conductive or semi-conductive polymer composites. However, since the discovery of CNTs in 1991 (Iijima, 1991), the excellent mechanical and physical properties as well as high aspect ratio have made CNTs one of the most preferred conductive fillers to develop multi-functional conductive polymer composites during the past decades. Due to the large disparity of electrical conductivity between the polymer matrix and CNTs, i.e., the electrical conductivity of CNTs is several orders of magnitude larger than that of all neat polymers (Ebbesen et al., 1996), a very small amount of CNTs added into polymers can remarkably improve the electrical conductivity without significantly reducing the intrinsic properties of the polymers. Compared to traditionally conductive or semi-conductive materials, such as silicon, metal, ceramics etc, CNT-polymer nanocomposites possess desirable conductive properties from conductive fillers while keeping the beneficial attributes of the polymers, such as flexibility, light weight, easy processability and good chemical and biological compatibility (Yu et al., 2009; Nambiar and Yeow, 2011; Shang et al., 2011). Such unique combination of the excellent electrical conductivity of CNTs and the beneficial features of polymers have made CNT-polymer conductive nanocomposites one of the most promising material candidates for a variety of applications, such as stretchable electronics, conductive coatings, electromagnetic shielding and solar cells (Yang et al., 2005; Berson et al., 2007 Yu et al., 2009; Shang et al., 2011), which cannot be achieved by the traditional rigid conductive or semi-conductive materials.

Electrical behaviors of CNT-polymer nanocomposites

In addition to the as-mentioned features of CNT-polymer nanocomposites, i.e., a very small amount of CNTs added into polymer can significantly improve the overall electrical conductivity of the composites, it has been confirmed by both experiments and simulations that the electrical conductivity demonstrates a percolation-like behavior as shown in Figure 1.1 (Kim et al., 2005), i.e., the electrical conductivity of the composites increase abruptly when the CNT volume fraction reaches a certain critical value, which is usually referred as percolation threshold (Kim et al., 2005; Gojny et al., 2006; Yan et al., 2007; Seidel and Lagoudas, 2009; Qunaies et al., 2013).

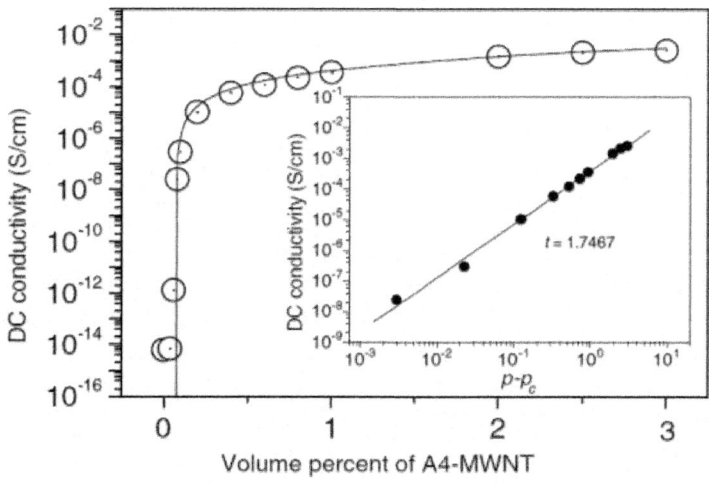

Figure 1.1: Variation of electrical conductivity of CNT-polymer composite with CNT volume fraction (Kim et al., 2005).

Explanations for the percolation-like behavior attribute to two conductivity mechanisms: electron hopping (or quantum tunneling) at the nanoscale and conductive networks at the microscale (Ounaies et al., 2003; Du et al., 2004; Chang et al., 2009; Zhang et al., 2009). As argued by Deng and Zheng (2008), the contribution of the electron hopping and the conductive networks to the electrical conductivity of the composites depends on the CNT concentration. From the perspective of quantum mechanics, electrons always have the probability of hopping intra-tube or from one CNT to another (as illustrated in Figure 1.2), but the probability is highly dependent on the separation distance between CNTs (Seidel and Lagoudas, 2009). When the CNT concentration in the composite is extremely low with larger separation distance between CNTs, independent and electron hopping governs the electrical conductivity of the composite. However, when the separation distance between CNTs decreases with the CNT concentration, some adjacent CNTs may be electrically connected resulting in microscale conductive networks Figure Figure 1.3

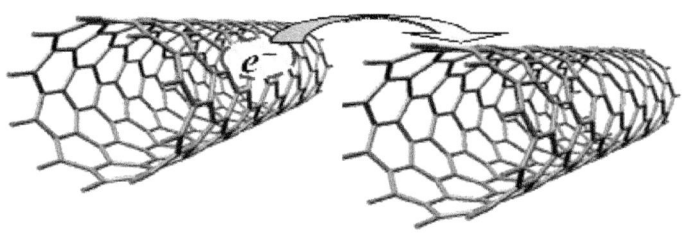

In addition to the dependency of the electrical conductivity of the composites on the CNT concentration and the conductivity mechanisms, it has been demonstrated that stretching can also significantly affect the electrical performance of the CNT-polymer composites (Park et at, 2008; Cheng et al., 2009; Hu et al., 2010; Miao et al., 2011; Miao et cll., 2011; Shang et at, 2011; Wang et at, 2011; Miao et at, 2012; Mayoral et al., 2013), which is of great importance for the application of the composites as stretchable electronics. Figure 1.4 (Park et al., 2008) shows an example of the stretching effects on the resistance of the composites for different CNT volume fractions. From the figure, it can be seen that the stretching increases the resistance of the composites, especially for lower CNT volume fraction (i.e., MWCNT 0.56 vol%). Therefore, to facilitate the full potential application as stretchable electronics it is naturally necessary to investigate the stretching effects on the electrical behavior of the composites. Furthermore, due to CNTs' large aspect ratio and low bending stiffness (Li et al., 2008), it is well observed and accepted that CNTs dispersed in polymers are usually not straight but rather have a certain degree of waviness. Such wavy feature of the CNTs is suggested to have considerable effects on the electrical behavior of the composites (Yi et al., 2004; Li et al., 2008; Takeda et al., 2011; Yu et al., 2013), i.e., increasing waviness of CNTs will decrease the electrical conductivity of the composites. Therefore, the consideration of CNT waviness effects is necessary and essential for the improvement on the prediction of the electrical conductivity of CNT-polymer composites.

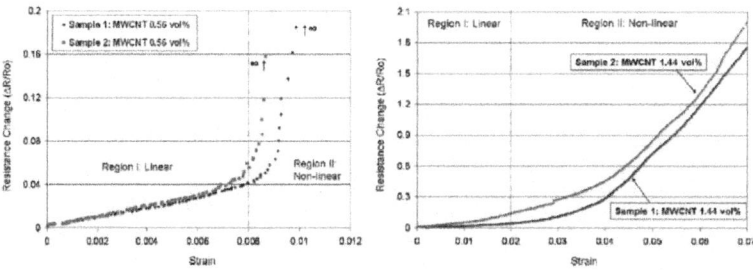

Figure 1.4: Stretching effects on the electrical property of CNT-polymer composites (Park et al., 2008).

Literature review

Although extensive explorations by experiments and simulations have been done on investigating the electrical conductivity of the composites, there is relatively less theoretical work in this topic. Therefore, this thesis will work on the investigation from the perspective of theoretical modeling and make qualitative predictions on the electrical conductivity of the composites. In this literature review section, existing studies on the investigation of the electrical conductivity of the nanocomposites involving the effects of the as-mentioned factors will be presented and discussed in the following sub-sections.

Studies on the electrical conductivity of CNT-polymer composites For the electrical conductivity of CNT-polymer composites, extensive experimental work has been found to investigate the variation of the electrical conductivity with CNT volume fraction. Sandler et al. (1999) dispersed catalytically-grown carbon nanotubes in an epoxy matrix and found that compared to carbon black the use of CNTs reduced the percolation threshold and increased the overall electrical conductivity of the composites. Ramasubramaniam and his co-workers (2003) fabricated homogeneous carbon nanotube/polymer composites and found that very low CNTs loading is required to achieve satisfactory electrical conductivity without compromising the beneficial properties of the polymer. Qunaies et al. (2003) and Gojny et al. (2006) evaluated the electrical conductivity of CNT reinforced polyimide composites and epoxy resin, respectively. Both of their experiments demonstrated that the electrical conductivity increased with the increase of CNT volume fraction and concluded that the electrical conductivity of the composites could be attributed to the formation of conductive pathways. There are some other experimental work (Barrau et al., 2003; Grunlan et al., 2004; Kim et al., 2005) that had observed the same phenomena, i.e., the addition of a very small amount of CNTs into polymers can significantly increases the electrical conductivity of the composites.

In addition to experimental work, there also exist modeling and simulation work on the investigation of the electrical conductivity of the composites. Regardless of the conductivity mechanisms, percolation theory, in which a power law equation is applied, had been widely adopted to predict the electrical conductivity of the composites after percolation (Kirkpatrick, 1973; Grimmett, 1999; Ramasubramaniam et al., 2003; McLachlan et al., 2005). Comparisons showed that the power law equation successfully predicted the trend of experimental data. However, such proposed power law equation is phenomenological and several parameters in this power law equation need to be fitted from existing experimental data. In addition, the power law equation cannot capture the variation of electrical conductivity prior to percolation and is unable to distinguish the two conductivity mechanisms. In addition to the percolation theory, Monte Carlo (MC) and Molecular Dynamics (MD) simulations have been considered as effective ways to study the CNT-polymer composites (Ma and Gao, 2008; Zhang and Yi, 2008; Lu et al., 2010). However, MD and MC simulations are numerically expensive for large-scale simulation and impractical to obtain the overall property of the composites. In addition, MC and MD simulations cannot provide any explicit formulation for material design and optimization. Alternatively, some micromechanics theory based models have also been extended to predict the overall electrical conductivity of CNT-polymer nanocomposites.

Deng and Zheng (2008) developed a simplified micromechanics model to evaluate the effective electrical conductivity for CNT composites by accounting for the percolation, conductive networks, conductivity anisotropy and non-straightness of CNTs. Based on Deng and Zheng's model, Takeda et al. (2011) considered the effect of electron hopping among CNTs to predict the overall electrical conductivity of the composites. However, both Deng and Takeda's work did not involve any prediction of the electrical conductivity before percolation. The influence of electron hopping and the formation of conductive networks on the electrical conductivity of CNT-polymer composites was also investigated by Seidel and Lagoudas (2009) using a Mori-Tanaka micromechanics model. In their work, electron hopping was assessed through an interphase layer surrounding the CNT, while the effect of conductive networks was captured by changing the CNT aspect ratio. The developed micromechanics model was successful in qualitatively identifying the potential causes for the low percolation concentrations. However, large discrepancy was observed between Seidel's predicted results and experimental data after the percolation. It should be mentioned that in Seidel's work (2009), the thickness and the electrical conductivity of the interphase layer were kept constant, and the two electrical conductivity mechanisms were considered separately in the simulations, i.e., either electron hopping or conductive networks solely dominated the electrical conductivity. From a statistical mechanics point of view, it is believed that some of CNTs in the polymer form conductive networks, while others contribute to the effective electrical conductivity of the composites through the electron hopping (Deng et al., 2008). The probability of the formation of conductive networks will increase with the increase of the CNT concentration.

In addition, the thickness and the electrical conductivity of the interphase layer accounting for the electron hopping will also vary with the CNT concentration. In order to more accurately predict the overall electrical conductivity of CNT-polymer composites, all these factors need to be incorporated into a micromechanics model.

Investigation on stretching effects

Although efforts have been devoted to investigating the electrical properties of CNTpolymer nanocomposites through experiments and simulations (Kim et al., 2005; Gojny et al., 2006; Li et al., 2007; Yan et al., 2007; Chang et al., 2009; Seidel et al., 2009; Takeda et al., 2011), most of the existing studies were focused on preparing composites with well dispersed conductive fillers or predicting the electrical behavior of the asreceived composites without considering stretching effects. However, for the potential application of the composites as stretchable electronics it is necessary to investigate the stretching effects upon the overall electrical conductivity of the composites in order to accurately predict and control the device performance. Existing experiments have demonstrated that stretching may significantly influence the electrical behavior of conductive polymer composites. For example, under a uni-axial stretching, Bao et al. (2011) experimentally examined the morphology and the electrical conductivity of carbon nanofibre composites before and after stretching and showed that the mechanical stretching could lead to decrease in the electrical conductivity of the composites due to breakdown of conductive networks.

Park et al., and Hu et al., (2010) demonstrated that an abrupt increase in resistance of CNT-polymer composites subjected to a uni-axial stretching. In addition, Miao and his co-workers (2011a, 2011b and 2012) examined the piezoresistive response of CNT-polymer composites by experiments and percolation theory and indicated that the stretching decreases the electrical conductivity of the composites. There are some other experimental results having the same trend (Das N. C. et al., 2002; Dang et al., 2007). However, other experimental results (Cheng et al., 2009; Shang et al., 2011; Wang et al., 2011) presented an opposite trend, i.e., stretching increased the electrical conductivity of CNT-polymer composites due to substantial alignment enhancement of CNTs along the stretching direction. In addition to uni-axial stretching, bi-axial stretching is another typical stretching mode for CNT-polymer composites. Shen et al. (2012) and Mayoral et al. (2013) experimentally investigated the bi-axial stretching effects on the electrical conductivity of CNT-polymer composites and obtained similar contradictive trends as that for uni-axial stretching. However, both of the work stated that the bi-axial stretching enables the CNTs to re-orientate along the two stretching directions in the polymer matrix. Such a stretching mode is expected to reduce the anisotropy of the electrical properties of the composites in the stretching plane compared to the uni-axial stretching case, altering the electrical behavior of the composites. In addition to experimental studies, efforts have also been devoted in investigating the stretching effects upon the electrical conductivity of the CNT-polymer composites theoretically and numerically. Taya et al. (1998) and Lin et al. (2010) applied the fibre percolation model and Monte Carlo method, respectively, to investigate the stretching/compression effects upon the electrical properties of fibre-filled composites and their results showed that the deformation could shift the percolation threshold of the composites. Ghazavizadeh et al. (2011) proposed a 3-D Monte Carlo model to evaluate the effect of mechanical loading on the electrical percolation threshold of CNT reinforced polymers and indicated that a percolating nanocomposite became non-percolating under a unidirectional stretching. A recent study by Tallman and Wang (2013) also indicated that stretching could increase the percolation threshold due to the CNTs' re-orientation. Although contradictory stretching effects have been observed in the literature, i.e., some studies found stretching decreases electrical conductivity while others found stretching increases electrical conductivity, it is suggested that there are three major expected changes occurred during the stretching, including composite volume expansion, reorientation of conductive fillers and change in conductive networks, which may contribute to the variation of the overall electrical properties of the polymer composites. It should be mentioned that there is limited work on theoretical modeling with the consideration of the stretching induced three changes as mentioned, especially on bi-axial stretching. Therefore, another objective of the thesis is to investigate the effects of the stretching induced volume expansion, CNT re-orientation and conductive network change on the overall electrical conductivity of the CNT–polymer composites following the previously developed micromechanics modeling work.

Modeling on CNT waviness effects

As cited in previous sub-sections, numerous works have been done on investigating the electrical properties of the composites, including the work on stretching effects. However, most of the work was focusing on the composites with straight conductive fillers. It is well accepted that for CNT dispersed composites, CNTs in the composites are usually not straight but rather have a certain degree of waviness mainly due to their large aspect ratio and low bending stiffness (Shaffer and Windle, 1999; Qian et al., 2000; Li et al., 2008). The waviness of the CNTs is suggested to have considerable effects on the electrical conductivity of the composites (Yi et al., 2004; Li et al.; 2008; Takeda et al.; 2011; Yu et al.; 2013). Therefore, the consideration of the CNT waviness effects is necessary and essential for the improvement on the prediction of the properties of CNTpolymer composites. Due to the difficulty in experimentally characterizing CNT waviness, most of the existing works on investigating the CNT waviness effects were focused on using modeling and simulation techniques. For example, assuming wavy CNTs with a sinusoidal shape, Yi et al. (2004), Berhan and Sastry (2007) and Fisher et al. (2003) considered the effect of the waviness on the percolation onset of CNT-polymer composites. It was observed that the waviness increases the percolation threshold of the composites. Shi et al. (2004) considered CNTs being in a helical shape and investigated the CNT waviness effects on the mechanical properties of the composites. Their investigation found that CNT waviness decreases the elastic modulus of the composites. Approximating wavy CNTs as elongated polygons, Li et al.'s (2008) computational simulation on the CNT waviness effects suggested that the waviness could increase the percolation threshold and decrease the electrical conductivity and elastic stiffness of the composites. Recently, assuming wavy CNTs with bow and sinusoidal shapes, respectively, Dastgerdi et al. (2013) and Yanase et al. (2013) investigated the effects of CNT waviness on the mechanical property of CNT reinforced polymer composites by micromechanics model and demonstrated that the waviness could significantly reduce the stiffening effect of the CNTs on the composites.

From the existing studies on the waviness effects, it can be seen that most of the studies were focusing on the mechanical properties and relatively less theoretical work has been found on the investigation of the CNT waviness effects upon the electrical conductivity of the composites, especially when the composites are under stretching. For complementary understanding on the electrical behavior of the composites, the CNT waviness effects are needed to be taken into account.

Objectives

Based on the introduction and literature review, it can be seen that developing a micromechanics model incorporating the two conductivity mechanisms, nanoscale electron hopping and microscale conductive networks, the stretching effects and the CNT waviness effects is necessary and essential to better understand and predict the electrical behaviors of the CNT-polymer composites and provide guidelines for the composites based stretchable electronics. Therefore, the objectives of the current work will focus on the following:

1. Develop a mixed micromechanics model for CNT-polymer composites with the incorporation of the nanoscale electron hopping and the microscale conductive networks.

2. Investigate the uni-axial stretching effects on the electrical conductivity of the composites by considering stretching induced changes in the composites structures.

3. Based on the consideration of the uni-axial stretching, extend the investigation to the bi-axial stretching effects upon the electrical conductivity of the composites.

4. Incorporate CNT waviness effects into the developed micromechanics model under uni-axial stretching

Thesis structure

General introduction, literature review, motivation and objectives of the current work are presented in Chapter 1. For the work on specific objectives, detailed introductions are presented in later Chapters. In Chapter 2, a mixed micromechanics model is developed to predict the overall electrical conductivity of CNT-polymer nanocomposites, in which the two electrical conductivity mechanisms, electron hopping and conductive networks, were incorporated. By incorporating stretching induced changes into the developed mixed micromechanics model, Chapter 3 investigates the uni-axial stretching effects on the electrical conductivity of the composites Based on the work of Chapter 3, the investigation on the stretching effects is extended in Chapter 4 to the bi-axial stretching case. The CNT waviness effects on the electrical conductivity of the composites under a uni-axial stretching are studied in

Chapter 5. At last, Chapter 6 summarizes the conclusions of the thesis and proposes future work on the modeling of the electrical behavior of CNT-polymer nanocomposites.

REFERENCES

1. Bao, S., Liang, G. and Tjong, S., 2011. Effect of mechanical stretching on electrical.conductivity and positive temperature coefficient characteristics of poly (vinylidene.fluoride)/carbon nanofiber composites prepared by non-solvent precipitation. Carbon 49,.1758–1768.

2. Berhan, L. and Sastry, A. M., 2007. Modeling percolation in high-aspect-ratio fiber.systems. II. The effect of waviness on the percolation onset. Phys. Rev. E 75, 041121.

3. Barrau, S., Demont, P., Perez, E., Peigney, A., Laurent, C. and Lacabanne, C., 2003. Effect.of palmitic acid on the electrical conductivity of carbon nanotube-epoxy resin composites..Macromolecules 36, 9678–9680.

4. Chang, L., Friedrich, K., Ye, L. and Toro, P., 2009. Evaluation and visualization of the.percolating networks in multi-wall carbon nanotube/ epoxy composites. J. Mater. Sci. 44,.4003–4012.

5. Cheng, Q., Bao, J., Park, J., Liang, Z., Zhang, C. and Wang, B., 2009. High Mechanical.Performance Composite Conductor: Multi-Walled Carbon Nanotube Sheet/Bismaleimide.Nanocomposites. Adv. Funct. Mater. 19, 3219–3225.

6. Dastgerdi, J. N., Marquis, G. and Salimi, M., 2013. The effect of nanotubes waviness on.mechanical properties of CNT/SMP composites. Compos. Sci. Technol. 86, 164–169.

7. Deng, F. and Zheng, Q. S., 2008. An analytical model of effective electrical conductivity of.carbon nanotube composites. Appl. Phys. Lett. 92, 071902..Du, F., Scogna, R. C., Zhou, W., Brand, S., Fischer, J. E. and Winey, K. I., 2004. Nanotube.networks in polymer nanocomposites: rheology and electrical conductivity..Macromolecules 37, 9048–9055.

8. Ebbesen, T. W., Lezec, H. J., Hiura, H., Bennett, J. W., Ghaemi, H. F. and Thio, T., 1996..Electrical conductivity of individual carbon nanotubes. Nature 382, 54–56.

9. Feller, J. F., Linossier, I. and Levesque, G., 2002. Conductive polymer composites (CPCs):.comparison of electrical properties of poly (ethylene-co-ethyl Acrylate)-carbon black with.Poly (butylene Terephthalate)/poly (ethylene-co-ethyl Acrylate)-carbon balck. Poly. Adv..Technol. 13, 714–724.

10. Fisher, F. T., Bradshaw, R. D. and Brinson, L. C., 2003. Fiber waviness in nanotubereinforced.polymer composites-1: Modulus predictions using effective nanotube properties..Compos. Sci. Technol. 63, 1689–1703.

11. Gojny, F. H., Wichmann, M. H. G., Fiedler, B., Kinloch, I. A., Bauhofer, W., Windle, A. H..and Schulte, K., 2006. Evaluation and identification of electrical and thermal conduction.mechanisms in carbon nanotube/epoxy composites. Polymer 47, 2036–2045.

12. Grimmett, G., 1999. Percolation. Springer, Berlin..Grunlan, J. C., Mehrabi, A. R., Bannon, M. V. and Bahr, J. L., 2004. Water-based singlewalled-nanotube-filled.polymer composite with an exceptionally low percolation threshold..2004 Adv. Mater. 16, 150–153

13. Hu, N., Karube, Y., Arai, M., Watanabe, T., Yan, C., Li, Y., Liu, Y. and Fukunaga, H.,.2010. Investigation on sensitivity of a polymer/carbon nanotube composite strain sensor..Carbon 48, 680–687.

14. Iijima, S., 1991. Helical Microtubules of Graphitic Carbon. Nature 354, 56–58..Kim, Y. J., Shin, T. S., Choi, H. D., Kwon, J. H., Chung, Y. C. and Yoon, H. G., 2005.

15. Electrical conductivity of chemically modified multiwalled carbon nanotube/epoxy.composites. Carbon 43, 23–30.

16. Kirkpatrick, S., 1973. Percolation and conduction. Rev. Mod. Phys. 45, 574–588..Li, C., Thostenson, E. T. and Chou, T. W., 2007. Dominant role of tunneling resistance in.the electrical conductivity of carbon nanotube–based composites. Appl. Phys. Lett. 91,223114.

17. Li, C., Thostenson, E. T. and Chou, T. W., 2008. Effect of nanotube waviness on the.electrical conductivity of carbon nanotube-based composites. Compos. Sci. Technol. 68,.1445–1452.

18. Lin, C., Wang, H. and Yang, W., 2010. Variable percolation threshold of composites with.fiber fillers under compression. J. Appl. Phys. 108, 013509.

19. Lu, W., Chou, T.-W. and Thostenson, E. T., 2010. A three-dimensional model of electrical.percolation thresholds in carbon nanotube-based composites. Appl. Phys. Lett. 96, 223106.

20. Ma, H. and Gao, X. L., 2008. A three-dimensional Monte Carlo model for electrically.conductive polymer matrix composites filled with curved fibers. Polymer 49, 4230–4238.

21. Mayoral, B., Hornsby, P. R., McNally, T., Schiller, T. L., Jack, K. and Martin, D. J., 2013..Quasi-solid state uniaxial and biaxial deformation of PET/MWCNT composites: structural.evolution, electrical and mechanical properties. RSC Adv. 3, 5162–5183.

22. McLachlan, D. S., Chiteme, C., Park, C., Wise, K. E., Lowther, S. E., Lillehei, P. T., Siochi,

23. E. J. and Harrison, J. S., 2005. AC and DC percolative conductivity of single wall carbon.nanotube polymer composites. J. Poly. Sci. Pt. B: Poly. Phys. 43, 3273–3287..Miao, Y., Chen, L., Lin, Y., Sammynaiken, R. and Zhang, W., 2011. On finding of high.piezoresistive response of carbon nanotube films without surfactants for in-plane strain.detection. J. Intell. Mater. Syst. Struct. 22, 2155–2159.

24. Miao, Y., Chen, L., Sammynaiken, R., Lin, Y. and Zhang, W., 2011. Note: Optimization of.piezoresistive response of pure carbon nanotubes networks as in-plane strain sensors. Rev..Sci. Instrum. 82, 126104.

25. Miao, Y., Yang, Q., Chen, L., Sammynaiken, R. and Zhang, W., 2012. Modelling of.piezoresistive response of carbon nanotube network based films under in-plane straining by.percolation theory. Appl. Phys. Lett. 101, 063120.

26. Narkis, M., Ram, A. and Flashner F., 1978. Electrical properties of carbon black filled.polyethylene. Poly. Eng. Sci. 18, 649–653.

27. Ounaies, Z., Park, C., Wise, K. E., Siochi, E. J. and Harrison, J. S., 2003. Electrical.properties of single wall carbon nanotube reinforced polyimide composites. Compos. Sci..Technol. 63, 1637–1646.

28. Park, M. and Kim, H., 2006. Evaporation-based method for fabricating conductive.MWCNT/PEO composite film and its application as strain sensor. Proc. 12th US-Japan.Conf. Compos. Mater. Michigan, US, pp 78–86

29. Park, M., Kim, H. and Youngblood, J. P., 2008. Strain-dependent electrical resistance of.multi-walled carbon nanotube/polymer composite films. Nanotechnology 19, 055705.

30. Qian, D., Dickey, E. C., Andrews, R. and Rantell, T., 2000. Load transfer and deformation.mechanisms in carbon nanotube-polystyrene composites. Appl. Phys. Lett. 76, 2868.

31. Ramasubramaniam, R., Chen, J. and Liu, H., 2003. Homogeneous carbon.nanotube/polymer composites for electrical applications. Appl. Phys. Lett. 83, 2928–2930.

32. Sandler, J., Shaffer, M. S. P., Prasse, T., Bauhofer, W., Schulte, K. and Windle, A. H., 1999.

33. Development of a dispersion process for carbon nanotubes in an epoxy matrix and the.resulting electrical properties. Polymer 40, 5967–5971.

34. Seidel, G. D. and Lagoudas, D. C., 2009. A Micromechanics Model for the Electrical

35. Conductivity of Nanotube-Polymer Nanocomposites. J. Compos. Mater. 43, 917–941.

36. Shaffer, M. S. P. and Windle, A. H., 1999. Fabrication and characterization of carbon.nanotube /poly(vinyl alcohol) composites. Adv. Mater. 11, 937–941.

37. Shang, S., Zeng, W. and Tao, X. M., 2011. High stretchable MWNTs/ polyurethane.conductive nanocomposites. J. Mater. Chem. 21, 7274–7280.

38. Shen, J., Champagne, M. F., Yang, Z., Yu, Q., Gendron, R. and Guo, S., 2012. The.development of a conductive carbon nanotube (CNT) network in CNT/polypropylene.composite films during biaxial stretching. Compos. Pt. A: Appl. Sci. Manuf. 43, 1448–1453.

39. Shi, D. L., Feng, X. Q., Huang, Y. G. Y., Hwang, K. C. and Gao, H. J., 2004. The effect of

40. nanotube waviness and agglomeration on the elastic property of carbon nanotubereinforced.composites. J. Eng. Mater.-Tran. ASME 126, 250–257.

41. Takeda, T., Shindo, Y., Kuronuma, Y. and Narita, F., 2011. Modeling and characterization.of the electrical conductivity of carbon nanotube-based polymer composites. Polymer 52,.3852–3856.

42. Tallman, T. and Wang, K. W., 2013. An arbitrary strains carbon nanotube composite.piezoresistivity model for finite element integration. Appl. Phys. Lett. 102.

43. Taya, M., Kim, W. and Ono, K., 1998. Piezoresistivity of a short fiber/ elastomer matrix.composite. Mech. Mater. 28, 53–59.

44. Tchoudakov, R., Breuer, O., Narkis, M. and Siegmann, A., 1997. Conductive polymer.blends with low carbon black loading: high impact polystyrene/thermoplastic elastomer.(styrene-isoprene-styrene). Poly. Eng. Sci. 37, 1928–1935.

45. Wang, X., Bradford, P. D., Liu, W., Zhao, H., Inoue, Y., Maria, J. P., Li, Q., Yuan, F. G..and Zhu, Y., 2011. Mechanical and electrical property improvement in CNT/Nylon.composites through drawing and stretching. Compos. Sci. Technol. 71, 1677–1683.

46. Yan, K. Y., Xue, Q. Z., Zheng, Q. B. and Hao, L. Z., 2007. The interface effect of the.effective electrical conductivity of carbon nanotube composites. Nanotechnology 18,.255705

47. Yanase, K., Moriyama, S. and Ju, J. W., 2013. Effects of CNT waviness on the effective.elastic responses of CNT-reinforced polymer composites. Acta Mech. 224, 1351–1364.

48. Yi, Y., Berhan, L. and Sastry, A., 2004. Statistical geometry of random fibrous networks,.revisited: Waviness, dimensionality, and percolation. J. Appl. Phys. 96, 1318–1327.

49. Yu, Y., Song, S., Bu, Z., Gu, X., Song, G. and Sun, L., 2013. Influence of filler waviness.and aspect ratio on the percolation threshold of carbon nanomaterials reinforced polymer.nanocomposites. J. Mater. Sci. 48, 5727–5732.

50. Yu, Y., Song, S. Q., Bu, Z. X., Gu, X. F., Song, G. B. and Sun, L., 2013. Influence of filler.waviness and aspect ratio on the percolation threshold of carbon nanomaterials reinforced.polymer nanocomposites. J. Mater. Sci. 48, 5727–5732.

51. Zhang, R., Dowden, A., Deng, H., Baxendale, M. and Peijs, T., 2009. Conductive network.formation in the melt of carbon nanotube/ thermoplastic polyurethane composite. Compos..Sci. Technol. 69, 1499–1504.

52. Zhang, T. and Yi, Y., 2008. Monte Carlo simulations of effective electrical conductivity.in short-fiber composites. J. Appl. Phys. 103, 014910.

CITATION

CHAPTER 1

Alexander L. Kalamkarov; Marcelo A. Savi; Micromechanical modeling and effective properties of the smart grid-reinforced composites; dx.doi.org/10.1590/S1678-58782012000500002

CHAPTER 2

Zhenkun Lei, Xuan Li, Fuyong Qin, and Wei Qiu, "Interfacial Micromechanics in Fibrous Composites: Design, Evaluation, and Models," The Scientific World Journal, vol. 2014, Article ID 282436, 9 pages, 2014. doi:10.1155/2014/282436

CHAPTER 3

Juergen M. Lackner, Wolfgang Waldhauser, Lukasz Major and Marcin Kot ; Tribology and Micromechanics of Chromium Nitride Based Multilayer Coatings on Soft and Hard Substrates; doi:10.3390/coatings4010121

CHAPTER 4

Ladevèze et al.: A micromechanics-based interface mesomodel for virtual testing of laminated composites. Advanced Modeling and Simulation in Engineering Sciences 2013 1:7.

CHAPTER 5

Leandro José da Silva Túlio Hallak Panzera; André Luis Christoforo[l]; Juan Carlos Campos Rubio Fabrizio Scarpa; Micromechanical analysis of hybrid composites reinforced with unidirectional natural fibres, silica microparticles and maleic anhydride; doi.org/10.1590/S1516-14392012005000134

CHAPTER 6

Xiaojun Zhu, Xuefeng Chen, Zhi Zhai, Zhibo Yang, Xiang Li, and Zhengjia He, "Strain Rate Dependent Deformation of a Polymer Matrix Composite with Different Microstructures Subjected to Off-Axis Loading," Mathematical Problems in Engineering, vol. 2014, Article ID 590787, 11 pages, 2014. doi:10.1155/2014/590787

CHAPTER 7

Andrew Ritchey, Joshua Dustin, Jonathan Gosse and R. Byron Pipes (2011). Self-Consistent Micromechanical Enhancement of Continuous Fiber Composites, Advances in Composite Materials - Ecodesign and Analysis, Dr. Brahim Attaf (Ed.), ISBN: 978-953-307-150-3, InTech, DOI: 10.5772/14319.

CHAPTER 9

Royan J D'Mello, Marianna Maiarù and Anthony M Waas; Effect of the curing process on the transverse tensile strength of fiber-reinforced polymer matrix lamina using micromechanics computations; DOI 10.1186/s40192-015-0035-y

INDEX

A

Air Force Office of Scientific Research (AFOSR) 19

B

Bright field (BF) 47

C

Carbon fiber reinforced epoxy composites (CFC) 43, 49
carbon nanotube (CNT) 151, 237, 240, 242
Carbon nanotubes (CNTs 273
Coatings. Chromium nitride (CrN) 43

D

Design of Experiment (DOE) 94
Design of Experiments (DOE) 89
Diamond-like carbon (DLC) 43, 48

E

Equivalent homogenous 131

F

Fiber-reinforced polymer matrix composites (FRPCs) 248
finite element (FE) 247

G

Generalized method of cells (GMC) 111

H

Homogenous medium 129, 131

I

Influence matrix 134, 139, 140, 147
Institute of Science and Technology on Smart Structures for Engineering (INCT-EIE). 19
Interfacial shear stress (ISS) 32

L

Laminate theory 133, 144

M

Micro-Raman spectroscopy (MRS) 24